The TRUTH *of the*
TECHNOLOGICAL WORLD

The TRUTH of the TECHNOLOGICAL WORLD

Essays on the Genealogy of Presence

FRIEDRICH A. KITTLER

With an Afterword by Hans Ulrich Gumbrecht
Translated by Erik Butler

STANFORD UNIVERSITY PRESS
Stanford, California

Stanford University Press
Stanford, California

English translation ©2013 by the Board of Trustees of the Leland Stanford Junior University. All rights reserved.

The Truth of the Technological World was originally published in German in 2013 under the title *Die Wahrheit der technischen Welt* ©Suhrkamp Verlag Berlin 2013.

This book has been published with the assistance

of the Hubert Burda Stiftung.

No part of this book may be reproduced or transmitted in any form or by any means, electronic or mechanical, including photocopying and recording, or in any information storage or retrieval system without the prior written permission of Stanford University Press.

Library of Congress Cataloging-in-Publication Data

Kittler, Friedrich A., author.
  [Wahrheit der technischen Welt. English]
  The truth of the technological world : essays on the genealogy of presence / Friedrich A. Kittler ; [edited and] with an afterword by Hans Ulrich Gumbrecht ; translated by Erik Butler.
    p. cm.
  "Originally published in German in 2013 under the title Die Wahrheit der technischen Welt."
  A collection of twenty-three essays which appeared between 1978 and 2010.
  Includes bibliographical references.
  ISBN 978-0-8047-9068-0 (cloth : alk. paper) —
  ISBN 978-0-8047-9254-7 (pbk. : alk. paper)
  1. Communication and technology—Philosophy. 2. Literature—History and criticism. 3. Communication—Philosophy. 4. Technology—Philosophy. I. Gumbrecht, Hans Ulrich, editor of compilation. II. Title.
P96.T42K584 2014
302.23—dc23   2014010076

    ISBN 978-0-8047-9262-2 (electronic)

Printed in the United States of America on acid-free, archival-quality paper.
Typeset at Stanford University Press in 10/14 Minion.

Contents

1. Poet, Mother, Child: On the Romantic Invention of Sexuality — 1
2. Nietzsche (1844–1900) — 17
3. Lullaby of Birdland — 31
4. The God of the Ears — 45
5. Flechsig, Schreber, Freud: An Information Network at the Turn of the Century — 57
6. Romanticism, Psychoanalysis, Film: A Story of Doubles — 69
7. Media and Drugs in Pynchon's Second World War — 84
8. *Heinrich von Ofterdingen* as Data Feed — 99
9. World-Breath: On Wagner's Media Technology — 122
10. The City Is a Medium — 138
11. Rock Music: A Misuse of Military Equipment — 152
12. Signal-to-Noise Ratio — 165
13. The Artificial Intelligence of World War: Alan Turing — 178
14. Unconditional Surrender — 195
15. Protected Mode — 209
16. There Is No Software — 219
17. *Il fiore delle truppe scelte* — 230
18. Eros and Aphrodite — 249
19. Homer and Writing — 259
20. The Alphabet of the Greeks: On the Archeology of Writing — 267
21. In the Wake of the *Odyssey* — 275
22. Martin Heidegger, Media, and the Gods of Greece: De-severance Heralds the Approach of the Gods — 290

Contents

23 Pathos and Ethos: An Aristotelian Observation — 303

24 Media History as the Event of Truth: On the Singularity of Friedrich A. Kittler's Works—An Afterword by Hans Ulrich Gumbrecht — 307

Notes — 331
Credits — 389

The TRUTH *of the*
TECHNOLOGICAL WORLD

1   Poet, Mother, Child: On the Romantic Invention of Sexuality

The Middle Ages had something called the *Clan*. Since the eighteenth century, the code for kinship has been called the *Family*. Clans were connected by the law of exogamy, which linked them and inscribed scions along the axes of generations and races [*Geschlechter*]. Families, on the other hand, introject norms and imagoes into offspring, thereby subverting binary sexual difference [*Geschlechterdifferenz*] and generating souls sexualized by incestuous desire.[1]

When Parzival is born, Wolfram von Eschenbach simply mentions that his mother and her ladies-in-waiting spread the legs of the infant. When they discern the *visselîn* (which translates into today's English as "willie"), they lavish affection on the child. Coded in terms of sex, the boy receives a phallic attribute that symbolically couples desire and power: now he is destined for exogamous alliances and knightly adventures. The clan is governed by the metaphor *visselîn* = *swert* ["sword"],[2] a figure running this way and that—which Freud took up to his own ends and confused with natural fact.

Instead of promoting the play of metaphor, Herzeloyde, out of love and fear, clothes the adventuresome boy in a fool's garb, so that its worldly echoes may bring him back to her.[3] She does so to no avail, however, for an *ars amandi* and law that are one and the same remove Parzival from the double bond with his mother. Condwiramurs (whose name says what it means—"to conduct love") initiates him into strictly exogamous eroticism—and as *amor de lonh* ("love from afar") at that. Taking the place of Parzival's father, old Gurnemanz prohibits the youth from appealing to childhood and motherly words at all, in order to inscribe him into the axis of succeeding generations. Finally, the boy's uncle on his mother's side—who (as in other cultures) wields greater symbolic power than a biological father precisely because he is not the child's actual sire—articulates, in the capacity of father confessor, debts of blood to relatives [*Verwandtenblutschuld*] and, as a genealogist, the alliances between two clans.

1

Parzival's innocence [*Tumbheit*] ends when the symbolic order, which Herzeloyde has kept silent, is voiced. And because Trevrizent tells Parzival of his expectant mother's dreams, which she never revealed to her son,[4] there is no unspoken remainder that might haunt the hero and open the way for psychology or psychoanalysis. The incestuous double bond vanishes without consequence.

The code governing the conjugal, nuclear family—which emerged in the seventeenth and eighteenth centuries in the intellectual bourgeoisie and became universal in the nineteenth—stands opposed to the code of the clan on every point. Now political, juridical, and economic power are no longer linked to kinship structures. The household becomes the family unit, which assumes all tasks of socializing a small number of children—who, moreover, are planned. Burdened with the responsibility of being more symbolic than ever, the biological father surrenders his preeminent position to the mother. She, in turn, as the new center of the family, takes the place of the nurses of old. (Paradoxically, then, an origin substitutes for a replacement.) Intimacy and education tie the few children in the family to parent imagoes and eclipse the law of exogamy (which Freud interpreted as incestuous itself, if by transference). In order to be able—indeed, in order to wish—to become mothers or fathers, Lessing's virgins dream of a Father and Goethe's youths dream of a Mother. The phantasm of the Family obscures exchange that occurs between many families (which culturalizes them).

In the process, infantile sexuality—which previously was just as public as it was unexamined—becomes worthy of mention in the first place. The nuclear family becomes a complex relay that produces the children's mobile and fragmentary sexualities through records [*Aufschreiben*] made from the standpoint of the conjugal norm. The separation between parents and the world of childhood enables loving mothers and fathers, pedagogues, and psychologists to store the children's declarations of love to the authors of their days. There results, especially for mothers, a microhistorical archive that drills family romances into children as their own "experiences." Children become individuals who interpret—instead of the accidents of birth and race—"developments" and origins "within" themselves according to the rules of "reflection" and hermeneutics.

This coupling—of sexuality that derives from cultural coding and of speech that, when it involves self-declaration and self-interpretation, goes by the name of "poetry"—is to be investigated by means of discourse analysis. Neither social psychology, which presupposes that the discourses in question have already emerged, nor psychoanalysis, which presupposes the sexualization of children, can analyze how such a link (and nothing else) is bound to texts (and nothing else). In terms of discourse analysis, Romantic poetry is the effect of a semiotechnics that made the conjugal family matrilineal around 1800. The recoding itself was enacted by Novalis's novel, *Heinrich von Ofterdingen*; the effects were articulated in the works of Clemens Brentano, Friedrich Schlegel, Ludwig Tieck, Achim von Arnim, and E.T.A. Hoffmann.

1. Matrilineal Recoding

Klingsohr's tale [*Märchen*] has the function of symbolizing the primary socialization that Heinrich's mother was supposed to narrate at the end of the novel.[5] In a reverse mirror image, it presents the constellation of figures in the work as a whole. Now the patrilineal pattern of initiation that occurs in the *Bildungsroman* is replaced by matrilineal sexualization. For this reason, the tale constitutes a discursive event. For the first time in literature, a family appears that *articulates* all the stirrings [*Regungen*] and regulations that occur between mother and child from "the cradle" (338) up to the consolidation of the Oedipus complex.

Thereby, the bourgeois family obeys a mandate. It must take over the task of cultural reproduction, for the era of dynastic alliances has come to an end. The bourgeois family unit occupies a position between an "afamilial" and barren underworld of archaic mothers, on the one hand, and a heavenly dynasty that has grown sterile, on the other. Dynasties do not produce; they combine: stars and figures—signs and signs. This play of alliances comes to a halt as soon as Arcturus, who "cannot be king alone" (308f.), loses his wife to the bourgeois family and his only daughter—for whom he cannot find a husband of equal birth (cf. 214f.)—to the slumber of death. The order of alliance literally falls apart in its hypergamy: to make known and put an end to Freya's unredeemed status, the ancient hero (a symbolic father) must break the phallic sword of the dynasty.

The end of the law that codifies bodies as signs and punishes transgressions of the code by the sword inaugurates the norm that sexualizes children and makes them into individuals. The bourgeois family does not combine and distribute signs. Instead, it produces: children and imagoes. What is at first a nuclear family—"the Father," "the Mother," and their son, "Eros"—is augmented by Sophia, who comes from heaven, the Scribe or Death (303), Ginnistan or

3

"Fantasy," and little Fable, whom the Father sires with Ginnistan. Initially, Ginnistan is only a nursemaid for the Son, who makes up for the Mother's lack of milk. Soon, however—and to put matters in Freudian terms—she becomes sensuality [*Sinnlichkeit*], to which the Mother opposes interiority [*Innerlichkeit*] and familial cohesion. Familial eroticism, that is, plays out between the weakness [*Mangel*] of infants (which makes them dependent on others), the inability [*Mangel*] of a mother to nurse, and paternal desire: it couples child care and eroticism. For this reason, the culturalization of children that it effects takes the form of love for the breast—and not of their own mother, but of a Mother (294).

Orality is followed by the mise-en-scène of the phallic-narcissistic stage. In keeping with a pedagogy tailored to children, Ginnistan makes the sword fragment that the Father has found—and the Scribe archived—into a toy.[6] The splinter becomes a magnetic snake that phallically extends to the North; that is, it rouses "Eros" for the future beloved, Freya. Eros himself, in this phallic game, suddenly becomes a youth. The phallus, then—which is synonymous with the name "Eros"—means becoming the object of desire for a/the Mother. This inducts the precocious youth into premature oedipality: into a round dance [*Reigen*] of heterosexual pairings that cycles through all combinations between Father and Son, Mother and Nurse. First, Ginnistan abducts Eros into the bedroom; however, she obeys a wave from Sophia and replaces sensuality with tenderness. The "quiet embrace" (295) between the Mother and Eros, which echoes an imaginary dyad, steers the desire of the Father back to Ginnistan, so that the agent prohibiting incest simultaneously affords an example of its transgression. And because the desire of speaking beings is the desire of the Other (Lacan), the example arouses a forbidden desire in the Son. On the orders of Sophia, the Mother and Ginnistan have to exchange forms so that he "will not be led into temptation" (296). Unlike the gesture of the wave, however, the prohibition is violated although—and because—it is articulated. Since "all barriers are there only to be overcome,"[7] they sexualize the Mother, who was "quietly embraced" previously. The act of uttering the prohibition creates, in the first place, what it declares unattainable: the imago ("gestalt") Mother.

Accordingly, the "Fantasy" of Mother, writ large, stages a play that steers the infantile wish that is "Eros" from the image of the nurturing-washing Mother—by way of a "forbidden thrill [*Rausch*]" (305)—toward the future image of amorous union with Freya. In this process, Ginnistan plays the part of all female imagoes. "Fantasy," then, is not merely the unconscious fantasy of

On the Romantic Invention of Sexuality

the author; it symbolizes the sexual rite of initiation itself under the conditions of the nuclear family.[8] The path to reproduction must be staged before the eyes of the speaking being; it does not follow instinct, but fantasy. The infant—whose senses and motor skills are still disorganized after a painful and premature birth—achieves the social identity function [*Einheitsfunktion*] of "I" only when others inscribe it with phantasms and present a deceptive image of integral corporality beforehand. The scenario of Ginnistan offers a historical variant of the mirror stage Lacan describes: her gaze and desire steer Eros's eyes onto the prefiguration [*Vor-bild*] of unity that he does not possess. He "thanks" her "with a thousand delights [*Entzücken*]" (300) for sexualization. Hereby, the Mother, Ginnistan, and Freya—as well as natal and "target" families—become confused.

The end of the tale consolidates the child's sexuality, which has been produced maternally: it constitutes the very basis of a new Golden Age. Unlike traditional fairy tales, which simply end with hierogamies, Klingsohr's narrative subordinates the couples—Eros and Freya, Arcturus and Sophia, and the Father and Ginnistan—to Motherly Love [*Mutterliebe*]. Because there is no room for Eros's mother among the couples, Sophia—the Heavenly Mother—promotes her to a position where, present in absence, she stands at the origin of the entire system; that is, the Mother becomes the Mother of All, including figures who have "other mothers." All the characters drink from her ashes in the baptismal ritual; after the fact, this inexhaustible beverage makes up for the Mother's lack of milk and for the pains the children experienced in the process of birth. With delight [*lustvoll*], they feel their *generatio continua* from the Mother, who "underlies" all marriages in the form of imaginary incest. The children's love for each other is love from and for the Mother.[9]

The Universal Mother [*Allmutter*]—continuously giving birth, heightening sensation, and producing phantasms of incest—takes the place of the Symbolic Father who formerly distributed his seed among the races [*Geschlechter*] and generations. Accordingly, the correlate of the Mother's ascendancy is the elimination of the Scribe (i.e., Death), the sole figure the tale fails to assign a place in the final tableau. His textual archive is done away with so that the incestuous nature of the new norm will remain a "secret" to the precise extent that it stimulates (ongoing) orality. Hereby, the Mother becomes the signified for all sounds that are made: "her presence" (315) is felt in the amorous whisperings of the endogamous couples. Orality and the poetry of discourse become one and the same.

2. The Voice of the Mother and the Poetic Individual

Matrilineal recoding follows and celebrates the rules of communication in a culture that "invents motherly love for infants."[10] The coupling of orality and poetry stems from a psycho-pedagogy that, since Locke and Rousseau, has prescribed that mothers themselves should nurse and speak to the being without language (*infans*) in their charge. At the end of Klingsohr's tale, the matrilineal and fatherless siblings/couples sing and whisper instead of performing a speech act that would promise loyalty, and the "milk-blue stream" (300) of the Mother herself replaces that of the Nurse. These narrative events take contemporary critiques of the unmotherly mothers of old literally:

> [They] fulfill these duties, and with exactness, but they do not go beyond them; they neither sing nor speak to the child; they do not seek to awaken its senses; they do not have the intention of developing the sensations it has through ... the incitements [*agaceries*] of maternal tenderness.[11]

The center of the nuclear family—the Mother—becomes the relay point for a new kind of productivity, which rouses the senses in threefold manner: to individual perception, to sexuality, and to aesthetics. That Romanticism considers poetic discourse to be individual expression and the bearer of elementary sensuality derives from the communicative matrix formed by a nursing, loving, and speaking mother and an infant. Drinking at Ginnistan's bosom, Fable gives thanks for the "unbreakable thread" that "seems to wind forth from her breast" (314) and makes a pure idiolect of poetry. Likewise, Brentano's Godwi nurses at the breast of his beloved as the "source of all sustenance and voluptuousness [*Nahrung und Wollust*]"—"all the power of the word, all the magic of poetry."[12]

Matrilineal recoding changes the status of literature. The poetic function posited by Roman Jakobson—previously a matter of the autonymy [sic] of cultural symbols—becomes phatic in nature. Accordingly, in *Heinrich von Ofterdingen*, the "secret word" (or signifier) *Mother* replaces "numbers and figures" (344) and in so doing opens communication between "lovers." As Heinrich Bosse observes:

> While to classical thought the institution of signs rendered possible human communication, it is now the very fact that man communicates with man which will define the signs.[13]

Just as the speech prescribed for new mothers, because it produces linguistic competence in the first place, shares no positive content, poetry itself becomes a play of sounds [*Lauten*]. That it "speaks in order to speak"[14]—as Novalis puts it elsewhere—brings back the intransitive quality of the initial situation of communication. Sounds melt with nature; noises murmur and whisper with the maternal voice, which induces harking [*horchen*] and not hearing [*hören*] in the infant. The matrix of motherly lullabies—which take the place of less complicated methods of quieting children—gives rise, at the border between speaking and sleeping, to a new lyricism that has existed ever since "Wanderer's Night Song," by Goethe.

To be sure, humanizing [*hominisierend*] speech in order to make (infants) speak had always occurred. Only now, however, was it bespoken—that is, discussed. Herder derived "the I" from learning to feel [*Empfindenlernen*] at the mother's breast, and "the knowing and feeling of the human soul" [*Erkennen und Empfinden der menschlichen Seele*] from acquisition of language in the infant.[15] Such psychologizing of discourse displaced the ontogenetic thresholds of what—and who—can be addressed [*Besprechbarkeit und Ansprechbarkeit*]. Rousseau, in turn, considered self-consciousness the effect of complete alphabetization,[16] and Brentano's traveling student even recalls how he read the first sounds from his mother's lips.[17] Bespeaking initial speech makes it worthy of mention in the first place. It opens space for the free play of little geniuses who arouse admiration, not by performing speech acts that are binding but through toying with sounds [*Lautspielen*] and infantile words.[18] Of course, it is mothers who protect and promote the dreams and dream narratives of their poetic children against the incursions of prosaic or evil fathers.[19]

With this displacement of the threshold of socialization, a parameter of discourse that is corporeal (and not digital) won power over mute bodies. Voice transformed into the *mythos* of a theory of lyric that discerned "the secret-filled depth of human spirit and poetry"[20] in its murmurings; likewise, it whispered originary truth to a linguistic science that explored Indo-European languages as a family—and investigated "language" in general (instead of letters as sounds). The celebration of the voice amounts to the rejection [*Verpönung*] of writing: the voice's presence and individuality deny the absence and the symbolism of the signifier. In Klingsohr's tale, Fable—who sings—unseats and replaces the Scribe (295, 308). Similarly, Brentano's *Chronika des fahrenden Schülers* begins with a mother who teaches her infant to sing and pray, and it ends with a siren whose book lures a youth far away, into erotic ruin.[21]

In poetry [*Poesie*], the poet [*Dichter*] becomes another. If, as Julia Kristeva has claimed, Western literature translated the conjunctive hierogamy of Oriental texts into a disjunction between the One and the Other—the speaking poet and the mute woman[22]—Romanticism marks the moment where the former becomes a childish individual, and the latter a mother. Henceforth, "the dear woman exists" as a "mother" who addresses her words when she talks; she does so, "as everyone knows," in order to "make the speaking being . . . speak."[23] Instead of being defined by the binary code of sex, the poet is defined by his matrilineal individuality. Klingsohr's tale depicts the poet in Heinrich as "little Fable"; that is, it does not portray him as her half-brother. This is also how the possibility of female poets arose: Goethe left the "*aristeia* of mothers"—the blind spot in *Dichtung und Wahrheit*—for Bettina Brentano to write.

If poetry repeats the voice that has sexualized its speaker, then its utterance already contains the eroticism invoked by what is uttered. If it *reproduces* what words merely *represent*, no word can reach where it originates. Poetry is an origin as omnipresent and as hidden as the Mother in Klingsohr's tale: a vocal shadow that the words cast yet never can express directly. Tracking the sexuality that inhabits it as a voice, poetic discourse generates the very thing it claims it cannot say. Such positive feedback between speaking and sexuality occurs in the chapter "Devotion and Jest" [*Treue und Scherz*] in Schlegel's *Lucinde*, where the eponymous character—who is called "a child," after all—is enjoined to "caress" a "motherly" beloved[24]; another instance is the eroticizing confession of incestuous sexuality that Medardus makes as a scribe in *The Devil's Elixirs*, by Hoffmann.[25]

3. Hermeneutics of the Origin and the Norm

According to standing ideas, sexual matters came to penetrate literary discourse to the extent that bourgeois society prohibited their expression. Foucault demonstrated that the opposite is the case. Sexuality is an effect of discourses. To affirm that its origin is unspeakable is to call forth discourses about it—which, because they are sexualized themselves, can never end. Sexuality, then, functions within a machinery that makes bodies speak and incorporates them into a new organization of power and knowledge. In contrast to cultures that let live and make die, our culture—and only our culture—has transformed into "society" [*Gesellschaft*]: it "makes live" and avoids killing [*macht das Leben und läßt das Töten*]. Planning conditions of and for life encompasses fields that did not pass into record under the law of Sword and Alliance. Moreover, it produces and stores knowledge that Aristotle deemed impossible: understanding what is individual [*das Wissen von Individuellem*]. Accordingly, "man" represents a recent invention in epistemological terms. "He" becomes a "subject" (in the double sense of the word) only through knowledge that declares "him" subject to the conditions of life governing "him" *and*, at the same time, the master who can recognize and change these conditions. Since 1800, literature and the human sciences have treated "phenomena of our being that actually turn out to be us, since they condition us—and we them—each in turn."[26]

The concept of sexuality represents one of many such instances of empirical-transcendental doubling. It relates bodies to a force of production that both precedes them and at the same time is derived from them. Without end, knowledge cycles between sexual origin, where the "human being" (in general) is produced, and the individual, whose origin seems to be unique. The dichotomy between law and transgression transforms into reciprocal reference between the norm and individual deviancy. This gives rise [*zu Wort bringen*] to new situations of communication and hermeneutics: on the one hand, rituals of confession and recollection, and on the other, analyses of the "Unconscious." These discursive events presume that sexuality voices the truth about us—which we cannot express when we articulate the truth about it, which it cannot speak itself.

Klingsohr's tale presents [*konstruiert*] this transformation of knowledge and power. It leads from juridico-political culture into the realm of familiality, sexuality, and productivity. The tale's incestuous norm involves transgressing the law of old, and it culminates in installing the human being on the throne. Eros ascends as "the new king" (314), yet his rule is paradoxical: he reigns only insofar as he is subject to a maternal origin which, for its part, only has "presence" to the extent that it comes to power in Eros. The individual *is* its history. The text reaches back to the cradle and forward to the Golden Age. Thereby, it transfers the ancient myth of the ages of the world [*Mythos der Weltalter*] into a logic of production: when the goal of the Romantic triad is achieved, human beings "dwell" (315) in temples; their sexual productivity is one with physical-chemical nature and organic life.

The tale performs the matrilineal recoding of characters/figures in simultaneous and transparent fashion. Thereby, it erects a dispositive that other works of Romanticism can cycle through in anamnestic and asymptotic ways. The maternal origin—which the tale names and at the same time places within the figures' interior lives [*Innerlichkeiten*]—becomes both the historically "sunken" *movens* and the goal for endless hermeneutic explorations. Following this shift from simultaneity into temporal profundity, the originary Family dwells within the Individual as its secret. Romantic works do not, like courtly romances, affirm genealogical identities through a succession of parents' and children's lives. Instead, they posit identity by means of an empirical-transcendental folding of the individual. As the process unfolds, however, it reveals just how much the sexualized family serves instances of power and knowledge.

Tieck's "Eckbert the Fair" offers a direct continuation of Klingsohr's tale. Both works transfer the conjugality of the fairy-tale form, which Klingsohr's predecessor and model, Goethe, had preserved, into endogamy. Whereas Novalis locates incest at the end of the narrative, as codification that occurs through Mother Sophia, Tieck makes it the unthinkable beginning of events, which is only (re)discovered later. Eckbert and Bertha have always already had the same father and been siblings—except that this fact is decoded only at the very end, by a witch, who is herself the vanishing point for all the childless couple's phantasms. The Witch is a Mother who can display both female and male traits, and therefore dominates the patrilinearity that the narrative preserves genealogically.

The same also holds on the level of events in the tale. A single witch replaces both foster parents to whom Bertha's father has given her, an illegitimate child. The dominant party is the foster father, who wants to raise Bertha only for work. Bertha, however—like the heroine of "The Elves"—flees into a fairy-tale world that the foster father's word(s) cannot reach. The world of childhood is one of the phantasms that derive from socialization in the nuclear family; here the distinction between adults and children[27] is reproduced in the wish to stay a child forever[28]—a matter that remains a phantasm because the children fall prey to an unsymbolized Mother. Just as Novalis equates childhood "development" that occurs without parental intervention and "education" that the father "has left entirely in the hands of the mother" (326), the Witch dominates the "small family circle" consisting of Bertha, the dog, and the bird. Accordingly, Bertha—their "daughter"—cycles through pre-oedipal sexualities. The animals, as "well-known friends,"[29] become narcissistic mirror images because a Mother coordinates [*inszeniert*] identification with them. Here differences are so slight that love can abruptly turn into paranoia. The bird—which lays eggs containing pearls and sings a song whose "words are constantly repeated" like dream poetry and lullabies[30]—displays both anal and oral traits.

Likewise, in Achim von Arnim's "Isabella of Egypt," the dyad between the parentless Bella and a witchlike foster mother produces narcissistic doublings such as the Golem Bella, anal beings like Bearskin [*Bärnhäuter*], and phallic ones like the gold-finding Mandrake [*Alraun*] (whose marriage concludes in thumb-sucking).³¹ These worlds—the grotesque one and the fairy-tale one—both *are* and *have* productivity. Bella's lover, a ruler under the conditions of early capitalism, prefers polymorphously perverse and productive sexualities to the love of, and marriage to, Bella. Similarly, in Tieck, the fairy-tale bird makes possible what Bertha "only dreamed of in childhood": to bestow (her father's) "wealth"³² on her foster parents—the measure by which they had evaluated her and found her lacking. Regression to the archaic Mother, then, is what enables the child to fulfill the mandate of productivity that the discourse of others has instilled [*einfleischte*].

Like her act of theft and her flight from the Witch's house, Bertha's narrative about events is subject to [*untersteht*] the discourse of others. Only for the sake of intimacy, whose norm is the Family, does Bertha tell parties other than Eckbert about her childhood. Beings possessed of interiority [*Innerlichkeiten*] who think that they "share themselves entirely [*sich ganz mitteilen*]" when they recall their origins embody the compulsion to repeat a situation of infantile communication: time and again, they speak about the family circle in order to integrate strangers *into* it as "friends."³³ At the same time, however—and in line with the operations of the mirror stage—narcissistic identification transforms into paranoia. Eckbert murders the man who has heard Bertha's confession, and he flees the party who has heard his own confession of killing because he fears the "misuse" of a "confidence [*Vertraulichkeit*]" that he himself has produced.³⁴ Communication that only intensifies feelings and reproduces the intimacy of nuclear families is just that paradoxical. In Novalis's novel, it entails eliminating a writer (the Scribe) for whom endogamy would still mean endogamy, and in Tieck's tale, it entails the murder of witnesses who might make the phatic speech of the endogamous couple into a public "text" capable of transmission.

The matter without precedent, however, is that hermeneutics of the Family addresses the very instance of power whose initial speech it interprets. Bertha's auditor mentions, in passing, a detail from childhood that escaped her: the name of the dog that had been her playmate. This item of inexplicable knowledge makes the man a member of the Family—indeed, it makes him the incarnation of the Witch. In the idiolectal name "Strohmian," the maternal point of origin [*der mütterliche Ursprung*] catches up with the girl who has fled and confessed. "A letter always arrives at its destination."³⁵ With a word that proves meaningless as a signifier, the Mother—in Romanticism—signals her status both of being the origin and of commanding speech. The phantasm is pathogenic and lethal: Bertha suffers a hysterical fit and dies.

The same thing befalls her brother and husband. The course of flight from confession and murder—which is meant to erase the traces of confession and murder—leads straightaway to the Other, whom Eckbert can neither murder nor flee because she gives chase and deals death herself. The Witch reveals that all parties who have heard the fugitives' confessions were incarnations of her, and that Eckbert and Bertha are siblings. Her genealogical discourse makes words fail Eckbert [*ihr genealogisches Wort macht Eckbert das Wort verwirken*]: mad and in the throes of death, he hears the voices of Mother Nature and his own phantasms melting into one. He could not have so much as "suspected [*ahnden*]"[36] incest, because language has always already commanded him. Indeed, it named him in the first place: "Eckbert" and "Bertha" are half homonymous.[37] "One is only ever in love with a name [*On n'est jamais amoureux que d'un nom*]."[38] Spellbound to their family through Christian and pet names, those who interpret them meet with death—death that occurs through words alone. A victorious Mother speaks first and last.

Matrilineal recoding, then, has the function of extracting [*entreissen*], from its products, the words it has beaten into [*einfleischen*] them. It is a machine that generates admissions and confessions—and, in so doing, generates the particular form of individuality which Romanticism deemed productive. When father confessor Trezvirent tells Parzival of a dream that was never revealed to him, he inscribes the youth into the Symbolic. Naming a forgotten [*entfallen*] name, however, performs the function of individuation because a family's memory [*Familiengedächtnis*] "spills" what it formerly declared secret. To ascribe meaning to the words and events of childhood to the extent that they are ("objectively") insignificant means making the family into the archive of criminological clues and sexological norms. It is not important whether the recollection of forgotten details from childhood affirms guilt or denies it.[39] It is itself a discursive event, and only the interiority that it has generated can call it a faculty [*Vermögen*] of its own. When interiority speaks, a culture speaks—one that accords the Family the production of all "meaning" to the same extent that other functions vanish.[40]

The matrilineal family becomes a relay for transmitting knowledge and power. The compulsion to confess—which ties Bertha to infantile sexuality, and sexuality to a mother—is no fairy tale. "Mademoiselle de Scudery," by

Hoffmann, continues Tieck's fairy tale in the framework of the institutions of Law and Psychology. The series of murders in Paris that undoes the holiest of bonds—that is, once more, that of the Family[41]—escapes the torture of the ancien régime. In contrast, what manages to get behind them is a speech act that answers for deeds forbidden by law. What escapes the established conception of truth are individual and unconscious motivations, which prohibit verdicts based on deeds alone, as well as productive aspects of criminals that promise future improvement and utility. Accordingly, the jurisprudence of Enlightened Absolutism decides to have the accused confess—without chains or witnesses—to a female writer who counts as a mother to him. When Mademoiselle de Scudery recognizes a child she once cradled, the psychology of crime is born.

The psychological account is itself familial. Once more, a mother has encoded what a mother in turn decodes. The goldsmith Cardillac—whose identity the accused man concealed, as if out of love for a father—has robbed his patrons and customers and stabbed them to death. He has done so in order to repeat a prenatal scene. Cardillac's mother, while pregnant with him, was seduced by the sight of jewels presented by a nobleman she had previously rejected—an embrace that lasted forever because death befell her lover. Now the son "embraces" and murders noblemen as they make their way to assignations with their mistresses. The newly minted pervert eliminates the libertine of the ancien régime because he unites criminality and productivity. Jewels, as the object of the mother's desire, entail fetishism of the same.

From childhood on, Cardillac has plied his trade/craft [Handwerk] as an art. The jewels the mother desired—as the phallus of a lover (and not of her husband)—led Cardillac to identify with her desire. Consequently, he embraces as a lethal mother. Matrilineal, then, are a craft that undoes borders between estates and a crime that does not occur simply for gain. The eccentric [Sonderling]—for whom the law makes no provisions—becomes the norm, and this entertains no relationship with repression whatsoever. The primal scene, perversions, and matrilineal art both *are* and *enable* juridical, psychological, and aesthetic forms of individuation. A culture that claims to be able to say how a "narrative" [Erzählung] told by a mother makes her child productive can optimize the choice of profession without invoking the order of estates. That said, it does well to have the mouths of "wise men" (as in *The Serapion Brethren*) offer instruction about the power of primal scenes—which it then confirms through the ears and writings [im Ohr und Dichten] of wise mother confessors.

4. Romantic Texts and Knowledge of the Soul

"The doctor is a second father confessor," one of Hoffmann's many personal physicians exclaims to a princess—who has reserved the sexual secret of her hysteria for priests. The alliance between the nobility and the church, whose statutes view bodies only in terms of blue blood and sinful flesh, yields to the alliance between family, psychology, and medicine, which investigates the "putty" [*Kitt*] sticking together "body and soul"[42]—the individual and sexuality. *The Devil's Elixirs* describes an endogamous family that brings forth eccentric souls [*Ausnahmeseelen*] and artists, revealing their—and its—productivity orally to "ingenious" psychiatrists and monks who cannot read genealogical texts.[43] Only in the newly established madhouse,[44] and not in the royal dungeon, can knowledge be obtained about knowledge that has been bought at the price of incest.

When literature becomes family hermeneutics—that is, when it investigates the sexualization of children and the hysterization of women in confessions, autobiographies, crime stories, and novels of the soul [*Seelenromanen*]—it has the same address as psychology. That makes psychoanalytic readings of Romantic texts possible, and tautological.

Displacing the threshold of addressability onto the mother-child dyad makes authors and characters "psychoanalyzable" in the first place: Freud's decodings of infantile sexuality begin exopoetically with Goethe's *Dichtung und Wahrheit* and endopoetically with Hoffmann's "Sandman." A fortiori the connection between author and characters becomes possible only when discourses [*Reden*] are referred to individuals and not to systems of symbols. In this manner, the appearance results that biographies explain texts—even though familial relations [*der Familiarismus*] in the one simply double those in the other.

Psychoanalysis inhabits the same space of discourse that invented and implemented the power of primary socialization. It is only on this basis—as is the case for Cardillac[45]—that text and interpretation coincide. Deciphering imagoes of the nuclear family in texts and the discourses that constitute them is merely a matter of rediscovering the sediments of codification that, around 1800, ascribed a meaning to the Family and especially to the Mother—a process that Freud considered "of paramount importance" for the "whole" of "later life."[46] At the same time, however, sexualization is subject to biotechnologies and forms of knowledge that made the Family into the "mother" of all imagoes in the first place. In *The Devil's Elixirs,* incestuous wishes—which are forgiven and then archived in monasteries—are aroused by portraits of the ancestral mother [*Ahnmutter*] that these same cloisters display. Likewise, when Heinrich's natal family is depicted in Klingsohr's tale, parental imagoes split between sires and scribes, sensuality and tenderness, only to be correlated, allegorically, to psychic faculties (338). It follows, then, that the multiplication of parental imagoes represents the stratagem of a kind of psychology that forms bodies through images and makes them into addressable souls. When Freud excavated such a process of image production from Hoffmann's "Sandman," he abandoned literary study along the lines of hermeneutics and empathy [*Einfühlung*]—but not the space of rhetorical invention [*Rede-Erfindungen*].

If pre-oedipal sexualization constitutes a program and the Oedipus complex represents a staging of "fantasy," then they are subject to a discourse [*einem Reden*] and not to a desire. In order to function, Romantic texts presume that objects of transference be spoken and heard; after mothers and psychologists, psychoanalysts join in. That hides the productivity of sexualizing discourse from exegetes. Psychoanalytic approaches to literature read Romantic texts as expressions of forbidden wishes and as compensation for social constraints. However, the joy that psychoanalysis has in such discoveries conceals a double blindness. An "individual" is assigned wishes that are actually technologies of socialization [*Sozialisationstechniken*]. Likewise, "society" is assigned prohibitions that are, in fact, obsolete. It is not the ancient law of the Symbolic Father—to which Freud reduced all forms of infantile sexuality—but rather the Norm that governs the texts. It contains positive figures that collaborate [*mitschreiben*] in the production of productivity [*Produktion von Produktion*] and extend invitations to enthrone the same fantasy that already wields power.

Finally, a trait of the psychoanalytic method of decoding is itself tautological. The search for conditions that constitute "the human being"—which at the same time this being makes—renews and prolongs the empirical-transcendental folding that has already occurred in Romantic texts. When Klingsohr's tale posits matrilineal sexualization for the public *Bildungsroman*—splitting and displacing family imagoes in the course of representing it—the work erects the hermeneutic dispositive that Freud's *Interpretation of Dreams* transferred into the scientific sphere. Even under the changed parameters that make the articulation of Romantic texts possible and disintegrate their transcendentalism—because writing has replaced the voice, the signifier the signified—interpretation remains a matter of the interplay between the latent and the manifest, the spoken and the unspoken, and "fantasy" and "reality."

15

Poet, Mother, Child

Yet discourses have no depth wherein their substance might lie [*in der ihre Sache läge*]. They are surfaces—the juxtaposition of familial coding, maternal memory, poetry, and psychology around 1800. Here, in intertextual space without shadow or shade, is where the philology that Nietzsche discovered could operate: the philology of rhetorical inventions.

2 Nietzsche (1844–1900)

> So you think you can tell Heaven from Hell.
> — Pink Floyd

The name of "literature"—and its theory—emerged alongside a public sphere that discoursed reasonably on culture, and alongside a philosophy that recognized an epistemological subject behind literary works. Nietzsche identified this constellation and brought about its disintegration: he withdrew fiction from philosophical judgments that concern truth, and he introduced a conception of the public sphere whose element is not reason but the production and consumption of media. For all that, the subversion he performed has affected literature itself (Artaud, Benn) more than its study.

Uncoupled from recognition/knowledge [*Einsicht*], literature entered relations with corporeality and power. Taking away the mandate of representing the ideas of Reason—or indeed, Absolute Spirit itself—meant passing beyond the borders that Kant and Hegel had imposed on both the productive energies of the body and on violence [*Gewalt*]. Nietzsche's literary-theoretical fragments articulate an aesthetics of production that recognizes no limits to creation and destruction. It replaces authorial psychology with the physiology of the artistically creative body, the theory of effects and affects in aesthetic education with the semiotics of sensory media, the philosophy of literary history with the genealogical analysis of discursive instances, and transcendental hermeneutics with philology.

1. Language, Fiction, Truth

Any project of philosophical aesthetics must, first and foremost, determine the relationship between philosophy and its object: art. Nietzsche did so by theorizing language as rhetoric. He placed literature and philosophy alongside each other on a field to which they both belong as forms of articulated language. This pragmatic-linguistic radicalization of Kant's critique of metaphysics undid the very distinction that had made it possible for philosophy to set the knowledge of things above literary discourse made to specific addressees—that is, it undid the difference between Concept and Metaphor. According to Nietzsche, all words are metaphors in a double, and literal, sense. First, they make nervous stimuli—which do not correspond to a thing but to a corporeal relation—into sounds; second, they transmit these sounds to an addressee.[1] The first instance of transfer has no priority over the second: the differentiation between stimuli is learned for the sake of others—indeed, consciousness itself is "only a means of communicability [*Mittel der Mitteilbarkeit*]" that has "developed in exchange" (*Nachlass* III 667). Rhetorical figures illustrate the matter clearly: a synecdoche like "sail" (instead of "ship") names a feature that stands out to communicating parties; it does not name the "thing itself" (*Rhetoric* § 3; *Collected Works* V 298f.).

As an "artistic transfer" (*Truth* § 1, III 315) from one medium to another, language expunges the ideas of Wholeness, Truth, and Authenticity. "There is no such thing as an unrhetorical 'naturalness' of language to which one might appeal. [...] *Language is rhetoric*, for it wishes only to transmit *doxa*, not *episteme*" (*Rhetoric* § 3; *Collected Works* V 298). The origin of linguistic rhetoric is not significant—indeed, it "originates" in an act of replacement; rather, its function is important. Rhetoric constitutes a form of elementary mnemotechnics. It operates as a machine for selection by setting up an environment that is memorable and ready-to-hand—one that, nevertheless (or for this reason), has no calculable utility. Rhetoric, which was a regional doctrine of art in antiquity, becomes universal; and "man," that "inscrutable animal [*das nicht festgestellte Tier*]" (*Beyond* III § 62, II 623), becomes one with the "drive to create metaphors" (*Truth* § 2, III 319). Nietzsche's effort to define [*bestimmen*] literature as language ultimately performs a reversal: language itself is literature—the fabrication of fictions.

Indeed, for Nietzsche, the scope of fiction extends so far that it changes status. That, in the final instance, deception means truth and simulation insight/knowledge follows from the passage of language to writing and concepts—which represent two further "technologies" of semiotic selection. By "jumping over" most words (*Beyond* V § 192, II 650), reading transforms verbal matter into "thoughts." Consequently, only the philologist still "reads words" at all (*Works and Letters* V 268). Thoughts and concepts—as "residues of metaphors" (*Truth* § 1, III 315)—subsume a verbal multiplicity, just as words subsume a swarm of sensations. In this way, the second selective operation, as if it were a primary function, erases the reference to the body that the voice has in speech. This accounts for Nietzsche's inimical relationship to writing (which separates him from his grammatological inheritors). Modern book culture rejects and eliminates embodied rhetorical techniques—what, *per antiphrasin,* we call "ancient literature" (*Greek Literature* III § 1; *Collected Works* V 209ff.). Accordingly, the modern *cogito,* in its state of disincarnate transparency, rests on something that remains unthought; its claim to knowledge is belief in grammar, whose tropes it parrots and forgets (*Nachlass* III 577).

It would appear, then, that Nietzsche's theory is still inscribed in the matrix of transcendental thinking: as the rehabilitation of language and rhetoric against Reason, which is hostile to them, philosophy would be the recollection of what thinking does not think, on the one hand, and the critique of this oblivion, on the other. Ever since Herder, the originary linguistic productivity of mankind has counted as the "unthought," which manifests itself in poetic speech and ultimately yields conceptual discourse.[2] However, Nietzsche leaves such transcendental anthropology behind in two ways.

First of all, production neither occurs in a mythical space where signs and referents are one, nor does it take place within a subject oblivious of what it has created (and creates). Instead, languages and fictions number among the many and disparate events of corporeal being. Their lack of "truth" does not lead theory to skepticism or positivism, but to Ariadne: "the path of the body" [*Leitfaden des Leibes*]."[3] Secondly, the deception and forgetting that are called "truth" and "insight/knowledge" are not sluggish figures whose aporias reflection might resolve. If the systems of signs necessary for life—instead of merely giving rise to interpretations [*Auslegungen*]—are already interpretations themselves, then no act of interpretation can reveal the "transcendental signified" underlying them.[4] Accordingly, Nietzsche's philosophy abandons the principle of critique and sides with the powers that inscribe and erase signs through the act of interpretation [*auslegend*]. It begins the ruse-filled game of naming *and* performing fictions—turning interpretation against interpretation, and rewriting the rhetoric of concepts as concepts of rhetoric. Regional concepts of literary theory (e.g., fiction, fable, interpretation) achieve the operative and strategic status of not just describing but also enacting "how the 'true world' finally became a fable" (*Twilight* IV, II 963).

Following the path of corporeality, philosophy becomes physiology, and by reinterpreting interpretations, it becomes genealogy.

2. On the Physiology of Aesthetic Media

The Birth of Tragedy from the Spirit of Music—the first and last closed "book" that Nietzsche wrote—names the link between physiology and genealogy in the title. A literary genre is declared born as a body. For all that, it takes two to conceive—and by extension, to give birth; here, matters differ from affairs of personal constitution. Physiological aesthetics disarticulates the unity of how both Art and Concept are conceived. A single origin is replaced by an "opposition that the shared word 'art' only seems to bridge," which is "tied" to aesthetics in the same way that "generation depends on the duality of the sexes." Inasmuch as it is sexual in nature, aesthetics cannot yield "logical insight" (§ 1, I 21). To express what is at issue, mythical names—"Apollo" and "Dionysos"—are required, as well as a physiological parallel: Nietzsche presents the opposition between the visual arts, on the one hand, and acoustic-gestural arts, on the other, as corresponding to the states of dreaming and intoxication. Dreams produce entoptic images that appear to the sleeper as defined shapes; intoxication produces sounds, rhythms, and dance figures, which emerge and vanish endlessly.

Following Schopenhauer, Nietzsche assigns dream to the realm of "representation," and he assigns intoxication to a desire that he and his forebear both call "will." The senses and the arts function neither as epistemological capacities that synthesize manifolds of perception, nor—as historians of art would have it—as canvases that imitate nature, nor, finally, as physiological filters that select relevant stimuli. The priority of ecstatic states over conscious perception activates specific modes of production:

> Apollonian intoxication keeps the eye stimulated above all, so that it receives the power of vision. Painters, sculptors, epic poets are visionaries *par excellence*. In the Dionysian state, on the other hand, the entire system of affects is roused and intensified, and so it discharges its means of expression all at once. (*Twilight* IX § 10, II 996)

Senses that are endogenously stimulated give rise, in dreams, to a hallucinatory "world of seeing"; in a state of intoxication, they produce a "world of hearing" (*Untimely* IV § 5, I 389). They form, in physiological but not in technical terms, media in the modern sense. Media escape the standards of knowledge: only materiality counts—the conditions of emission and reception, and the frequency of signs. In the Apollonian state, "the extreme calm of certain intoxicating sensations" creates the illusion that the images are autonomous, detached from the body that produces them (*Nachlass* III 785); in the Dionysian state,

Nietzsche (1844–1900)

the tempo of semiosis increases until all signs are eclipsed by the nonsignifying body.

From its inception, modern aesthetics has traversed this double meaning. That is, ever since Baumgarten, who coined the term, the doctrine of the beautiful has also been a matter of the senses. Nietzsche, therefore, as Heidegger demonstrated,[5] was continuing a tradition. In contrast to his predecessors, however, he cancelled the senses' reference to knowledge/insight, which had hierarchized them and placed their point of culmination in the eye's immaterial receptivity. When sensory media operate autonomously, sight loses its priority. Translated into the opposition between the Apollonian and the Dionysian, the pairing of the beautiful and the sublime changes status. Whereas Kant had declared that the beautiful can be taken in, and that the sublime defies any such efforts, Apollonian *opsis* forms only one part of a process of sign production whose paradigm is acoustic and gestural. Nietzsche's integration of the Dionysian into the theory of art puts an end to the reign of representation.

The matter is evident in Nietzsche's relationship to Schopenhauer. The equation between music and will, on the one hand, and the coupling of the other arts and representation, on the other, had prompted the latter to affirm that music is the "representation" and "imitation of a model [*Nachbild eines Vorbildes*] that itself cannot be immediately pictured."[6] But if one seeks only incitement to dance in music, one escapes the aporias of aesthetics conceived in terms of *mimesis*: "Aesthetics is nothing but applied physiology [...]. And so I wonder: what is it that my entire body wants of music in general? For no soul exists" (*Nietzsche contra Wagner* II, II 1041). The end of representation also puts an end to aesthetic psychology.[7] Dreams and intoxication reduce the "soul" to a "spiritualized eye, ear, etc." (*Works and Letters* II 255). Thereby, both the representations that occur and the subject who experiences them disappear as well. That is to say, the two concepts that sustained aesthetic discourse in the nineteenth century vanish:

> The whole opposition (which even Schopenhauer still uses to divide the arts as if it were a criterion of value) between the subjective and the objective does not belong to aesthetics at all . . . since the subject . . . can only be conceived as the enemy of art, not as its origin. (*Birth of Tragedy* § 5, I 40)

For a discourse of the media-producing body, the subject itself becomes a mere "medium." Physiology, instead of humanizing the arts, equates their seeming masters—human beings—with "images and artistic projections" that refer to a producer within consciousness [*diesseits des Bewußtseins*]. Such decentering of the subject—which amounts to an appearance [*Scheinbild*] produced by scattered affective tensions—displaces the method of aesthetics and the site of art.

Access to arts that are produced by a subject cannot occur by means of reflection:

> Our whole understanding of art [*Kunstwissen*] is fundamentally altogether illusory because, when we know, we are not one and identical with that entity that affords itself, as the sole creator and spectator of that comedy of art, an eternal pleasure. (ibid.)

Aesthetic "knowledge" derives from fixing borders that the body has always already transgressed when, in one, it produces *and* enjoys media. Aesthetics had been defined as a judgment of taste (Kant) or as "contemplative observation" that does not seek to "call forth" works but rather "to recognize scientifically, what art is" (Hegel).[8] Nietzsche deprived such "public" conceptions of knowledge and education of their franchise [*entzieht . . . das Wort*]. He marked the displacement that, historically, led to the mediated public sphere. Not for nothing is *The Birth of Tragedy* dedicated to Wagner, whose medial *Gesamtkunstwerk* "no longer speaks the educated language of a caste" (*Untimely* IV § 10, I 428). Nor is it for nothing that talk of the Apollonian—which is "fundamentally nothing more than an image of light cast on a dark wall" (*Birth* § 9, I 55)—sounds like a theory of film avant la lettre.

Nietzsche's decentering of consciousness refers the theory of art to the relationship between culture and bodies. Unconscious production provides its historical a priori—the site from which Nietzsche and psychoanalysis advanced their claims.[9] Freud formulated, on the model of the dream, how unconscious desire and the law of culture [*Kulturgesetz*] achieve compromises in the rhetorical complexity of texts. Literary fantasy animates—with replacements and sublimations—a scenario whose only rule is the universal law declared when familial associations [*Familienverbände*] were founded. Accordingly, the Oedipus complex permits works to be inscribed within a representational scheme—that is, to be interpreted textually and in terms of content; it also enables one to analyze the author individually—that is, to locate him in the conflict between the normal and the neurotic.

Nietzsche, however—in notes he made late in life—also formulated the Apollonian on the model of intoxication. Intoxication does not yield representation—a scenario—and it rejects hermeneutics. Because the dream uncouples desire and corporeal motorics, it forms an open system: "psycho-motoric induction" (*Nachlass* III 754) carries it from body to body. Accordingly, it exceeds—and not just endopsychically—"all family life [*Familientum*] and its venerable statutes" (*Birth* § 2, I 27); it openly injures the norms of the public sphere and communication. Correlated with psychosis and conspiracy,[10] art undoes the opposition between the normal and the pathological. It proceeds from collective and forbidden bodily techniques [*Körpertechniken*]: the sexual and alcoholic practices of Dionysian revelers, the narcotic activities of initiates at Eleusis, and the St.-Vitus dances performed during medieval epidemics (*Birth* § 1, I 24). For this reason, transgression—both as the praxis and as the contents of art (*Birth* § 9, I 55–60)—belongs to the way culture itself functions.

3. On the Genealogy of Literature

Genealogy, for Nietzsche, names the process of reading history as series of prohibitions and transgressions, struggles and tensions.[11] *The Birth of Tragedy* is the result—and deployment—of combat in and about discourse. In the struggle between the Dionysian and the Apollonian, sound and image, and words and meanings, the unity of literature vanishes along with the unity of its medium. Here, Nietzsche inscribes, into discourse, the split that linguistics will later make between "signifier" and "signified." Unlike Saussure's taxonomy, however, his position "sides" with the signifiers, stressing the innumerable and suprasegmental elements of language: intonation, rhythm, speed of delivery. All that "fades away" [*verklingt*] when conceptuality emerges, literature expresses [*bringt zur Sprache*] (*Collected Works* III 229).

The medial definition of literature subverts both the signified, understood as the integral meaning of words, and the idealistic poetics of semantic "content"—which vanish in the immortal parodies of the Faustian idea and Wagnerian materialism that Nietzsche stages. Literature means taking up communication [*Kommunikationaufnahme*]; consequently, it is regulated by bodily performances [*Redemomenten*]. Nietzsche accepts the classical triad of genres—epic, lyric, and drama. He rejects, however, the dialectic between subject and object involving normalized acts of narration, self-expression, and dialogue (Hegel).[12] Instead, processes of assuming-power [*Bemächtigungsprozesse*]—whereby the Apollonian and the Dionysian take the stage in a literal sense [*das Wort im Wortsinn ergreifen*]—constitute the trinity.

In Homeric epic, the Apollonian overcame pre-Greek states of ecstasy and, on the dismembered bodies of Titans, erected an Olympus of illusion and images. Epic poetry stands as a "monument of a victory" and does not represent the naïve beginnings of literature—as Schiller held (*Birth* § 3, I 31f.). Greek lyric, in turn, heralds the return of Oriental cults. Sound conquers image, and "desire [*Begierde*]" (§ 5, I 36) runs through all registers "from the whispering of inclination to the bellowing of madness" (§ 6, I 43). Such suprasegmental registers of the voice designate neither a subject nor a name, but rather the Dionysian body.

Nietzsche does not simply assign the two genres to Apollo and Dionysos; instead, they exist in a play of difference that subverts dichotomies.[13] Epic images are bounded only because of the counterweight provided by what is measureless; conversely, lyrical *melos* finds expression only after translation of "dream scenes" (§ 5, I 37) that occur neither in images nor in concepts. When Heidegger conceives works of art as reciprocal [*gegenwendig*] relations between world and earth,[14] he continues this nondialectical tension that Nietzsche posited: works are beautiful in keeping with forces that are not reconciled so much as made to bend under a yoke.

The third genre exercises the greatest force by harnessing vision and intoxication. Nietzsche—in a move that scandalized philological contemporaries such as Ulrich von Wilamowitz-Moellendorf—derived Attic drama from the dance, music, and dithyrambs of Dionysiac revelers. His claim, that the Doric word *drama* does "not mean 'to do' at all" but rather refers to a hieratic event (*Case* § 9, II 921), contests Aristotle's definition of tragedy on every point. Drama, according to Nietzsche, is *mimesis* only in the archaic sense of the word: as dance,[15] it does not imitate action but rather *is* action. What seems to be represented—the mythos of heroes—is hallucinated by a chorus that techniques of inducing ecstasy have made productive. The duality of protagonist and antagonist incarnates the sole hero of dithyrambs: the god they celebrate is "Zagreus"—"dismembered body" (*Birth of Tragedy* § 10, I 61).

Nietzsche's genealogy of drama interprets neither content nor form; it describes the "that" of its emergence. The community of worship is producer and spectator in one: ecstasy transports its members into the chorus, and it exalts the chorus into the god whom the transfigured community in turn beholds. This circular process does away with the poetics of effect and affect [*Wirkungspoetik*] as a separate matter. Tragedy does not purify one of affects (Aristotle), nor does it ennoble them into compassion (Lessing):

> One can disprove this theory in the most cold-blooded fashion: namely by measuring, by means of a dynamometer, the effect of a tragic emotion; and one gets, as

a result, what only the absolute mendacity of a systematist can misrecognize: that tragedy is a tonic [*tonicum*]. (*Nachlass* III 829)

The experience of the audience [*Rezeption*] is a single affirmation of productivity [*Produktion*], which, in tragedy, "still includes the pleasure of destruction within itself" (*Twilight* X § 5, II 1032). Only when such pleasure requires legitimation do poetics of effect/affect arise. Their emergence—which fixes the borders between the author and the public, between hero and actor—Nietzsche describes as a scene occurring between the last tragedian and the first dialectician. The fact that the author Euripides wrote under the "censorship" of Euripides qua "first great reader" (*Greek Literature* III § 1; *Collected Works* V 218)—who stood, in turn, under the "censorship" of the spectator Socrates—subjected tragedy to a philosophy that equated "true" pleasure and knowledge, to a psychology that calculated the effects of art, and to a poetics of "content" that presumed the existence of a text. As a result, language representing concepts took over [*Das Wort als Begriff ergreift das Wort*]. Socratic dialogue and Platonic discourse put an end to tragedy.

Genealogy, then, describes the emergence *and* the decline of Greek literature. It places it within a force field where the death of tragedy coincides with the birth of science. Accordingly, Nietzschean genealogy reads the first philosophical poetics only as polemical gestures [*Kampfschriften*]. Instead of practicing science, genealogy uses tragedy methodologically to pose the "problem of science"—upon which science itself cannot reflect (*Birth*, "Attempt at Self-Criticism," § 2, I 10). Thereby, Nietzsche issues a succinct rejoinder to the "end of art" announced by Hegel: philosophical discourse, which declared the matter a truth, in fact made the end occur by announcing it.

4. The Type of the Artist and the Production of Signs

In terms of overall design, *The Birth of Tragedy* remained within the discursive space of the nineteenth century: it discussed literature in terms of the system of all the arts—its foundation [*Stiftung*] in Greece and subsequent historical evolution. When the book closed, however, Nietzsche changed his approach to genealogy and physiology. Now he started with details.

In this perspective, the distinction between "truth" and "fiction," instead of being a matter settled once and for all, turns into an endless and open struggle. "Culture" names the various means of drilling [*einfleischen*] a soul and spirit into bodies, which subject these bodies to the conditions of truthfulness and sincerity in discourse [*Wahrhaftigkeits- und Ernsthaftigkeitsbedingungen des Sprechens*] (*Genealogy* II § 1f., II 799–801). Instead of legitimating such rules by way of a theory of speech acts, however, genealogy focuses on the violence with which they are inscribed. An operative conception of writing proves necessary inasmuch as genealogy entertains inimical relations with alphabetized interiority and exteriority. The question is not what acts of speech say, but rather who programs them.[16] And the answer does not concern individuals, but rather power formations [*Herrschaftsgebilde*]. Discourses are symptoms—or as Nietzsche puts it, "semiotics"—that reveal the origin, condition, and power of speakers.

As ever, poets represent an ambivalent type. They participate in the bloody task of making bodies hear and obey. Verses provide an instrument that fixes speech mnemotechnically, steers bodies rhythmically, and guards against disturbances in channels of discourse. Inasmuch as hexameter—according to legend—saw the light of day in Delphi (*Gay Science* II § 84, II 94), poets are "valets [*Kammerdiener*]" of priestly morality (*Gay Science* I § 1, II 34). Those who actually speak when poets open their mouths are the *others* who invented the categories that—through the autonym "true" and the heteronym "mendacious"—permit power to be mastered (*Genealogy* I § 5, II 776). At the same time, however, the valets are tricksters. The fact that the rules of discourse object less to what is untrue in lies than to what is harmful in them (*Truth* § 1, III 311) admits the possibility of fiction; this, in turn, yields the pleasure of lying at the price of "interiorization" (*Nachlass* III 418).

Deception needed to be sufficiently drilled in, over the course of generations, so that, ultimately, it became a dominant instinct, an end in itself. The fabrications of poets betray their origins in the lower orders, where one survives by means of mimicry and breaks with the idea of "character."[17] Poetry derived from the pressure for "truth"—just as the flourishing of the arts in Greece stemmed from slavery (*Greek State* III 277). Literature comes into being when the "slave intellect, that master of dissimulation, is permitted to celebrate its Saturnalia." It is transgression, which speaks "in nothing but forbidden metaphors" and constitutes a kind of parody that "dashes apart, throws this way and that, and ironically reassembles" the "scaffolding [*Bretterwerk*] of concepts" (*Truth* § 2, III 321). In the slave—whose work determines culture, and whose transgressions determine its festivities—the artist has his model. As much is evident in the Greek word *techne*, which makes no distinction between art and craft [*Handwerk*] (*Greek State* III 277).

Modern theories of production—which celebrate the "dignity of labor" in economic terms (ibid.) and the autarchy of works in aesthetic ones—only mask this slavery. "Taken into service" by an alliance between the state and sciences (*Untimely* III § 6, I 330), literature became "propaganda for reforms of a social and political nature" from the eighteenth century on. "The Author" and his "Oeuvre"—which has the task of "generating interest" (*Nachlass* III 509) for his person—both enact and *are* the educational system that invented the interpretive essay (*Future* II, III 201), as well as the public that consumed [*auffängt*] literary works in a critical and historical fashion (*Untimely Meditations* II § 5, I 242). Nietzsche opposes the "fabrications" (e.g., "Author" [*Nietzsche contra Wagner* IX § 1, II 1056f.]) from which modern literature arose—and not just the way that literature has been viewed in terms of social milieu (as was the case for Sainte-Beuve and Hippolyte Taine).[18]

Accordingly, he describes the way literature functions in a "history of 'education' [*Bildung*]" that is, in fact, a "history of narcotics" (*Gay Science* II § 86, II 96). In modern times, two complementary social types have emerged ensuring that work and leisure will remain separate: the Romantic artist, who produces sedatives instead of stimulants, and the philologist, who teaches the young how to "cram—the first precondition for robotically performing duties in the future (as a civil servant, husband, bureaucratic slave, newspaper reader, soldier)" (*Nachlass* III 630).

The poetics of "authorship" and "oeuvre" possess an erotic charge: only a sense of shame makes them conceal production as if it were the act of conception itself (*Greek State* III 277). In fact, "only one kind of power" exists, and it "is one and the same in art and the sexual act" (*Nachlass* III 924). By introducing sexuality, Nietzsche banishes *theologumena* from the aesthetics of production. Art is not *creatio ex nihilo*, but rather erotic invention. For this reason, it is anything but imaginary:

> We would err if we rested at its power to lie: it does more than simply make images [*imaginieren*]: it displaces values themselves. And it does not just move the "feeling" of value: this lover is worthier and stronger. Among animals, such a condition brings forth new weapons, pigments, colors, and forms: above all, new movements, new rhythms, new calls, and seductions. It is no different among human beings. (*Nachlass* III 752)

Eroticism and art are not restricted to being vehicles of expression or aims; their "power of transformation" (ibid.) produces expression and objectives in the first place. Bataille, a reader of Nietzsche, coined the term "expenditure" for this process/event. A "generous [*abgebende*] and overflowing fullness of bodily vigor" constitutes the "aesthetic state" (*Nachlass* III 535), which is, "so to speak, bred into a 'person' in the artist" (*Nachlass* III 715).

The positivity of creation [*Schaffen*]—which refuses to be reduced to fantasy—occurs as semiosis. The materiality of signs links erotology and medial aesthetics. If signs are not based on signifieds or referents, nothing and no one prescribes what all can be a sign or the sign of a sign. The artist stands in for this unlimitedness. His vigor involves the "extreme acuity of certain senses: so that they understand—and create—a wholly different language of signs, the same one that seems to be associated with certain nervous illnesses" (*Nachlass* III 716). Accordingly, all the arts are languages, and all languages are media that are their own message, since only the "excessive wealth of the means of communication [*Überreichtum an Mitteilungsmitteln*]" defines them. Through two complementary artistic capacities—positing signs in what is, as yet, uninscribed [*im Zeichenlosen*] and reading them there—"languages have their source [*Entstehungsherd*]: languages of sound, as well as languages of gestures and gazes" (*Nachlass* III 753). Being an artist is a function of physiological force, for force (i.e., "will-to-power") involves working with differences and producing them, "where otherwise, for a normal person, all distinction is lacking" (*Nachlass* III 784). Distinction, in turn, represents the necessary, determinate quality of a data set [*Zeichenmenge*] to signify when signs no longer simply represent something else. During the age when physiologists (Helmholtz, Fechner) identified threshold values for sensory perception, Nietzsche described the production of meaning [*Sinnenproduktion*] in terms of differences and intensities.

For artists, both creating [*setzen*] and reading signs represent unavoidable and coordinated matters. They cannot *not* communicate, and they cannot *not* interpret: "Wanting to say all that is capable of signifying [*Das Redenwollen alles dessen, was Zeichen zu geben weiß*]" and "needing to imitate, which already occurs when signs are sensed and represented [*das Nachahmen-Müssen, das einen Zustand nach Zeichen schon errät und darstellt*]" (*Nachlass* III 716), generate a positive feedback loop between affects and signs. Such is the effect of art:

> All distinct matters, all nuances, insofar as they recall the extreme heightening of force that intoxication produces, retroactively awaken this feeling of intoxication—the effect of works of art is the *arousal of the state of artistic creation*, i.e., intoxication. (*Nachlass* III 784)

Instead of idealistically mediating production, works, and reception through consciousness, Nietzsche short-circuits bodies and signs. Moreover, he holds that encounters with art bypass meaning and understanding and follow the signs themselves: "One hears with the muscles, one even reads with muscles" (*Nachlass* III 754). Literature constitutes a "mosaic of words, where every word, as a sound, as a locus, as a concept, exudes its power to the right and the left over the whole"; therefore, it also forms an economy where a "minimal extent and number of signs" achieves a "maximum ... of energy" (*Twilight* X § 1, II 1027).

Like the works they produce, the arts themselves are correlations of signs. First, they exist only because of semiotic processes that have preceded them. Contra theories declaring that the ends identified in rule-based poetics constitute the actual origins of literature, Nietzsche objects: "Every mature art has a fullness of convention at its basis: to this extent it is a language" (*Nachlass* III 754). Secondly, different arts—for example, lyric and music—are correlated by acts of instituting signs [*Zeichenstiftungen*]. Whereas *The Birth of Tragedy* called music an "immediate language" that "speaks directly to interiority and comes from interiority," genealogy holds that it was music's "ancient connection with poetry" that inscribed "so much symbolism" in the first place (*Human* I § 215, I 573). As a corollary, the genealogical perspective holds that ancient, quantitative verse was founded in the optical medium of dance steps—whereas modern, qualitative verse is based in signified content (letter to C. Fuchs, at the end of August 1888; III 1314f.). The arts, as they are conventionally understood, are historically variable and conventional connections between bodies of signs without any "immediacy [*Unmittelbarkeit*]" (*Human* II 2 § 168, I 940).

Finally, the production of signs collapses the cultural distinction between producers and consumers [*Rezipienten*]. When artists layer and connect semiotic systems, they act as interpreters; conversely, interpreters act as artists, too. Failing an *Urtext* to which all interpretations would have to refer, "forcing [*Vergewaltigen*], adjusting, abbreviating, omitting, filling-in, inventing [*Ausdichten*], falsifying"—all different terms that parody the notion of essence—become "the *essence* of all interpretation" (*Genealogy of Morals* III § 24, II 890). "That unspeakably more lies in what the things are called than in what they are" demonstrates the identity of interpreters and "creators" (*Gay Science* II § 58, II 77f.). Thereby, the very notions of tradition and oeuvre undergo a change. For Nietzsche, the erstwhile philologist, literary tradition amounts to a series of misunderstandings and falsifications,[19] and interpretation yields a strategy that—like all strategies—relies on two tactics: disciplining subjects [*Untergebene*] and combating opponents (*Daybreak* I § 84, I 1067f.).[20]

Such a subversive interpretation of the act of interpretation has a recursive impact on the praxis of the newly conceived philosopher. Healed of the philological deficiency of his ancestors, he does not equate understanding with actions performed by a subject because "interpretation itself is a form of the will-to-power" (*Nachlass* III 487). The answer to the question, "Who is interpreting [*wer legt aus*]?" must be: "our affects" (*Nachlass* III 480). For all that, however, Nietzsche gauges affects only in terms of the intensity and complexity that their semiotic practices create. Their measure is aesthetic. Art-creating affect—which makes "existence eternally justified" (*Birth* § 5, I 40)—operates without reducing complexity: "To depict frightful and questionable things is itself already an instinct of power and the majesty of the artist: he does not fear them.... Art affirms [*bejaht*]" (*Nachlass* III 784). If, then, the difference between pleasure and pain exists without "fixed norms" (*Nachlass* III 873), pleasure turns into a variable that stands open for inventing and naming [*Bezeichnen*]. Alternately—as the difference between grades of minimal displeasure [*minimaler Unlustreize*]—it constitutes a sign itself. With that, art and pleasure escape the aporias of aesthetic systems that, up to Adorno, have claimed to be able to identify what pleasure is and can only accept fiction, cruelty, and death in dialectical mediation. Art takes its pleasure [*hat ihre Lust*] in the "that" of "showing." Tragic pleasure admits no negation and no opposite. It lies in the creation of signs itself, which never does not occur.

3   Lullaby of Birdland

For Mimi

1.

When we entered the highest chamber, he said: "Long ago, with my servant, I spent eight summer days in this room, and I wrote a little poem on the wall. I should like to see the poem again. If the day is noted beneath, when this occurred, please be so good as to record it for me." Straightaway I led him to the southern window in the room; there, on the left, it stood:

Über allen Gipfeln	*Over all the peaks*
Ist Ruh,	*It is calm,*
In allen Wipfeln	*In all the tree-tops*
Spürest du	*You feel*
Kaum einen Hauch;	*Hardly a breath;*
Die Vögelein schweigen im Walde.	*Birds are quiet in the woods:*
Warte nur, balde	*Just wait; soon*
Ruhest du auch.	*You will rest, too.*
D. 7 September 1780, Goethe	

Goethe read these few lines, and tears flowed down his cheeks. Very slowly, he drew his snow-white handkerchief from his dark brown coat, dried his tears, and spoke in a gentle, mournful tone: "Yes, just wait, soon you, too, will rest!" He fell quiet for half a minute, looked once more through the window into the gloomy spruce forest. Then, he turned to me and said: "Well, let us go!"

Thus Christian Mahr reports how Goethe, on the evening before his last birthday, visited, one more time, the hunting lodge on the Kickelhahn mountain near Ilmenau.[1] The scene is not just historical. It made history, too—literary history. Here an author, near the end of his life, ceremoniously archives his beginnings. To the letter, Goethe follows the rules that in the years around 1800 produced the new, author-based kind of text called "literature."

These rules were also formulated in Goethe's *Bildungsroman*.[2] There one reads of Wilhelm Meister's relationship to the poems of his youth:

> Up to now he had carefully preserved everything that had flowed from his pen since his mind began to develop. His writings were tied up in bundles which he had hoped to take with him on his journey. [. . .]
>
> When we open a letter that we once wrote and sealed on a particular occasion but which never reached the friend it was sent to, and was returned to us, we have a strange feeling as we break the seal, our own seal, and converse with our different self as with a third person. Just such a feeling it was that gripped our hero.³

In the same spirit—as the archivist of his own authorship—the eighty-one-year-old poet ascended the Kickelhahn. "The old inscription was recognized,"⁴ Goethe notes in his journal apropos of the last journey he has made. The journey fetches back messages to the sender—messages that, unlike letters, are not fulfilled when they reach their addressees. They are "literature" in the new sense of the word, and that means that they remain the property of their author forevermore.⁵ That said, what is new is the division of labor. Whereas the aspiring poet Wilhelm Meister—in order to establish [*statuieren*] his authorship "since his mind began to develop"—must gather and order "the papers in chronological sequence,"⁶ Goethe, the old man, can build on the goodness of a geologist: Mahr takes note for him when a text was written—whose youthful author had already dated and signed it in anticipation.

However, something strange occurs. Just as the archivist Meister experiences a "strange feeling," the autobiographer and "clerk [*Kanzlist*] of his own interior"⁷—which Goethe has become in his old age—experiences a stream of tears that puts an end to the literary anagnorisis. Once more, rereading one's own writings becomes a conversation "with our different self." The reader lends his voice to what is written; he repeats and affirms what "Wanderer's Night Song" says. Thereby, he himself enters the chain of beings to whom the verses promise rest: first, the mountains and birds, then the writer, and finally, after fifty-one years, the reader, "too." Through the flow of tears, the archiving of the text becomes its return: everything—the view of the summits and the spruce forest, the self-address, the silence at the end—it all happens once more, just as the faded pencil lines at the southern window have described, and prescribed.

No one cries at his own words—if only because there is no such thing as "words of one's own." Only when someone else has written them does one read and cry. What literary scholars call the "lyrical I" does not exist at all. If the reader is promised calm, then this occurs for a "you." Fifty-one years ago, for the writer, it was no different.

The statement, "I rest," is a pragmatic paradox. No mouth can voice it, because sleep and death exclude speaking—just as speaking excludes sleep and death. The only exception to the rule is no exception: when the magic of animal magnetism enables the dead Mister Valdemar—in Poe's tale of the same name—to hold on to language and to answer the question about his condition by declaring, "I am dead," he promptly dissolves, in the blink of an eye, into a stinking mass "for which no language has a name."[8] For this mass, the word "corpse" is still a euphemism.

Absence occurs only in speech, but no speech occurs in absence. The verses on the Kickelhahn speak of this law. They are discourse about the site that excludes speech—and the site that speech excludes. "Wanderer's Night Song" does not mean that at the poem's end, "even the most restless being—man—must rest."[9] Instead, it means, simply, and without humanist add-ons, that the end of spoken speaking beings is at hand [*daß es mit den gesprochenen sprechenden Wesen zu Ende geht*]. The one who says that mountains, trees, and animals are mute will fall silent himself—and that means: he will become one with them.

Because it is discourse about discourse and discourse about the end of discourse, the text refers all its parameters to speaking—both to the speaker and to what is spoken about. Naming a final "breath" in the treetops means turning the work into a metaphor of respiration and voice, which are the Real of language and make it the sibling of sleep. To say that crepuscular birds "are quiet [*schweigen*]" means hearing their song as discourse [*ihr Singen wie ein Reden hören*], because (as Heidegger observed) "only in true speech is it possible actually to say nothing [*nur im echten Reden ist eigentliches Schweigen möglich*]."[10] Therefore, the poem invokes an acoustic twilight in which the voices of nature and speaking, sounds [*Laute*] and words, become indistinguishable. The last word—a vanishing [*verhallend*] "too"—explicitly puts an end to the difference between them. Sounds and speaking melt together in the moment when both cease. At its end, the poem performs what it speaks about: what is uttered and the utterance coincide. Any speech that dreams away its difference from sound(s) must end.

Therefore, it is Another who speaks. Where the text passes from the sounds of nature to the speaker who hears them, there appears, in his stead, a subject of utterance [*Äußerung*] that is a "you" for the implicit speaker-I. A nameless voice enters into play, without which the poem could not exist at all: the voice of a promise [*Zuspruch*], which calls the unspeakable end of speech a "rest."

Emil Staiger once observed that one would destroy "Wanderer's Night Song" by replacing "you feel" with "you notice."[11] One would destroy it even more effectively by replacing "you" with "I." For the promise of the Other—a fact to which Goethe's tears bear witness—is the discursive event in "Wanderer's Night Song." Because no one can perform the paradoxical speech act of naming his own absence in absence, spoken beings are absolutely reliant on alien discourses [*sind die gesprochenen Wesen auf fremde Reden schlechthin angewiesen*]. The law holds all the more for words such as "rest" [*Ruhe*], "sleep," and "death": they derive from the discourse of the Other. Neither deixis nor introspection could ever have dreamed them up.

If this oneiric (and therefore universal) law governs the absent party—that he "was already dead and just did not know it yet"[12]—then words, which is to say the appearance [*der Schein*] of knowledge about this law, belong to the Other alone. The nameless voice that surfaces at the end of "Wanderer's Night Song" articulates the unarticulated and speaks the unspeakable—and not because it knows but simply because it says anything at all. That falling-silent will be rest and not disappearance [*Vergehen*] (which is what the verses promise), whether or not a return will occur (the question posed by the tears), and the affirmation that the one who rests will be the same as the one who wakes (the comfort offered by the "you")—the words that are voiced can say all this only because it was once promised [*zugesprochen*] to them. Accordingly, the simple act of utterance grants the nameless voice the same calm [*Beruhigung*] of which it speaks. A guarantee that itself has no guarantee—because there is no Other of the Other[13]—ferries stupefied bodies over Absence.

To be sure, everyone takes up words [*jeder nimmt die Wörter in den Mund*] that name the body and its absences. The same hand also added the last two lines to "Wanderer's Night Song." But since "the subject . . . receives from the Other even the message he himself sends,"[14] they are repeated [*nachgesprochen*] and derive their power only from repetition [*Nachgesprochensein*]. There is further evidence of this occurrence in Goethe's works. Werther, in love for Lotte, speaks in the same way that "Wanderer's Night Song" does when it follows the words of a nameless voice:

> Yesterday as I was leaving she gave me her hand and said, "Adieu, dear Werther."—Dear Werther! It was the first time she had ever called me dear and it pierced me through and through. I repeated it to myself a hundred times and last night, getting ready for bed and muttering all sorts of things, I suddenly said, "Goodnight *dear Werther*"—and laughed out loud to hear it.[15]

This nocturnal exchange between an I and its double derives all its power from a promise. It rests on the symbolic gift of the Other, who alone can bid goodnight. The "third person" with whom Wilhelm Meister compares "our different self," then, is anything but a simple allegory; it rules the "self" itself—the very self that turns out to be imaginary. It is just that difficult to find rest on one's own. Only the pledge of Lotte's words makes Werther "dear" to himself. He—who is not lying with her body—falls asleep to the echo of her speech instead. In dreams, the hypnagogic discourse of the Other fulfills the desire of a love that was always already the wish to be loved. Werther's love for Lotte, after all, is "constituted not by [his] biological dependence, but by [his] dependence on her love, that is, by [his] desire for her desire."[16] And precisely on this point—as the novel declares explicitly—Lotte is the "image [*Ebenbild*]" of her mother.[17] The lonesome wanderer on the Kickelhahn and the lonesome sleeper in Wahlheim: both experience a literal pacification [*Stillung*] when the hypnagogic voice of the Mother returns. They get what Werther himself says he "needed" from the beginning: a "lullaby."[18]

2.

> Indeed, the children's nurses know the virtues of lilies in the nursery, heavenly theriacs, sedatives [*Requies Nicolai*], garlic potations, and opium—and when nothing else is to be done, humming and rocking.

The bitter scorn of an anonymous reform pedagogue says it: Western lands did not always employ such gentle methods of inducing sleep as "Wanderer's Night Song" takes for granted. At the end of the eighteenth century, when the gaze of the new anthropological sciences [*Menschenwissenschaften*] discovered where infants lay, it saw that naked violence surrounded them. Nurses and attendants calmed the cries of children by means that no friend of mankind could approve. Venerable methods of making children sleepy and putting them to bed included drugs, such as spook about in disreputable medicines (which is to say, once they shed their masks, opium), and a method called *Steckwickeln*, which involved placing the child on a board the length of its body, swaddling it tightly, and making a kind of mummy.[19]

Finally, methods involved the cradle, of which another reformer declared:

> Far more mistaken is the generally prevalent custom among the peasantry ... to force the children to sleep: one seeks to achieve this through constant, reckless rocking, by swinging and shaking, through bearing up and down and vigorous singing; these methods are sooner suited ... to produce, at most, a passing daze, which gives the first inducement to stupidity and idiocy.[20]

Lullaby of Birdland

Received means of putting children to bed and getting them to sleep did not acknowledge the soul of the child. The infant was treated as a body like any other. Practices did not take into account a relationship that, ever since the Age of Goethe, has given rise to innumerable celebrations and hymns: the interaction between mother and child. This is why, in the eyes of the pedagogical and psychological reformers who at the end of the eighteenth century discovered the infant as the main task of all cultural work,[21] prevalent methods all led "to stupidity and idiocy"—such are the attributes of a simple [*schlicht*] body when the gaze of psychologists assesses it. The children of nascent European states needed a soul, and so reform was a straightforward matter of declaring mothers "irreplaceable"[22] and putting them where for centuries nursemaids and attendants had stood.

With that, all the methods of pacification and putting-to-rest changed. The cradle fell into disuse. Twenty-five years after being lulled to sleep in "an oversized cradle of walnut inlaid with ivory and ebony," Goethe shared with his mother "that such rocking cribs have fallen out of fashion altogether"[23] because of the new freedom accorded to children. A great campaign of enlightenment against swaddling babies was initiated. Drugs were replaced by the gentle voice of the Mother.

A mother's gentle voice is a multipurpose instrument. It "records over" [*überspielt*] what is already there and puts harmony where differences otherwise tear at Occidental schemes of knowledge—the senses and the intellect, instinct and art, bodily discipline and cultivation of the soul [*Körpertechniken und Seelenherstellung*]. Johann Heinrich Pestalozzi, who not only founded the modern elementary school system [*Volksschulwesen*] but also "elevated" the "relationship between mother and child to the prototype of pedagogical relations in general,"[24] explicitly declared as much. Inasmuch as the new rules of infant care are taken to heart—that is, nursemaids and attendants are excluded—"the child first hears" (and hears only) the voice of its mother[25]: "The first feeling of a sound's relationship with the object that produced it is the feeling of connection between your voice and you, Mother!" (317).

This rule of an originary and indelible inscription is to be taken to heart by the mother and promptly applied:

> Bring forth tones yourself, clap, strike, stamp, speak, sing—in short, make sounds so that [the child] is happy and clings to you, so that it loves you; may lofty grace flow from your lips: please him through your voice, too, as no one else does, and do not believe that you need any particular art for this. The sweetness of the speech that flows from your heart is infinitely more valuable for the education [*Bildung*] of your

child than any art of song—wherein, in any event, you will always stand behind the nightingale. (319f.)

First of all, the maternal voice—whose "grace" [*Anmuth*] and sweetness [*Lieblichkeit*] inspire love in return—stands in marked contrast to the "vigorous singing" of nurses and maids. It has an immediate effect on the body of the child. This voice *is* nature and yields nature [*ist Natur und geht auf Natur*]. For this reason, the nightingale alone is its standard—and it offers the model for all birdsong:

> Mother! with whom I speak—as soon as the child recognizes your voice as yours, then the circle of what it understands enlarges, and it gradually recognizes the connection between the song of birds and the bird, barking and the dog, whirring and the spinning wheel. (318)

At the same time, however, a mother's voice represents that unique and paradoxical aspect of nature, which by itself and without undergoing any alteration, can make its own transition to art, education, and culture:

> Your instinct does not force you just to babble [*vorlallen*] notes to cheer and distract [your child]; this same instinct compels you to speak before, and to, him, and to pronounce words to him, even though you know that he does not yet associate a thought [*Begriff*] with them. (268)

Instinct makes the mother speak; that is, it makes her pass beyond instincts. Bodily desire [*Körperlust*] makes the infant hear; that is, it induces the child to receive concepts that will go beyond its body and articulate it. In this way, reformers' message of *antiphysis* glides miraculously from instinct to instinct. All violence seems to have been excluded from the acquisition of language. Indeed, all efforts aim for precisely such banishment. The words that motherly instinct instills in the instinct of the child represent the exact opposite of inherited educational practices [*überliefertes Bildungsgut*]. Whereas school makes the child "parrot whole sentences to itself and to the teacher in a language that it has not learned and that is not at all the language it speaks daily" (321), maternal instruction begins with what is nearest to the child and altogether everyday—with the field of perception and the child's own body. Pestalozzi's *Book of Mothers*, his guide for learning to notice children and speaking to them, begins by "teaching the mother to show her child the outer parts of his body and to name them."[26] Articulation is linked to deixis, which takes away all power from the sovereign willfulness [*Willkür*] with which a given culture articulates—that is to say, dismembers—bodies.[27]

But where no deixis reaches—in the field of symbolic relations that discloses objects and their demonstrability in the first place[28]—the voice that rouses both love and love-in-return remains, even after the acquisition of language and forevermore, a pure *melos* that refers to nothing yet means everything: Love itself. In this capacity, the maternal voice proves altogether irreplaceable. Only because it has heard the maternal pledge can the child find points of entry and names for all the absences that no deixis can reveal[29]—and without which it cannot achieve bourgeois individuality. "Without belief in" the Mother, there is "no belief in human nature"—"no belief in God, much less belief in the likeness of God and Man, Jesus Christ" (311). Therefore, the mother's gentle voice functions as the perfect replacement for the opium that nurses used to dispense: whoever has heard it once is addicted for life.

Accordingly, the new technology for instilling a soul into children involved disclosing a field in which speech and the sounds of nature transformed into each other without remainder [*ungeschieden*]. All at once, the fact that the act of hearing is infantile in a literal sense provided the basic precondition for theories of language and practices of language transmission [*Sprachüberlieferungspraktiken*].[30] Only the voice of the mother could fulfill this role, because it is half "breath" [*Athem*], through which the child learns to "feel" [*Empfinden*],[31] and half articulation, through which it acquires language. In this way, a sensibility emerges that excludes "stupidity," and a capacity for articulation that excludes "idiocy." "Respiratory erogeneity" (and its partial object, voice) is not "little studied"[32] at all—instead, it is "implemented" [*eingesetzt*] in a wholly explicit manner.

Because it begins in the intermediate space between nature and culture, breath and language, and sound and speech, culturization through the maternal voice remained equally far removed from the bodily interventions of nursemaids and from the experts [*Verständigen*] at schools—just as the knowledge instilled by teachers and domestic attendants fell subject, in a single breath, to Pestalozzi's critique. The drugs and swaddling of the nursemaids did not turn the crying of children (whom they pacified either by ruse or by force) into words—they simply "anesthetized" (Pfeufer). The grammars and encyclopedias that old-fashioned schools, addicted to rote copying, had drummed into students had always already severed the relationship to voices and crying—they simply instructed. In contrast, the historical innovation of the maternal voice established a relationship between the Real and the Symbolic of language, which freed the Imaginary itself: the Soul.

3.

> The first concern of nature for the weakness of my sex is concern for its rest. The first maternal care, the beginning of all motherly worries and the center of them all, is concern to calm the infant. Long, long before the mother spends even a moment teaching the child any kind of understanding [*Einsicht*], she spends whole days in movement and interrupts her h[oly] sleep for many nights to assure his rest. Long, long before she seeks traces of reason in him, she grabs for [*haschet nach*] traces of his love. Long, long before she thinks of directing the use of his senses, she already forms the same with great art into aptitudes and habits that assure his rest [*Ruhe*]. In this way, exalted Nature shows with all her might: *rest* is what is most necessary for the human child.

So begins Pestalozzi's fragment on the fundaments of education.[33] The "rest" that the mother assures and offers by providing for all that the infant lacks—in every sense of the word—is the same as the needs for which nature provides to every human being. It is no wonder that Goethe recognized the old inscription in and of nature precisely when celebrating his birthday: its message repeats the beginning itself—"the beginning of all motherly worries," which coincides with the birth of the child. It is also no wonder, then, that he repeated the end of the verses "in a gentle, mournful tone": their *melos* repeats the delicate voice that both promises rest in what has been uttered [*im Geäusserten*] and already *is* this rest in the act of utterance [*im Äussern*].

The new "fundaments of education" that reform pedagogy and reform psychology laid also represent the fundaments of the new poetic lyricism that, around 1800, found its voice—in the most literal sense. Lyric abandoned the grounding [*Boden*] of writing and, as the echo and resonance of an originary voice, became a voice in its own right. It forgot received regulations of language based on literacy, which had bound poems to the rhetorical arts, to the storehouse of erudition [*Tresor der Wissens*], and to poetic norms [*Normen der Verslehre*]. No traditional metrical scheme governs the lines that Goethe wrote on the Kickelhahn, and no *topic* underlies and authenticates their equation of sleep and death. The peace the poem announces may be the same as on any other night or, alternately, the last one ever—only the false profundity of literary scholarship ignores the first reading (in order to have yet another text by Goethe on the "final things"). The alternative corresponds exactly to the charge placed on the maternal voice: of mediating between all forms of absence, both everyday and religious, through the fullness of its presence.

"Wanderer's Night Song" is bare of all knowledge: just as the mother brings language nearer to her child through birdsong and the sounds of nature, the poem concerns only the immediate environment—where, once more, the voices of birds speak. For this reason, its lyricism escapes conceptuality and passes over to where language and the self-expression of nature become one and the same.

An acoustic twilight encompasses and defines the new poems around 1800. Goethe:

Rausche, Fluß, das Tal entlang,	Rush, river, down the valley
Ohne Rast und Ruh,	Without rest or repose,
Rausche, flüstre meinem Sang	Rush, whisper my song
Melodien zu.	Its melodies.[34]

Eichendorff:

O wunderbarer Nachtgesang:	O wondrous song of night:
Von fern im Land der Ströme Gang,	Course, from afar in the land of streams,
Leis Schauern in den dunklen Bäumen.	Soft shuddering in the dark trees.[35]

Brentano, begging for a "whispering lullaby":

Singt ein Lied so süß gelinde,	Sing a song as sweetly mild,
Wie die Quellen auf den Kieseln,	As the wells over the pebbles,
Wie die Bienen um die Linde	As the bees around the linden
Summen, murmeln, flüstern, rieseln.	Hum, murmur, whisper, bubble.[36]

And finally, the lines that betrayed the secret of all the murmuring and rushing sounds of nature to prosaic paper in the first place:

Da lieg ich nun des Nachts im Wald.	There, by night in the forest I lie.
Ein Wächterhorn von ferne schallt,	From afar, a watchman's horn sounds,
Das Rauschen, das den Wald durchzieht,	The rushing blowing through the woods
Klingt wie der Mutter Wiegenlied.	Seems like a mother's lullaby.

The texts declare it themselves—and more precisely than it ever could be said by the literary scholars and philosophers who owe them the insight that the new lyrical "language in its distance from meaning . . . imitates *Rauschen* and solitary nature."[37] The imitation of the sounds of nature is the imitation of the single discourse [*Rede*] that, ever since, has been called both "nature" and "speech" simultaneously, for it induces calm, and nothing else [*schlechthin beruhigt*].

"Der Mutter Wiegenlied" (The Mother's Lullaby) represents the matrix of

Romantic lyric, period—and therefore the matrix of Goethe's "Night Song" as well. Despite all doubts that scholars of German letters have expressed, a Silesian lullaby—as Adalbert Kuhn already observed in 1843—is both the source and the plaintext [*Klartext*] of Goethe's most famous poem:

Schlaf, Kindlein, balde!	Sleep, little one, soon!
Die Vögelein fliegen im Walde;	The birds now fly in the woods;
Sie fliegen den Wald wohl auf und nieder,	They are flying here and there,
Und bringen dem Kindlein die Ruh' bald wieder.	And soon will bring rest back to you,
Schlaf, Kindlein, schlaf!	Sleep, little one, sleep![38]

Literary scholars avoid these unambiguous words—which eliminate all doubt as to who is speaking in Goethe's verses—in two ways running in opposite directions. The discovery of the relationship between Mother and Child disappears either in the supposedly timeless truth of the "soul" or in a history of historical drama [*Haupt- und Staatsaktionen*].

In order to explain that around 1800 lullabies suddenly represented a fitting subject for literature, an essay entitled "Zum Erlebnisgehalt des Wiegenliedes" (The Experiential Content of the Lullaby) simply invokes "the primordial unity of mother and child as the origin of all yearning, first and last, and thereby the origin of all religious and artistic creation."[39] Such psychological metaphysics denies the fact that the Middle Ages and Renaissance did not produce literary lullabies—as well as the fact that even in the eighteenth century the uncommon word *Wiegenlied* could (in keeping with venerable rhetorical tradition) mean a poem dedicated to parents on the occasion of a child's birth.[40] This occurs through reference to old Christian cradlesongs [*Krippenlieder*][41]—whose replacement by literary lullabies is, in fact, the matter that needs explaining. Conversely, discussions such as the one presented in an essay simply entitled "Critical Reading" (which expressly calls for "Wanderer's Night Song" to be viewed sociohistorically) invoke a certain ill mood prevailing between Goethe and Carl August, Grand Duke of Saxe-Weimar-Eisenach, three or four days before the poem was composed. Accordingly, "Wanderer's Night Song" is supposed to express Goethe's "doubt at his success in shaping his life" along the lines of "the Enlightenment ideals in Weimar society."[42]

So close to each other do the passion for political enlightenment and the passion for ignorance dwell among academics. Whether they recur to timeless givens of the soul or to decisive social and historical events [*Haupt- und Staatsaktionen*], both ways of failing to understand [*Verkennungen*] the lullaby are just as comforting and deceptive as the poem is itself: they close ears and eyes to the fact that speaking itself is pure exteriority. "[A] symbolic world . . . with this same world . . . the machine is built."[43]

The machines of speech do not just have a history—they also make history. The psychological-pedagogical techniques of culturization that were bestowed on Central Europe around 1800 changed the parameters of literary influence and effect [*Wirkung*]. When lyric became "The Mother's Lullaby," it was no longer limited to the speech acts performed by poems composed according to the *ars poetica* of old: celebration, lament, praise, and delight. Such forms of literature all presume the capacity for articulation of speakers and hearers. "The Mother's Lullaby," on the other hand, undermines this very precondition. It has effects on levels that concern the mute body; its parameters are *melos*, sound, and breath. Speech issues only to extinguish in an infinite paradox. The definition that Gotthilf Heinrich Schubert gave to lyrical meter in 1814 is as comprehensive as it is unheard of today: its effect is "calming, partially soporific, and conducts the soul into the realm of dark feelings and dreaming."[44]

This definition is made to order, as it were, as a commentary on "Wanderer's Night Song." Therefore, it could be integrated without any problem at all into German literary history. One hundred fifty years later, in an exemplary analysis of Brentano's "Der Spinnerin Lied" (The Spinner's Song), Richard Alewyn wrote that the poem "glides effortlessly into the ear, so free of violence that one's attention is sooner lulled to sleep than exerted." All the same, lyric does not exercise such wonders for the form-immanent reason that, to Alewyn's joy, it combines "simple words and short sentences strung together without syntactic extravagance" and does not "exert the intellect or feeling."[45] Instead, it does so because "Der Spinnerin Lied" is part of Brentano's *Chronika eines fahrenden Schülers*: here a mother sings to her son in order to make the speechless infant sleep and, by this means, establish [*stiften*] a childhood memory that will prove unforgettable because it occurs so early on.[46] In altogether literal fashion, the lyricism of 1800 emerged from the short circuit between the mouths of mothers and the ears of children.

Unlike professional exegetes, lovers and technicians first noticed this connection. Bettina Brentano, more enamored of Goethe than any other reader, wrote of his poems' effects:

> It is Goethe who sends these lightning flashes through me, then looks at me healingly as though my sorrow pained him, swathing my soul again in the soft wrappings from which it had freed itself, that it may find peace in slumber, and slumbering

thrive, in the glory of night, in the sun; and to the air by which I am rocked, he confides me. I do not want to feel differently in regard to him than in this poem; it is the cradle in which I feel myself surrounded by sympathy, nearer his care, catching the tears of his love, on which I thrive.[47]

Then, finally, there is Wagner, who according to Nietzsche brewed up a whole array of opiates and sedatives of the will—that is, transferred all the imaginary effects of Romantic poetry into the Real of technology. Wagner wrote of his *Lied* "Dors mon enfant":

> It was so successful that, when I had tried it over softly several times on the piano, my wife, who was in bed, called out to me that it was heavenly for sending one to sleep.[48]

If the hypnagogic voice of the mother provides the model for the new poetry and its effects, these poems—notwithstanding even the most up-to-date theories of lyric—do not involve matters of expression. The parameter of "expression" refers an utterance to its speaker. Hypnagogic effects, in contrast, occur in the addressee.

Rarely does Lacan's law—that "style," on no count whatsoever, is "the man," but rather "the man one addresses"[49]—hold as strictly as it does for the literary lullaby. In its distance from meaning [*Bedeutungsferne*], the lullaby sounds for an infant that harks yet does not hear. The idealistic philosophical aesthetics of the Age of Goethe—which considered lyric to be self-expression—discerned the matter as little as do the linguistic aesthetics of our own time, which declare lyric to be "egocentric" or "inner" speech.[50] Both definitions remain stuck in the discursive space opened by the invention of the "soul." It is the characteristic ambiguity of the psychogenic maternal voice that it seems to introduce, simply through making a sound, the child into speech that can be celebrated as the child's own discourse—and, in its highest form, as the lyric of a genius. From this ruse of history derives the interiority that seems to speak in the poems—an interiority that theories ascribe to them yet again.

The verses scribbled on the wooden wall of the Kickelhahn inaugurated a new epoch of lyric because they spoke—and speak—simultaneously of the end and of the origins of speech. Of the end: after the breath in the treetops has ceased and the birds fall silent, breathing itself will rest—the breath that is articulated respiration, that is, the voice. Of the origins: the lines that do without a title or a marker of genre record, instead, the day of their creation and the name of their creator—and so they stand apart, on the strength of this signature alone, once and for all, from the fading murmur without an author that they are, and they stand apart from the fading murmur without an addressee that they create [*dichten*]. Discourse about the end of discourse and discourse about the origin of discourse are just that complementary.[51]

The phantasm of the author as the master from whom discourse is to be born and to whom it should forever belong emerged at the same point in history when the "Language of Mankind" [*Menschensprache*] became the "Mother Tongue" [*Muttersprache*].[52] Goethe—to celebrate his last birthday—fetches back, into discourse and presence, the same lines that speak of how sounding and speaking end in calm and absence. He himself, as the exemplary Author, does what the godhead does in *Torquato Tasso*: he makes speech possible even where "man falls silent [*der Mensch verstummt*]." The verses on the Kickelhahn—as the product and document of authorial biography and chronology, such as it has existed since Goethe's quasi-divine gesture transformed into his *Collected Works*—have, so far, survived the fading and disappearance of their original inscription.[53]

Since then, other technologies have emerged to manage the effects, and decay, of discourse. When the question, "Who is speaking?" is asked, tautology—the fact that the one who is speaking is always the (same) one speaking—vanishes. Now, "Wanderer's Night Song," a title given later on, no longer conceals the fact that neither the "wanderer" nor the author offers the words of promise and rest [*das Wort des Zuspruchs und der Stillung*]. Instead, a historical figure of the Other does so. Since then, other sounds have grown audible, too. "Birdland" was no land of winged creatures, but a bar in New York. "The Bird" was the name of an alto saxophone player. And when he played "Lullaby of Birdland," it was a signal, not a lullaby.

4 The God of the Ears

In Gedanken an Rochus und die Insel 12

The Greeks had a god who dwelled in the acoustic realm. When shepherds dreamed and the quiet of noon turned, Pan suddenly droned in the ears of all.

Pan, a curve of auditory space, had always been closer to the Great Goddess than her despairing paramours, who chased her only in the field of vision. Full of envy, Actaeon describes the hunt:

> At the time it seemed to me I saw, up there on the rock, the back of old Pan, who was also lying in wait for her. But from afar one might have taken him for a stone, or for some stunted old tree trunk. Then he was no longer discernible, though his pipes still rang out in the air. He had become melody. He had passed into the sighing of the wind, where she was sweating, where breathed the fragrance of her underarms and lower body, when undressed.[1]

"To look at a room or a landscape"—to say nothing more of goddesses—"I must move my eyes around from one part to another. When I hear, however, I gather sound simultaneously from every direction at once: I am at the center of my auditory world, which envelops me. . . . You can immerse yourself in hearing, in sound. There is no way to immerse yourself similarly in sight."[2]

The great god Pan, it has been said, is dead. But gods of the ears cannot fade away. They return behind the mask of our amplifiers and sound systems. They return as a rock song.

> *Pink Floyd: "Brain Damage"*
> The lunatic is on the grass
> The lunatic is on the grass
> Remembering games and daisy chains and laughs
> Got to keep the loonies on the path
> The lunatic is in the hall
> The lunatics are in my hall
> The paper holds their folded faces to the floor

The God of the Ears

    And every day the paperboy brings more
    And if the dam breaks open many years too soon
    And if there is no room upon the hill
    And if your head explodes with dark forebodings too
    I'll see you on the dark side of the moon
    The lunatic is in my head
    The lunatic is in my head
    You raise the blade, you make the change
    You re-arrange me 'til I'm sane
    You lock the door
    And throw away the key
    There's someone in my head but it's not me.
    And if the cloud bursts, thunder in your ear
    You shout and no one seems to hear
    And if the band you're in starts playing different tunes
    I'll see you on the dark side of the moon. (Lyrics and music: Roger Waters)

The Dark Side of the Moon (Harvest LP I C 072–05–259)—from the year it appeared, in 1973, until 1979—sold eight million copies[3]; more recent figures indicate eleven million. Books and the number of printings they go through become laughable when streams of sound feed into streams of money. "Brain Damage" needs no further description. The damage has already been done.

And yet it all began so simply. In the 1960s, Roger Waters, Nick Mason, and Richard Wright were three architecture students roaming about with guitars, performing old Chuck Berry tunes in the suburban concert halls of England. The first name of their group, long since forgotten, was "The Architectural Abdabs." Then, one spring day in 1965, a lead guitarist and singer joined them. This individual created "Pink Floyd"—the name and the sound. Using distorted amplifiers, the mixing board as a fifth instrument, sounds that circled through space, and all one can do by combining electrical optics and the technology of low frequencies, Syd Barrett—with eyes like two black holes—opened up "Astronomy Domine," the domain of astronomy, for rock 'n' roll.

The star above the London underground shone for barely two years. Everyone knows Andy Warhol's dictum that in the age of electronic media everyone will be famous—for fifteen minutes. At Barrett's last appearances, when concerts were not canceled altogether, his left hand hung at his side; the right one simply played one and the same open string, over and over[4]: the monotony of Chinese torture as the beginning and end of music. Then the man who discovered Pink Floyd disappeared from the scene altogether, somewhere in the diagnostic no-man's-land between LSD psychosis and schizophrenia. Pink Floyd found a replacement guitarist and the formula for worldwide success.

Yet even with seven-figure LP sales, it still holds that the streams of money flowing through the machine of capital are fed by a decoded, deterritorialized stream of madness, whose immediate realization is electric.[5]

For six years, Pink Floyd kept silent about the act of exclusion that had made the group's continued existence possible. "Brain Damage"—a song about outside and inside, exclusion, inclusion, and their sublation—reveals the truth. At the beginning, everything is still in order [*stimmt noch alles*]. There, indoors, stands the property owner—key in hand and up to date on current nonsense thanks to the newspaper. Here, on the lawn, presumably the beautiful lawn of southern English residences and Gottfried Benn's reveries,[6] lurks the madman (or the madmen). At least that is how a law would have it—one that territorializes and prescribes that madmen must remain on fixed pathways (above all, *outside*). Such is the law of architects.[7] The blockade that gave it concrete form would be built in 1980–81 by Waters—the erstwhile architecture student—as a gigantic wall running through Earl's Court and the Halls of Westphalia.

Yet nothing ever goes as smoothly in the acoustic realm as it does in show business. After all, "ears, in the field of the unconscious, are the only opening it is impossible to close."[8] From the grass, down the hallway, and into the head: the unyielding progress of madness passes through ears that cannot defend themselves. At the end of the song—whether it be called "Brain Damage" or "The Wall"—the dam is broken, and one's head explodes; what remains is only screaming without a receiver [*ohne Empfang*]. No word, no wall, and no dam between outside and inside can withstand *sound*, because sound is both what is unrecordable [*das Unaufschreibbare*] in music and its immediate technical implementation [*unmittelbar ihre Technik*].

Foucault wrote *The History of Madness in the Age of Reason*. Bataille wrote *The Story of the Eye*. To Roger Waters, the lyricist of "Brain Damage," we owe a short story about ears and madness in the Age of Media.

When the serial inventor Thomas Edison—following an idea hatched by Charles Cros—built the first gramophone, the reproduction offered only the shadow of the recording. Even horn-shaped pipes [*Schalltrichter*] for amplification could scarcely make vibrations that were mechanically recorded and reproduced louder than the original. It was not just because Edison was almost deaf that he had to scream into his phonograph on that fateful day, 6 December 1877.[9] Only in futuristic fantasies written by contemporary Symbolists did the

"Wizard of Menlo Park" connect his phonographs to a speaker—many speakers—which permitted him to fetch, into his laboratory, the dances of children on the grass outside.[10]

In actual fact, citizens and emperors at the turn of the century were more interested in voices than in the ritornello that puts voices and identities into motion. In 1897, the poet laureate of Wilhelmine Germany was granted acoustic immortality. After lengthy excurses about how voices—unlike faces—cannot lie (and therefore represent first-rate sources for psychological study), he spoke a lovely closing couplet into the recording device:

> Hear, then, from the sound of this saying / The soul of Ernst von Wildenbruch [*Vernehmt denn aus dem Klang von diesem Spruch / Die Seele von Ernst von Wildenbruch*].[11]

From sound to saying, from saying to soul: Wildenbruch was desperate to equate the Real (his recorded, but mortal voice) with the Symbolic (the discourse articulated in his verses), and in turn to equate the Symbolic with the Imaginary (the poetic spirit of creation within him). Praise the Lord: ever since, sound technicians have steered the opposite course. Time and pure research [*Grundlagenforschung*] have assured that the breath of the soul has drowned in sound and decibels.

Only for as long as records were mechanically cut and played did the human voice govern them, and, given their wretched frequency range between 200 and barely 2000 hertz, this is no wonder. Only after a world war—the first one—had, in its push for innovation, established the principle of amplification, could Edison's mechanical apparatus be electrified. For the first time, the frequency spectrum and sound dynamics of orchestras were present in record grooves and loudspeaker coils. In Respighi's *Pini di Roma* (1926), a single nightingale—electrically preserved and amplified—held its own against Toscanini's entire philharmonic.[12]

To perfect the magic, all that was required was another world war. Now the drive for innovation gave German engineers the tape recorder and their British counterparts the hi-fi record (which made even the subtlest differences of timbre between German and British submarine motors audible; naturally, the audience was restricted to aspiring officers in the Royal Air Force[13]). Soon, with magnetic tape—the booty of war—in hand, America's dormant recording industry (which had discharged other duties between 1942 and 1945) set a new standard: for the first time, it was possible to manipulate acoustics in the space between the production of recordings and their reproduction.

Before long, British industry also recognized that technologies developed to locate U-boats could have peaceful applications as well. In 1957, Electrical and Mechanical Industries (EMI)—not coincidentally the label that subsequently held Pink Floyd under contract—offered consumers the first stereo record.[14] Since then, the two ears with which human beings happen to be equipped have not represented the variable moods of nature, but rather a source of revenue: they are used to locate individual voices and/or instruments between the two speakers of a living room. If ever they should fail to do so, it is only because the sound engineer was too clever. In 1959, when John Culshaw produced Georg Solti's wonderfully distorted *Rheingold*, each and every god and goddess found an audible location on the stereo soundscape. The voice of the great technician Alberich, however—when he invisibly, and drastically, demonstrated the powers of the magic cap [*Tarnkappe*] to his brother—emerged from all possible corners at once.[15] What was a special effect for Culshaw, Syd Barrett made the rule. Legend has it that when recording, he twisted the knobs on the mixing board back and forth as wildly as if the two stereo channels were themselves an instrument . . .

Since that heroic age, things have proceeded as if by explosion. So-called reproduction has transformed into the production of sounds, and the vow of "high fidelity" has degraded into a palliative formula compared to what technology actually makes possible. Today it is only for commercial reasons, and not for technical ones, if the standards of radio and recordings are limited to soundscapes and do not simulate real—or even absolute—sound spaces instead. Where money and madness are at play, all restrictions collapse. The proof was offered by none other than Barrett. With his Azimuth Coordinator, he gave Pink Floyd a technical advantage over competing groups. As the name indicates, the Azimuth Coordinator is a system that enables one to bring sonic events—tracks and layers within the mass of sound—directly to the listener's ear at will and in variable positions within all three dimensions of space. "Brain Damage" sings the device's praises.

The song has three beginnings. Three times in a row, sound reproduction performs a historical step forward.

"The lunatic is on the grass . . ." Children playing and laughter—that is, precisely what the Edison of Symbolist science fiction wanted to listen in on—come from outside and enter the house, muted by walls and deprived of spatial coordinates through the distance they have traveled. (Likewise, a part of "Wish You Were Here," which uses the equalizer to cut the high and low ends and

The God of the Ears

then transfers the music to a single track, simulates a simple transistor radio.[16]) Verse One, then, reproduces the impoverished age of monaural reproduction in acoustic quotation.

"The lunatic is in the hall / The lunatics are in my hall . . ." Step by step, line by line, the monaural distance (and with it, abstraction) comes to an end. The hallway, which when repeated becomes "my" own, has a defined reference to the spatial coordinates of the hearer/speaker himself. The hall is close enough for him to distinguish, by hearing, between right and left—and close enough, too, for him to make out the approach of many lunatics. This is exactly how the acoustically constructed ladder functions at the unforgettable end of "Grantchester Meadows," when steps march from left to right—from the surface of the vinyl directly into space, and thus into the ears of listeners. Verse Two, then, is the age of high fidelity and stereophony.

"The lunatic is in my head / The lunatic is in my head . . ." In other words: brain damage has occurred; an Azimuth Coordinator is at work. If sounds—ones directed anywhere in the space of hearing—can pop up in front or back, right or left, and above or below, then the space of everyday orientation vanishes into thin air. Thus, the explosion of acoustic media transforms into an implosion that, immediately and without distance, befalls the center of perception itself. The head—not just the metaphorical seat of "thought," but also the nervous switchboard of actual fact—becomes one with arriving information, which is not just a matter of so-called objectivity, but of *sound*. The final sequence of "Brain Damage" is pervaded by the notes made by a synthesizer, as if to demonstrate the thesis that synthesizers have long since replaced the synthetic judgments of philosophers.[17] A tone generator that can steer and program sounds in all parameters—frequency, phasing, harmonic tone, and amplitude—transports the conditions of possibility for so-called experience into the realm of total physiological simulation.

Accordingly, "The Story of the Ear in the Age of Technical Explodability" has always already been *The History of Madness*. Brain-damage music makes everything true that once spooked about in heads and madhouses as so many dark premonitions. According to the information a psychiatric lexicon provides, "in comparison to other senses, the sense of hearing is most often affected by hallucinations."[18] The scale of such so-called acouasms extends from white noise to hissing, water drops, and whispers, and on to speech and screams—all of which madness either perceives or produces. It seems as if this psychiatric reference work had meant to compile a list of effects used by Pink Floyd. White noise occurs in "One of These Days," hissing in "Echoes," water drops in "Alan's Psychedelic Breakfast," and screams in "Careful with That Axe, Eugene." Whispers, in turn, are everywhere and all over the place ...

What is surprising, given the extent of so much clairaudience, is that psychiatrists find it at all surprising that in today's world acouasms no longer come from susurrating devils or howling witches, but rather from radio stations or radar antennas.[19] The insane seem to be better informed than their doctors. They announce that madness—instead of being the metaphorical rambling of radio stations in the brain—is, on the contrary, a metaphor for technology [*Techniken*] itself. Because madness always meets up with the most up-to-date testing standards, its antennas register the current state of data processing with historical precision.

It is only under cultural conditions commanding that discourses be heard as individual speech acts and the like that discourses involving conditions of transmission [*Diskurskanalbedingungen*]—static and hissing, stereophony and echo—sound crazy. When speech acts are fundamentally mass-media acts—anonymous and collective events[20]—madness is the truth and vice versa. A press statement that EMI (another mass-media act) issued when Pink Floyd's suggestive track "Let's Roll Another One" was banned from the airwaves,[21] illustrates the point nicely. "The Pink Floyd," English journalists learned, "does not know what people mean by psychedelic pop and are not trying to create hallucinatory effects on their audiences."[22]

Even if Barrett's glorious Azimuth Coordinator had not already assured that ambulances had delivered Pink Floyd listeners to the hospital after experiencing fits of dizziness, statements of this kind are a foolproof method to drive people insane. To say that one does not intend to do something means saying how easy it would be. After all, it is impossible to close one's ears. Mass-media acts lie and are themselves crazy [*spinnen*]. This chagrins only philosophers—everyone else hears it gladly. The request voiced to an unknown god or engineer in the song "If" (over the same synthesizer washes as in "Brain Damage") cannot be fulfilled: "And if I go insane, please, don't put your wires into my brain."

Brain damage is unavoidable. The antennas at which the fear of madness—in both senses of "of"—trembles have long since invaded brains, and psychiatrists have not even noticed as much. The broadcast occurs on all frequencies, from long wave to VHF. Waters sings the verses of "Brain Damage" in solo voice over a thin expanse of sound that simulates the innocence of acoustic guitars.

The refrains, in contrast, are "bells" of sound—countless tracks stacked on top of each other, which turn inside out in the listener's ears and head as they resonate. The verses are spoken by an "I." Initially, they are spoken to the lunatic outside, but by the end—after the Azimuth Coordinator has engineered immediate proximity—they address the speaker. The refrains, although formulated as "if"-sentences, offer a response—a discourse of the Other, who puts the stanzas back the other way around and brings them down to earth [*Kopf auf die Füße*]. Now Barrett returns and performs what they have told him [*was sie ihm zugesprochen haben*]. "You make the change / You re-arrange me 'til I'm sane."

Such healing and reconfiguration occurs quite simply and concretely through arrangements and recording technology. In the first German radio play made for dummy head [*Kunstkopfhörspiel*]—after all, dummy-head technology is just an Azimuth Coordinator for a private, domestic setting—all voices and sounds were recorded with stereo microphones, except for one, which was intended to represent both a computer output and an input for madness. In elegant fashion, the radio play delivered what its title promised: *Destruction*. If among the innumerable voices that can be located in three-dimensional space, one—and only one—pops up without coordinates, it is unfailingly localized in the listener's imploding head. Under conditions of perfected spatial simulation, there is no longer any need for the ruses employed by Alberich and Culshaw. The most harmless and old-fashioned recording technology drives the heroes and hearers of a *Kunstkopfhörspiel* mad.

"Brain Damage" functions no differently. The third verse, about someone in my head who is "not me," is mixed with laughter. This laughter does not only transform all fear of antennas extending into the brain into a great Nietzschean affirmation. Additionally—because it was recorded monaurally, in a crafty exception to the rule—it *is* the wire in the brain itself.

At the very beginning of the record, the first audible sentences drown in laughter when a triumphant voice announces that it has always been mad and knows it, too. When the voice returns at the end of the record and the panicked laughter explodes in the listener's head, Pink Floyd's madman celebrates victory over his backing band. There are two kinds of music, then. The one is a quotation (and not a recollection) of voice and nature; the other—to use the words of Ingeborg Bachmann—is a song from beyond humanity [*von jenseits der Menschen*].[23] "I've always been mad, I know I've been mad."

Brain damage means that the *other* kind of music will triumph. "Radio is far superior to nature; it is more comprehensive and can be varied."[24] Nothing and no one limits what electronic media enable. Without fear of madness, one can always make other kinds of music, too. "Pretty, but a little out of date," Barrett is supposed to have mumbled when, years after being kicked out of the band, he arrived at Abbey Road Studios and listened in on his former group. "Brain Damage" makes this murmur into a promise filled with laughter. When Pink Floyd plays the other kind of music, their madman will return. "And if the band you're in starts playing different tunes / I'll see you on the dark side of the moon." Or, in French translation: "*Des dieux nouveaux, les mêmes, gonflent déjà l'Océan futur.*"[25]

Nietzsche—who could only have known another kind of music through the works of Wagner—once dreamed of "a deeper, mightier, and perhaps more perverse [*bösere*] and mysterious music, a super-German music, which does not fade, pale, and die away at the sight of the blue, wanton sea and the Mediterranean clearness of sky"; this music would be "a super-European music which holds its own in presence of the brown sunsets of the desert."[26] Precisely this is what the madman of "Brain Damage" relocates to the Dark Side of the Moon, where other musics meet. The sunset, in turn, is what Pink Floyd stood before at their legendary concert in Pompeii: they remained motionless at the shore for hours until, just as the flaming red sphere touched the surface of the sea, they began playing to the strike of a gong.

It is no accident, then, that *Dark Side of the Moon* was performed when the London Planetarium opened. Only the mightier—and perhaps more perverse—music of the twentieth century has its antennas in the domain of astronomy. Classical European instruction in musical harmony [*Tonsatz*] sought to master the incessant noise [*Rauschen*] all around by means of form and binary coding (major/minor, consonance/dissonance, etc.). Romantic music was—and stayed—a process of decoding such oppositions: a *Song of the Earth* [*Lied von der Erde*] that exploded all triadic changes of harmony [*Dreiklangsharmonik*] like "rotten junk" [*morschen Tand*] when the word "earth" was heard. For all that, however, twentieth-century music also left the earth, our lifeworld, behind. Cosmic rays and neurological energies—powers, that is, which lie both within and beyond humanity—form its two poles.[27] When a short circuit occurs, they are unleashed.

The cover art of *Dark Side* could not express the matter any more clearly. The design team for Pink Floyd—with the telling name of "Hipgnosis"—devised (lest any doubt remain . . .) a beam of light against a black background, splitting into spectral colors before coming back together in a line that forms an electrocardiogram: an oscillogram of the heartbeats with which *Dark Side* begins and ends. In this way, electronic technology finally caught up with the premonitions that, since time immemorial, had short-circuited the raving brains of lunatics, the moon, and the stars.

Indeed, brain-damaged listeners become moonstruck. One reads verses and forgets verses. Pink Floyd, on the other hand, sticks in one's head—for the "I of today [*ich von heute*], who learns more from newspapers than from philosophies, who stands closer to journalism than to the Bible, for whom a good hit song contains more of the age than a motet [*dem ein Schlager von Klasse mehr Jahrhundert enthält als eine Motette*]."[28] Even if a voice at the end of "Brain Damage" mumbles, "I can't think of anything to say," even if books are laughable, and even if descriptions of music get lost in space [*hinterm Mond sind*], something to write is always there—simply because there is something that never stops transmitting [*einfach weil etwas nicht aufhört, sich {ein}zuschreiben*]. After all, "Brain Damage" does not sing of love or anything like that—it is a single, positive circuit of feedback between *sound* and listeners' ears. The music [*Klänge*] announces what it does. And that outstrips all the effects that Old Europe promised itself with the Book of Books or the words of immortal poets.

The simple secret of all poetic lyric is to tear words from oblivion [*Vergängnis*]. When the Greeks invented the hexameter, they had nothing else in mind. "The rhythmic tick-tock"[29] was meant to make certain discourses inescapable for human ears—and amplify them, over great distance, for the ears of the gods. (The ones are so forgetful, and the others so hard of hearing.)

Nietzsche—who rediscovered this technique for channeling discourse—offered positive philological proof right away: Greek rhythm does not measure syllables according to the meaning of words (as occurs in modern lyric), but simply in terms of acoustic length or brevity. For this reason—and for this reason alone—ancient lyric is bound to a foot in a literal sense: the foot of dancing bodies. In contrast, modern European languages determine emphasis and verse rhythm on the basis of the meaning of words.[30] Consequently, the music of lyric has vanished along with bodily memory. One can no longer tell how texts are to be sung or danced. It remains a matter of chance whether, after the fact, they are set to the mnemotechnical medium of music.[31]

Perhaps this is the reason why Romantic lyric—more directly than all other genres of poetry—has been tied to the experience and psychology of those who write it. It became possible in the Imaginary, before any composition had occurred, to breathe [*einhauchen*] an inner music even into verses that were read silently. Because phantasmagoric voices whispered between the lines (for male readers, the voice of the Mother, and for female ones that of the Author), poetry remained stuck in enamored consciousness [*im verliebten Gedächtnis*]. *Classical Forget-Me-Not* [*Klassisches Vergißmeinnicht*] was the name of a tiny book of poems by Goethe. Only under the conditions of high capitalism—when consumers got bored with this kind of psychology and came to prefer harder drugs—did lyric adopt the cold medium of writing for its mnemotechnics. Baudelaire's *Fleurs du Mal* begins with a direct address to the reader that tells the whole story—from yawning ennui up to the hookah.

Modern lyric: a treat by and for fetishists of letters at a time when, all around, letters and notes—the sole symbolic means of recording sound in Old Europe—were being replaced by electric media. "High" and "popular" culture [*E- und U-Kultur*] . . .

Not for nothing, then, was Wildenbruch moved when he spoke his verses about phonography into the phonograph. On that day, the bell tolled for lyric as it had long existed, beloved by so many. "What is poetry for in the age of mechanical reproduction?" The media are too generous to restrict their storage capacity to the sounds, sayings, and soul of a Wildenbruch. Mnemotechnic aids like Authorship and Individuality have become superfluous now that record grooves and magnetic tape can capture [*bannen*] *sound*—the Unrecordable itself. In popular culture, the ancient connection between words and music has returned after millennia, no longer via the feet of verses and dancers but as an inscription in the Real.[32] Pink Floyd sticks in one's head, simply because people no longer need to be reminded, simply because machines themselves *are* the mind. And with that, it also becomes possible to store—beyond words and melodies—the colorations of instruments [*Instrumentalfarben*], sonic spaces [*Klangräume*], and indeed, even abyssal stochastic noise [*die abgründige Stochastik des Rauschens*].

Respighi's little nightingale has had a storied career. The mad laughter of "Brain Damage" and the beatific sounds of a summer day in "Grantchester Meadows" are not just hymned—they are also, and at the same time, audible themselves. A meadow near Cambridge, with all its noises, grounds the song, which conjures it up once more. What books or scores could only hint at through awkward games [*vertracktes Spiel*] (songs in character [*Rollenlied*], changes of perspective, imitations of nature [*Naturzitat*]) has become an event in absolute sonic space. And so they return: the calm of noon, the field of a meadow, the laughter of a god.

And ever since rock groups, instead of performing numbers prefabricated by lyricists, composers, and arrangers at the behest of record companies, have themselves placed parameter on parameter, track over track, and words on top of instruments in studios—that is, since the LPs of the 1960s—custodians of the law have been eliminated from the space of *sound*. "There's someone in my head but it's not me." Only throwbacks like copyright—which did not arise in the Age of Goethe by chance—still force one to name lyricists and composers (as if any such parties existed in the space of *sound*). It would be much more fitting to list the circuit diagrams of the facilities and the model numbers of the synthesizers employed (as occurs on the cover of *Dark Side*).

And so, for now, a few things still are operative:

> The famous personalization of power is like a territoriality that accompanies the deterritorialization of the machine, as its other side. [...] One sometimes has the impression that the flows of capital would willingly dispatch themselves to the moon, if the capitalist state were not there to bring them back to earth.[33]

"I'll see you on the dark side of the moon." Who can tell the difference between what belongs on the Moon and what belongs down on Earth? "So you think you can tell Heaven from Hell," go the mocking lyrics in "Wish You Were Here." And the final words on *Dark Side of the Moon*—which, whispered, are hardly audible among the heartbeats with which the album fades out—say the same thing: "There's no dark side of the moon, really. As a matter of fact, it's all dark."

Even a heart connected to contact microphones and oscilloscopes falls silent. And when all differences fade away—along with loud and soft, bright and dark, and Heaven and Hell—then another space draws closer, which other cultures call *satori*. For this reason, one should not take [*hören*] the explosion of media of our times as theoretically as its prophets did. According to Marshall McLuhan, the message of synthesizers is simply "synthesizer." But if—against vast darkness—there is no Dark Side of the Moon at all, then electronic media may in fact be announcing the advent of much darker figures. Waters: "The medium is not the message, Marshall . . . is it? I mean, it's all in the lap of the fucking gods . . ." (pause for laughter).[34]

5 Flechsig, Schreber, Freud: An Information Network at the Turn of the Century

In memoriam G.J.

Madness only seems to be a marginal phenomenon. As soon as access is secured to archives that have held apart the present day and history, power and the past, one regularly finds—after the customary thirty years have elapsed—that what seemed marginal has resulted from a politics of knowledge. Too late to be able to do anything, the owl of Minerva recognizes that the exclusion of madness from the culture in question has served to hide its constitutive place in the system. What this culture deemed alien, borderline, and intolerable belatedly assumes a place as one of its constitutive elements [*Formen*].[1] That is no accident. According to Foucault, these constitutive elements are historically specified rules of speaking and writing, of discourse administration and discourse networking.

"It would be worthwhile," Lacan wrote decades ago,

[to note] the places in social space that our culture has assigned [the insane], especially as regards their relegation to the social services relating to language, for it is not unlikely that we find here one of the factors that consign such subjects to the effects of the breakdown produced by the symbolic discordances characteristic of the complex structures of civilization.[2]

What follows is an attempt to demonstrate Lacan's supposition empirically—and on the basis of a case that, from Freud to Lacan, has counted as the very paradigm of psychosis. For all that, however, the case has hardly ever been viewed as a symbolic discordance within our own culture. *Memoirs of My Nervous Illness* by Daniel Paul Schreber, this most celebrated work of all mad, German books—or German books by the mad—bears witness to a breach in the order of discourse only if one does not, for the umpteenth time, "psychiatrize" or psychoanalyze it. What the paranoiac Schreber wrote, what his psychiatrist Flechsig wrote, and what his psychoanalyst Freud wrote: this whole mass of writing must remain a mere pile of paper. Discourse administration is to be conducted with media-technical precision or not at all. The information network Flechsig/Schreber/Freud consists only of dusty books from 1882 to 1911. What they record, however, is the fact that dusty books—the fundamental medium of power in the Europe of old—lost their monopoly around 1900.

1.

Schreber's *Memoirs*, which were composed the same year that *The Interpretation of Dreams* was published, appeared in 1903—in a private edition commissioned by the resident of an insane asylum. The "*main motive in publishing this book,*" the author affirmed, is to "*offer my person as object of scientific observation for the judgment of experts.*"[3] Freud came just in time when, in 1910, he cashed this blank check. The following year—the year of Schreber's death—"Psychoanalytic Notes upon an Autobiographical Account of a Case of Paranoia" was printed. Psychoanalysis, however, could not, and cannot, involve the scientific observation of bodies. Rather, it interprets paranoia as a psychic conflict first and foremost—in the case of Schreber, as homosexuality that a father famous for therapeutic, pedagogical nurseries [*ein heilpädagogischer Schrebergartenerfinder von Vater*] had prompted in his son, a lawyer and judge.

All that would remain of the *Memoirs*, then, would be yet another Oedipus complex—were it not for the fact that Schreber wrote everything down. In contrast to what happened in the "talking cures" to which Freud subjected the neurotics who came to his practice, in the Schreber case "the subject of analysis was not actually a person but a book produced by him."[4] This did not occur only because Schreber resided in the oldest insane asylum in Germany, far away from Vienna.[5] Rather, it happened because even the inmates of madhouses achieve theoretical dignity when they write books. Freud testified that Schreber's 516 pages of autobiographically described paranoia display "a remarkable resemblance" to the "theory" of paranoia itself—as if his own seventy-six pages of psychoanalytic observations were altogether superfluous. Indeed, at the end of his discussion, Freud feels obliged to appeal to a psychiatrist friend, who would be able, if need be, to swear that the father of psychoanalysis already had a theory of psychosis before reading Schreber. What is at stake, then, involves more than Oedipal complexes and what therapy can achieve [*Heilbarkeiten*]. Freud and Schreber dispute far weightier matters: intellectual property, scientific priority, and the riddle "whether there is more delusion in my theory than

I should like to admit, or whether there is more truth in Schreber's delusion than other people are as yet prepared to believe."[6]

That is no surprise. Psychosis always forms a tangent to the politics of knowledge. In order to enable qualified observations on his body, Schreber's *Memoirs* describe, with neurological precision, all the nervous tracks [*Bahnen*] that connect [*verschalten*] the discourse of a malignant God,[7] over millions of kilometers, to linguistic centers in his own brain. In Freud's estimation, these same "sunrays, nervous fibers, and spermatozoa," correspond exactly to the "libidinal cathexes" that distinguish neuroses and psychoses. Madness and theory entertain a relationship of solidarity, then. Already in "Project for a Scientific Psychology" (1895), Freud had described "souls" as sequential circuitry [*Schaltwerk*], where neurons, bound and unbound, set up channels, circumvent inhibitions, occupy mental representations, and so on. The "psychic apparatus" (Freud's apt coining) consists of neuroelectrical data flow; before his hysterical patients forced him into the talking cure, Freud had been a neurophysiologist himself. Accordingly—and up to the end of his life—he insisted that the hypothetical apparatus he had posited possessed an anatomical substrate. Alas, talking cures involve "nothing . . . but an interchange of words," unlike procedures in laboratories; therefore, this substrate, this "Real," must remain "unrecognizable."[8]

Today it has been altogether forgotten that Freud intended to "establish psychology on foundations similar to those of any other science."[9] His theory was based on all the revolutionary findings the natural sciences had made about human beings in his day. After Paul Broca, Franz S. Exner, Jean-Martin Charcot, and Paul Emil Flechsig, scalpels and microscopes had dissolved the life of the soul—and discourse, in particular—into matters of neurophysiology. In younger days, Freud had researched the localizations of individual nervous circuits whose networking, in everyday terms, is called "language." The talking cure could not measure up to such standards without assistance.

Yet psychoanalysis did not need Habermasian cures of its own "scientific self-misunderstanding," but rather brains offering conclusive proof. For this purpose, the patients on the couch in the Berggasse were out of the question. After all—and as, of all people, a contemporary "physiologist of art" [*Kunstphysiologe*] recognized—these patients were all suffering from normality itself. For Georg Hirth, it was just as elementary as it was "incomprehensible" that "the wholly healthy human being and the healthy animal *feels absolutely nothing*" of its own central nervous system—that "great factory"; "indeed," he marveled, it "seems not to have a sense that the organ performing all of this even exists."[10]

Exceptions to such a "law of the non-sensation of brain activity"[11] occur only in psychoses. Around 1900, anyone who—like Schreber (or a few years before him, the insane Doctor Gehrmann[12])—exhaustively[13] described how a mad deity occupies the nervous conduits of his sensory and speaking organs represented a scientific marvel. Schreber was just as foretold as he was unhoped-for. Freud remarked that what is recounted in *Memoirs of My Nervous Illness* "sounds almost like endopsychic perceptions of the processes whose existence" he himself considered "the basis of the explanation of paranoia."[14] And that, in terms of method, is a necessary tautology. Only because Schreber perceived, in his own body or brain, what psychoanalysis considered hypothetical and held at the margins of theory, did this theory not go mad itself. That is also why Freud was spared the fate that Schreber feared (and met). According to the standards of the hardest natural sciences of the day, there could not *not* be a psychic apparatus affording endopsychic perception even to the most delirious mental patients.

Just turn your brain inside out, and psychoanalysis has the Real: what it deems irreplaceable yet cannot find.

Schreber's brain offered proof positive for Freud's theory. Brain and theory fit together like a lock and key. The only question that remained was what locksmith had built them and matched them to each other. Yet Freud duly avoided addressing the matter, and he would sooner risk losing his case of scientific priority against the brilliant jurist Schreber. For if (as is always said when inventions are disputed) Schreber and Freud did not discover the endopsychic perception of the soul apparatus "independently"—that is, if madness and theory were "drawing from the same source"[15]—then Schreber's brain would lose its value as scientific proof. According to Winnicott, psychoanalysis has no concept of intellectual property: it simply carries knowledge onward from patients' mouths. Yet if these sources—the patients—are already caught up in information networks, then matters prove altogether mindless [*geistlos*]. In such a case, and to follow the lead of Daniel Paul Schreber, the only possible help might come from paranoia.

2.

The brain that Schreber dissected autobiographically neither fell from the heavens nor into a no-man's-land. It belonged to the university nerve clinic in Leipzig. More specifically, it was the property of the director, Professor Paul Emil Flechsig, M.D.—"Paul Prince-of-Hell," as his patient Schreber (exploiting the similarity of their Christian names) put it with due precision.[16]

Flechsig—and no lesser luminary than Freud himself said so—had led German psychiatry into "a new epoch."[17] He put an end to the concept of madness that had united the poets and thinkers of the Age of Goethe and its alienists. The condition of possibility for Mignon and the Harpist, for Orestes and Serapion, had been the fact that their deviancy [*Störung*] inhabited language. That is why Johann Christian August Heinroth—Flechsig's sole predecessor at Leipzig—had derived mental illness from moral trespasses, which he treated with "psychic cures." But it is also why there "gaped," between Flechsig and his predecessor, "a gulf, no less deep and far than the one between medieval [and modern] medicine."[18] In his inaugural lecture, delivered in 1882, Flechsig expressed only scorn for psychic cures and moral trespasses. Notwithstanding its own verbal nature, the lecture disregarded language—to say nothing of "spirit." Flechsig, an anatomist by training, only acknowledged reality [*Reales*], especially where psychosis was concerned: circumscribed and describable localities in the physiology of the brain. Whenever speaking of "mental illness" [*Geisteskrankheit*], he added the words "so-called." And because "there are no independent maladies [*Erkrankungen*] of the soul without the same for the body,"[19] he preferred the more exact term of "nervous illness" [*Nervenkrankheit*]. That is why, with due juridical precision, Flechsig's patient—in the very title (*Denkwürdigkeiten eines Nervenkranken*) and at the end of the book as well—does not contest "the presence of a mental illness in the sense of a nervous illness"[20]; he would, however, dispute its existence in the sense that a Heinroth, Hoffbauer, or Hegel would have attached to it.

Under historical conditions that reduced language and spirit to epiphenomena of a neuroelectrical data flow, Germany's universities had to adapt. Lecture and seminar rooms were joined by laboratories, and the heads of traditional departments were joined by principal investigators.[21] This is precisely what occurred in Leipzig when King Albert of Saxony gave the newly appointed Flechsig a psychiatric and nerve clinic "with all the modern requisites."[22] The last survivors of German Idealism perceived such changes as an "attack on the foundations of state and religion."[23] However, a researcher who preached neurophysiological materialism "down from the altar" of the Leipzig university church could count on the rewards offered by "canny strategists" and politicians of knowledge [*Wissenspolitiker*].

An autobiographical item of note for Flechsig (and not Schreber) is that he once showed the king the "brain map" he devised. (This is the same image in front of which Flechsig poses in the photo that appeared in his *Festschrift*.) The doctor recalls:

> The canny strategist immediately noticed the similarity between the brain's pathways and a network of rail lines. Despite the newness of the matter, he understood right away the enormous complication and the difficulty of unraveling it—especially since, in the course of explanation, I pointed out that the total length of the cerebral fibers, placed end to end, would presumably exceed the circumference of the Kingdom of Saxony significantly. This impressed the king so much that later, at dinner at court, he called to me across the table and loudly asked: "How many kilometers do the brain lines run?"[24]

This strategic question would become central for combat when modern, rapid-fire weaponry was introduced.[25] The new nerve clinic at Leipzig undertook the matter, too, and used everything at its disposal: technical apparatuses, staff, and madmen alike. Here Flechsig researched the connection between particular forms of aphasia and the curvature of the brain in specific sites, the different locations of centers of perception and centers of association—that is, the material substrate of talking cures. Discoveries named after him include centers of sight and hearing, as well as the frontal tract of the brain: "primary Flechsig optical radiation," "Flechsig acoustical radiation," and the "Flechsig cortical bridge" [*die primäre Sehstrahlung Flechsig, die Hörstrahlung Flechsig, die temporale Großhirnrinden-Brückenbahn Flechsig*].[26] Most importantly, he discovered the material substrate for Lacan's mirror stage: the fact that it is only after the completed myelogenesis of sensory nerves that "unified perception of the body is possible" for infants.[27]

Unfortunately, however, university nerve clinics were there for people, too. Patients brought in for treatment had little interest in the length, in kilometers, of their nerve fibers. Here Flechsig's problems began—relative ones for diagnosis, and absolute ones for therapy. On the one hand, his iron rule held that "the analysis of the human mind that has fallen ill [*des kranken Menschengeistes*] is a physical problem, above all," and that "any kind of metaphysics" would amount to "a narcotic."[28] On the other hand, the canny anatomist knew that the physics of the brain could be determined "on living parties only by way of more or less conjectural conclusions [*zusammengesetzte Schlüsse*]." The "protected position of the brain" simply "entails as much."[29] Thus, Flechsig's entire psychiatry pushed on, toward a royal road of diagnosis that was at the same time a therapeutic dead end: "post-mortem findings."[30]

No sooner said than done. In 1884 and 1893, Daniel Paul Schreber—first, after unsuccessfully running for a seat in the Reichstag, then as the newly appointed Senate President of the Court of High Appeals in Dresden—was brought to Flechsig. His simple wish was to sleep, or as Schreber himself put it in his magnificently bureaucratic German, the human right to have a "not-thinking-of-anything-thought [*Nichtsdenkungsgedanken*]."[31] Accordingly, the case history of his first visit records the misuse of sleeping aids and "great hypochondria,"[32] an understandable condition given that the patient had sought, in vain, to participate in discourse governing the State. The second case history, on the other hand—after Schreber's appointment to Senate President—registers hallucinations and manifest paranoia. Time and again, Schreber wrote to his physicians, "If you want to kill me, do it right away."[33] This prompted Flechsig to declare the patient "dangerous to himself and others."[34] In making this determination, he left unanswered an "Open Letter" that Schreber later placed at the beginning of his *Memoirs*. The letter simply asks whether the "most honorable Privy Councilor" had not only failed to treat him, but also "[used] a patient in your care *as an object for scientific experiments*."[35] The madman, then—in accordance with all the programmatic pronouncements his doctor had made—declared that paranoia represents an effect of knowledge politics: a psychiatrist who experiments on the nerves of a patient is literally committing "soul murder."[36]

Medical procedure and diagnosis met up. The insane are always subject to the most up-to-date standards; accordingly, they note the current state of data processing with due historical precision. Even if Flechsig had not prescribed a dose of 0.3 grams of opium three times daily, his experimental neurological psychiatry would have made any hypochondriac paranoid. And even if his area of specialization had not concerned matters of legal accountability, Schreber—Saxony's second-highest-ranking judge—would have fallen victim to "an attack on the foundations of state and religion" as a matter of course at the Leipzig clinic. Here accountability and linguistic competence, morality and spirit, were over and done. Whereas judges exercise power through the verdicts they pronounce, psychiatrists—ever since Flechsig—have done so through the nerves they dissect. Reversing the morality of civil servants and speech acts, their motto might be: "How to do things without words."

Schreber was lucid enough to recognize this power-grab for what it was—and what it meant: his own powerlessness. The last scion of a great race of civil servants, he wrote what disadvantages had bedeviled the Schrebers of late: the "choice of those professions which would lead to closer relations with God such as that of a nerve specialist" were denied to them."[37] Accordingly, the accomplished jurist tried his hand at neurology in an amateur capacity. Schreber's *Memoirs* provide supplements to Emil Kraepelin's *Compendium der Psychiatrie*. In a more general sense, they were written to enable observations on the author's body while he was still alive. Alternately—and "short of this"—Schreber could only harbor "hope that at some future time such peculiarities of my nervous system will be discovered by *dissection of my body*, which will provide stringent proof. I am informed that it is extremely difficult to make such observations on the living body."[38]

That was prophetic, for in 1911 Schreber's corpse was, in fact, dissected.[39] However, it is also what, in telecommunications, is called "cleartext." Although no mention is made of the informant's name, there can be no doubt about the data feed. Faced with the difficulties posed by living brains in an age without electroencephalograms, Flechsig had declared postmortem findings the *via regia* of psychiatry. Moreover—and with "remarkable eloquence"—Flechsig had conducted "long interviews" with Schreber about the treatment of psychosis.[40] As the addressee of private lectures, the patient could foresee his own dissection, and so he tried to do something about it. Schreber wrote so that Flechsig might make an exception and investigate a nervous system while the subject was still alive. Psychotics are the subject itself of science; what they write, then, is preventative soul murder. The *Memoirs* exist at the exact location of a murder and a corpse—which even Roberto Calasso overlooked in his remarkable work of human science fiction about Flechsig, Schreber, and Freud. Here the textual corpus supplements a body—that is, something real [*ein Reales*] that Freud considered "unrecognizable" and both Flechsig and Schreber deemed "entirely impossible."

The fact that the *Memoirs* exist—as well as what exists within them—follows from this state of affairs, where "power alone counts."[41] Schreber's so-called visions of madness, instead of offering evidence of endopsychic perception (as Freud wished), simply repeat psychiatric discourse. They dabble in the language of science[42] to avoid getting cut by its blade. This—and nothing else—is what the most thought-provoking and fundamental component of the *Memoirs* means: Schreber's theorem that God is persecuting him.

> The above picture of the nature of God ... differs markedly in some respects from ... Christian views.... It seems to me that a comparison between the two can only favor the former. God was not *omniscient* and *omnipresent* in the sense that He *continuously* saw inside every living person, perceived every feeling of his nerves, that is to say at all times "tried his heart and reins." But there was no need for this because after death the nerves of human beings with all the impressions they had received during life lay bare before God's eye.[43]

This image of God is as sharply focused as Flechsig's *Festschrift* photo. Everything (including the attack on religion) follows the script of the doctor's lecture at the university church, "Brain and Soul" [*Gehirn und Seele*]. "God Flechsig"[44] did not supervise—as the psychologists of the Age of Goethe had done—the stirrings of sentiment in living people. Rather, as a good neurologist, he waited for postmortem results. He could do so because everyday language, upon which Heinroth had relied both for diagnosis and therapy, had been replaced by nerve language. All the data the doctor needed was offered by localized cerebral engrams, which could be read only after death. Autopsy made it possible to "identify the lawful relations of dependency between mental disturbances and cerebral anomalies"[45]—which Flechsig had postulated in 1882, twenty years before his patient did the same.[46]

Consequently, this same "nerve language"[47] (which, incidentally, represents one of many "expressions ... of ... a medical nature" in the *Memoirs* that "*would not have occurred to me*"[48]) also forms the news channel running between the neurologist God and his paranoiac victim. Schreber suffers from voices "of which," as Hirth observed, "a healthy person does not, as a rule, become aware." "*From without,* incessantly, and without any respite," God causes Schreber's "nerves to vibrate in the way which corresponds to the use of the words concerned, but the real organs of speech (lips, tongue, teeth, etc.) are either not set in motion at all or only coincidentally."[49] That is a reasonably exact definition not of hallucinated voices, but of the very processes of innervation to which all aphasia researchers, from Broca to Flechsig, traced discourse—likewise, it provides the basis for Saussure's linguistics.[50] Schreber, as if he were himself a researcher of Flechsig's stature, describes the effects of the nerve language: "a natural sensitivity for similarity of sounds" is preserved, but the "meaning" of the spoken and repeated words becomes incomprehensible.[51]

The question is, Just how does Flechsig manage to direct the linguistic centers in Schreber's brain from his distant divine location—Cassiopeia or Orion? A doctor who has declared to his king that the "total length of cerebral fibers, placed end to end," outstrips the dimensions of the Kingdom of Saxony should not be surprised if the nerve language reveals its business secrets to his patient. Schreber's nerves have managed to master the "difficulty of unraveling" that Flechsig identified. They do not rest bundled in his brain, but rather, "placed end to end," bridge the millions of kilometers between his body and God's, thereby enabling the flow of information in both directions. Nerves of this kind provide the answer to King Albert's strategic question. Moreover, and just like the anatomical features named after the doctor, they are entitled to bear the honorific title of "rays."[52] Whereas Albert could only see the information network called "brain" on Flechsig's diagram, Schreber's "mind's eye" could perceive endopsychically—he "saw" nerves "as long drawn-out filaments approaching my head from some vast distant spot on the horizon."[53]

A God wired like that can drive one crazy, to be sure. To achieve this end, *He* simply needed to drive out Schreber's old-fashioned bureaucratic belief in intellectual property. Whenever the patient thought he had had an original thought (for example, while reading a newspaper or playing the piano), neurological examination and data-checking [*Nervenmessung und Nervenspeicher*] revealed that the thought had already existed beforehand. And when that did not do the trick, God jammed transmissions [*wird . . . Störsender*]. He fed pure nonsense into Schreber's nervous system, which the putative master of his own speech was then supposed to give open "verbal expression"—as if it were his own.[54]

Madness, then, is technological, and God—quite unlike what Christians believe—is a deity of information channels such as Marconi and Siemens built. Since, as Schreber surmises, "it is presumably a phenomenon like telephoning,"[55] it is all over, not just for intellectual property but also for words and books, sermons and Bibles. The same holds for judgments—whether made by courts in Saxony or tribunals on high. What runs down the wires is the Real of the current century: electrical data flow. Eighty years before Pink Floyd, Daniel Paul Schreber might have called out to the unknown God or Engineer: "And if I go insane, please, don't put your wires into my brain."

3.

Because psychotics perform social functions [*verwalten Sozialleistungen*] involving language, symbolic discordance—this trademark of complex levels of civilization—affects them the most. Nothing more need be said about the acid test [*Zerreißprobe*] involving bureaucratic and nerve languages, accountability,

An Information Network at the Turn of the Century

soundness of mind [*Zurechnungsfähigkeiten*], and postmortem findings. Possessing a culture that—as C. P. Snow put it—consists of two cultures, means that the one must necessarily signify madness to the other.

Psychoanalysis has never said anything about this. Freud did not make a single mention of the fact that Schreber's delirious nerve language was the same as his physician's neurological discourse.[56] One who declared Flechsig the hero of a new epoch of psychiatry—and, in turn, received praise from the great Flechsig for his research on aphasia[57]—could not afford to notice as much. However, to found new scientific disciplines, it takes, first of all, cliques [*Seilschaften*],[58] and secondly, victims. Iatrogenic (i.e., professional) psychosis made it advisable for Freud to interpret Schreber's persecution by Flechsig as repressed homosexuality. Thus, the innumerable pages of the *Memoirs* about—or addressed to—Flechsig were transformed into metaphors for Schreber's biological father, who is only mentioned in one short passage of the text.[59] In this way, Freud inaugurated the literature on Schreber—which today is too vast to take in in its entirety—that traces all the son's woes to the father's admittedly ruthless methods of child rearing; the orthopedic devices that Schreber père invented are deemed to represent "the real background" for a God who "deals only with corpses [*der den Menschen nur als Leiche kennt*]."[60]

And so it happens that Flechsig's autopsies are just as forgotten today as they are omnipresent. Freud's compensation [*Ersatzbildung*] for knowledge politics involves the "primal father" Schreber and two brothers—Freud himself and Schreber fils—narcissistically competing for intellectual property. The fact that all this intellectual property derives from a theory of nerves antedating the theory of the libido—and belonging to a certain Prof. Dr. Flechsig—remains successfully repressed. Freud would rather believe in the endopsychic perception of nerve fibers than see them in the delirium of professorial *Festschrift* photographs. The unrecognized Real at the theoretical margin of psychoanalysis is the flow of information. Schreber and Freud both prolonged a discourse that threatened to make discourse itself superfluous. That makes the madness of the one as paradoxical as the theory of the other—hence their "remarkable resemblance."

Heroically, psychoanalysis cleaved to language at a time when the biotechnologies of a Flechsig and the media technologies of an Edison evacuated the power of the Word.[61] Freud, unlike his contemporaries, wrote what made itself heard [*zu Wort gekommen ist*] in talking cures. No science proceeds more literally than psychoanalysis.

Flechsig, Schreber, Freud

Heroically, Schreber wrote memoirs even when a neurologist God sought to eliminate all thinking on his part. Even when Flechsig's experiments or "miracles" "pull[ed]" all "the nerves out of [Schreber's] head,"[62] the author pressed on. "For all miracles are powerless to prevent the expression of ideas in writing."[63]

6   Romanticism, Psychoanalysis, Film: A Story of Doubles

On a winter's night in 1828, a Romantic poet—by no means the greatest in their ranks—encountered the spirit of poetry itself. Adalbert von Chamisso, the boon companion in Berlin of E.T.A. Hoffmann, Karl Wilhelm Contessa, Julius Eduard Hitzig, and Friedrich de la Motte Fouqué, had looked a bit too deep into his cups with his Serapion Brethren. The usual "riotous activity" went on until midnight. Then the "weary reveler" "stole away"[1] homeward through the streets of the metropolis, followed by the echo of his lonesome steps.

Not always—indeed, never, according to Freud[2]—is "home" [*Heim*] the opposite of *unheimlich*. Arriving at the windows of his own house, Chamisso saw, or hallucinated, a light burning in the study. He "turned to stone" out of fear and tarried at the door. Only after a bold decision to put an end to the effects of alcohol did he go inside. Thereupon, he saw what the echo already had given him to hear: that he had a double.

The double is the spirit of poetry. While the company of Romantics sat at the "clinking of glasses," seeking to induce inspiration in a relatively professional manner so that poems like Chamisso's "Apparition" might arise, another apparition had long since taken a seat where poets write. The light Chamisso saw in his study, then, was no Romantic delirium, but rather the condition of work for his double. Accordingly, Chamisso's question—"Who are you, ghost [*Spuk*]?"—elicits a wholly justified question in return: "Who is it that disturbs me at such a ghostly hour [*Geisterstunde*]?" For a double who has spent the entire evening reading or writing—at any rate, engaged in some kind of authorial activity at the desk—a "weary reveler" must indeed seem to be some kind of ghost.

Now the roles have been reversed and—as Lacan's theorem of the mirror stage and sibling transitivism might have predicted—a duel becomes likely. The poet and his double cross their blades as words, or more precisely, ter-

69

cets, are exchanged. Everything proceeds as if the two enemy brothers were not Chamisso and Chamisso but rather Sosia and Mercury. Their combat involves "squaring" a "circle that threatens madness"—the impossible matter of proving that one is Chamisso. Simply because, in 1828, passport photos, files of fingerprints, anthropometric tabulations, and data banks did not exist, the two duelists must remain in the verbal or poetic sphere. The impossibility of positive proof of identity gives way to the agreement that each party will provide a self-definition and wait for the effect. First Chamisso, then the double, declares who/what he is.

What occurs to Chamisso is a conventional fabrication [*Dichtung*]—words that prove surprising only because a mouth reeking of alcohol pronounces them. He says, "I am such a one as only aspires to what is beautiful, good, and true." What occurs to the double is novel and to the point, especially given that he is seated at a writer's desk. He says, "I am a cowardly, mendacious wretch."

Such insolence at the outer limit of poetry is barely possible to hold in tercets—and accordingly, produces a devastating effect. Chamisso manages only to mumble that his double Chamisso is the true Chamisso—then he finds himself outside again in the Berlin night, exposed as a fraud and in tears. This time it is forever: the tercets—and "Apparition"—are over.

Only in 1914, eighty-six years later, did the story continue. Now it no longer occurred in verse, but as scientific prose. Otto Rank, Freud's specialist or adjutant for literary history, dug up—in addition to many other cases—Chamisso's encounter with his double. Consequently, alcoholic episodes of Romanticism became matters of scientific necessity for the twentieth century. The proof of identity that Chamisso had failed to offer, Rank now provided. The first insight pronounced by the new science of psychoanalysis is that only writers who are haunted by "neurological and mental illnesses" are also haunted by doubles.[3] The second insight held that what contemporary readers of Chamisso—so long as they did not take what they read for a moral metaphor—had to consider incredible or fantastic is literally true. Freud's theory of narcissism, both for patients in the present as well as in the case of dead writers, can account for the psychic mechanism that "creates such internal splitting and projection" as occurs with Chamisso's double. The duel between what is beautiful, true, and good, on the one hand, and cowardly mendacity, on the other, is a reality in the Unconscious. It measures, "as Freud has demonstrated, . . . the distance between the ego-ideal and the attained reality."[4] Half a century after his death, then, Chamisso received confirmation in writing of who he was. The doppelgänger, instead of resulting from seeing double through teary eyes or being a poetically moralizing metaphor, was "the phantom of our own ego."

Here—leaving Kittler aside—I am quoting Rank, who is quoting E.T.A. Hoffmann,[5] who is quoting a certain Clara. And that means: when psychoanalysis verifies the fantastic, precisely because it transfers poetry into science, certain fundamental assumptions remain unexamined. They involve, in the first place, the basic assumptions of Hoffmann and the literary epoch that produced the phantasm of the double. Secondly, they concern the basic assumptions of Clara and the philosophy that saw to the empirical-transcendental doubling of "man." Rank speaks of Goethe, Fichte, Jean Paul, Hoffmann—his historical memory reaches back exactly a century. He never asks, however, why it is since then, and only since then, that doubles have inhabited the page. Even if all psychoanalyses—that is, dissections—of Romantic fantasies achieve a resolution, a remainder persists. This is the simple fact that doubles appear at writing desks.[6]

One finds proof of this circumstance quickly inasmuch as there is no longer any need to pore over all the books in question. It is enough simply to reread Rank's study. Rank registered all the ghosts at desks—he just did not unmask them.

Guy de Maupassant sat, "one afternoon in 1889,"

> at the desk in his study. His servant had strict orders never to enter while his master was working. Suddenly, it seemed to Maupassant as if someone had opened the door. Turning around he sees, to his extreme astonishment, *his own self entering*, who sits down opposite him and rests his head on his hand. Everything Maupassant writes is dictated to him. When the author finished his work and arose, the hallucination disappeared.[7]

What had happened only under the influence of alcohol in 1828 became autobiographical reality in 1889. Naturalism and psychoanalysis are synchronous. As if to clarify the genesis of "Lui" and "Horla"—his two novellas featuring doubles—Maupassant performed psychiatry on himself. He told of a hallucinated dictator at his desk; just as quickly, the entity entered the archives of contemporary psychiatry and, through them, Rank. All the "sciences of the soul" were content. But no one asked why the double appears at a desk, of all places.

And yet the answer is found in the works of Goethe himself. As is well known, *Wilhelm Meister's Apprenticeship* features a baroness who dresses the hero in the gown of a count and installs him in the latter's study to provide a gallant surprise for the man's wife. Whenever the aspiring poet and citizen has performed roles on stage or recited love verses, he has played to the Countess "alone"; she, for her part, "could not take her eyes off him."⁸ Here, a love that is as secret as it is literary finally can take wing through the doppelgänger trick. Bearing all the attributes of his rival, Meister sits in the Count's chambers. A lamp, representing the forefront of technological development in 1783, shines on him and the "book" in his hands. Education [*Bildung*] is just that easy to stage. However, instead of the poet(ry)-loving Countess—for whom the *tableau vivant* has been arranged—the Count steps into the room, only to suffer a shock that will stay with him for life. He will never learn that his double was not a warning from God, but rather a ploy. The Countess would sooner leave him to religious delusions than admit her failed rendezvous.

The effect on the Count is a misrecognition that befalls psychoanalysts even today. To see doubles as a "phantom of our own ego," one must, as a matter of principle, conceal the strategies by means of which crafty others have produced the phantom. Whether these others are schemers like the Baroness or poets like Goethe is unimportant. In either case, they clothe their hero with the attributes of his paternal rival—the Baroness does so in the castle, the poet does so on paper. If the Count is to believe that his double is standing before him, this must first be believed . . . by Goethe's readers. Except for the words affirming the optical identity of the two men [*Mannsbilder*], nothing guarantees this at all. Words can accomplish this more readily, the emptier they are. Wisely, the entire novel does not contain a physical description of the hero. Wilhelm Meister remains as blank as the sketch of a silhouette.

"There are no individuals. All individuals are also *genera*," decreed Goethe⁹— precisely the individual, that is, whom Germanists have celebrated for inventing the individual in literature. But as the title of Manfred Frank's book (*Das individuelle Allgemeine*) indicates, the individual of 1800 was an individual generality or a generic individuality; that is, not an individual at all. The reason is perfectly obvious: it involves the technological conditions of the time. Meister and his Count, Goethe and his readers: they could all believe in doubles simply because words do not designate singularities. Not even the word *Doppelgänger* itself. And media other than words did not exist in Classical-Romantic times.

The poor depressive Count must have sensed as much. Otherwise, he would not that same evening have sent for Meister to reconstruct his shock. Once more, the aspiring poet has a book placed in his hands—this time not to play a count converted to reading Goethe, but simply to read aloud. Naturally, Meister trembles in fear that his mask may have been seen through. But precisely this trembling in his voice is, "thank goodness, . . . appropriate to the content of the story"; consequently, the Count praises the "expressiveness with which Wilhelm had been reading."[10] It is hardly possible to state more clearly that Classical-Romantic doubles derive from books as such. Whoever, like Meister, uses reading and reciting fundamentally as a means of identification, wins the love of a countess and the praise of a count.

Notwithstanding the legends surrounding poets, then, the fact that words do not designate singularities is not their weakness, but their craft. Identification can click into place in the empty spaces—such is the new prescription for legibility of the time. It holds both for the story that Meister reads aloud and for the one that his readers read. Daniel Jenisch, who wrote the first interpretation of *Wilhelm Meister* in 1797, already said as much: the doppelgänger episode in the novel simply serves to program the reader to read for identification. "The most conspicuous feature of *Meister's Apprenticeship*"—that is, "what makes this novel a work from Goethe's hand"—is the history-making innovation of introducing a hero like you and me. Meister stands neither above nor below his readers: he has no "particular qualities" that might separate him from us. Because, around 1800, individuals are not recorded, he has only "general qualities of human nature."[11] In other words, Meister's singular characteristic is that he has no characteristics and is simply the double of his readers. The logical consequence, then, is that all Germans have the obligation to read Goethe. The novel tells the very "*story of us all*; in this Wilhelm Meister we discern—like the Count, when he sees the disguised adventurer on the sofa—our own self; we, however, do not [...] react with petrified terror, but with pleasant amazement at the magic power of the enchanted mirror that the poet holds before us."[12]

Enchanted mirrors from other lands and times showed goddesses or demons. In Germany during the age of Classicism, they reflected the blank face [*Schafsgesicht*] of citizens who confused what they experienced in life with what they read. What *Wilhelm Meister* teaches—following Friedrich Schlegel[13]—can mean life only for people who have always already been taken in by words. And so long as, in a best-case scenario, the *laterna magica* competed with the magic mirror of poetry, the trick was not difficult. As Novalis put it, "If one reads properly, a real, visible world in our inside unfolds according to the words."[14] Individual letters are skipped over and the book forgotten until, somewhere between the lines, a hallucination appears—the pure signified of the printed signs. In other words, Classical-Romantic doubles emerged on the school bench, where one learned how to read properly.

Musset's "Nuit de décembre," a long poem that Rank esteems so highly—

which every two stanzas (or years) confronts the poet with his double—begins with a passage that Rank withheld:

Du temps que j'étais écolier,	A schoolboy, I my vigils kept
Je restais un soir à veiller	One night, while my companion slept;
Dans notre salle solitaire.	Into the lonely room forlorn
Devant ma table vint s'asseoir	There came and sat a little lad,
Un pauvre enfant vêtu de noir,	Poor, and in somber garments clad,
Qui me ressemblait comme un frère.	As like me as my brother born.[15]

The poor "lad ... in somber garments clad" has been produced neither by narcissism nor by an ego, and neither death nor immortality is his message. Everything happens much more easily than psychoanalysis ever dreamt. The child clothed in black is "poor" only as the victim of the general push for literacy that gripped Central Europe around 1800. It has been this way ever since new methods of teaching children to read sweetened and sensualized the alphabet—ever since people no longer experienced letters as violent, foreign bodies—ever since, that is, they could also believe that printed letters were addressed to them. Lacan called it *alphabêtise*. Baudelaire, as if to decode the ghosts of Chamisso and Musset, began his volume of poems with the address, "*Hypocrite lecteur,—mon semblable,—mon frère!*"

That is plain speech [*Klartext*], and it draws the curtain down on poetry. None of Baudelaire's heirs in *l'art pour l'art* ever brought up the mendacity involved in writing for mendacious readers again. Books ceased to pretend that letters are harmless vehicles delivering hallucinations to our inner eye—especially the delusion that such a thing as "interiority" or the "self" exists. This double disappeared along with what is true, beautiful, and good.

The form that emerges from the depth of mirrors in our day is very different. It has nothing to do with literacy or poetry. In 1900, Ernst Mach described how he saw a stranger on the omnibus and thought to himself, "What a shabby-looking schoolmaster that man is!" In actual practice, the great physicist and theorist of perception needed a few milliseconds to recognize that the stranger was, in fact, his own mirror image. And Freud, who retold Mach's uncanny encounter, was able to provide a parallel case of his own. He "was sitting alone in [the] *wagon-lit* compartment when a more than usually violent jolt of the train swung back the doors of the adjoining washing-cabinet, and an elderly gentleman in a dressing gown" entered. Freud "thoroughly disliked" this party.[16] Mirror images of one's self in the glass of the door to the toilet seem made to order, for they demonstrate the double meaning of the familiar and the uncanny [*heimlich/unheimlich*] and remind even the father of psychoanalysis of his bodily functions.

If doubles spooked about in omnibuses and express trains, of all places, this occurred for good reason. And if the double by the name of "self"—this poetic-philosophical phantasm—stemmed from the general alphabetization of Central Europe, then the shabby forms before Mach or Freud were the products of the general motorization taking place in these same lands. *The Analysis of Sensations* says nothing of this development—nor does "The Uncanny." And yet, mobile reflective surfaces, gliding panoramas, and the countless doubles called "commuters" [*Verkehrsteilnehmer*] have existed only since railroads and the internal combustion engine were invented. The same Mallarmé who put an end to reading and readability advised engineers to move engines to the back of vehicles. Then happy passengers might feast their eyes, without any interference, on the "magical" spectacle of perspectives gliding by through "bow windows." Mallarmé called his "invention" the "vision of a commuter with taste"—the automobile as a traveling camera.[17]

Such was the vision of a writer who systematically sealed off his own medium—writing—against effects of hallucinations and doppelgängers. When asked, in a survey, about illustrated books, Mallarmé replied with a categorical no. The inquiry prompted a counterquestion on his part: "Why not go to a cinematographer right away, whose sequence of images will replace the text and pictures in many a book to great advantage?"[18] That is plain talk, too. Since 1895, there has been a split between an imageless cult of letters called "serious literature," on the one hand, and on the other, any number of technical media that motorize images—like railroads or film. Literature no longer even tried to compete with the wonders of the entertainment industry. It handed its magic mirror over to the machines.

This and only this accounts for the horror and dismay experienced by Professors Mach and Freud, when, for a few milliseconds, the old-fashioned medium of the book had to yield to the film of so-called reality. Silent films implement, in technological positivity, what psychoanalysis can only think: an Unconscious that has no words and is not recognized by His Majesty the Ego.

The very dumbness [*Dummheit*] of film makes it an advantageous replacement for many a book—and Romantic ones, especially. Film can store bodies that, as everyone knows, are just as dumb. When, in the last comedy of Romanticism—Georg Büchner's *Leonce and Lena*—King Peter of Popo ordered a search for his fugitive son, the policemen of the Grand Duchy of Hessen were

Romanticism, Psychoanalysis, Film

not to be envied. They had only the "warrant [*Steckbrief*], the personal description [*Signalement*], and the certificate" of a human being in general: "Walks on two feet, has two arms, a mouth moreover, one nose, two eyes, two ears. Distinguishing characteristics: a highly dangerous individual."[19] That far, and no further, did poetic art go in the matter of recording bodies—as far as the individual generalities of a sketch à la Wilhelm Meister. In contrast, film—like criminology and psychoanalysis, too—numbers among the technologies for securing evidence that, as Carlo Ginzburg has noted,[20] optimize the control of fugitive bodies.

One finds ready proof in all the dumb, crazy, mongoloid, and hysterical bodies that early silent films parade. Every single one of them presents the shadow of the body of the party filmed —in other words, a doppelgänger. A camera pan would suffice for King Peter to have the unmistakable, unforgeable certificate of authenticity for his son—Leonce storming through nature as a Romantic actor. Parties who believe that printed words mean them have simply been misled. And anyone who is filmed has already been caught and handed over—even if this occurs simply through mobile mirrors, as in Freud's case. On film, all actions look dumber; on sound recordings—which suppress bone conduction from the larynx to the ear—voices have no soul; and on passport photos, everyone looks like a criminal. This is not the case because media lie, but because they dismember the narcissism underlying a unified conception of the body.

Media enact a historical escalation of violence, and they force those affected into total mobilization. The first theorist of the uncanny seems to have intuited more of this than his critic, Freud. Already in 1906, Ernst Jentsch compared the panic provoked by automata or doppelgängers to the collapse of a "defensive position"—the "lack of cover in the events" of a "war" that, in his estimation, "never ends."[21]

As is well known, in 1917 UFA—the German feature film company—was founded under the auspices of the Office for Image and Film of the General Staff [*Bild-und-Film-Amt im Großen Generalstab*]. This occurred at the behest of First Quartermaster General and Infantry General Erich Ludendorff.[22] It is little surprise, then, that the media war has never ended. In Vietnam, elite units like the U.S. Marine Infantry were only prepared for attack and death on condition that NBC, CBS, or ABC had a camera team at the site of deployment.[23] The very fact that a body was torn apart by a Vietcong grenade made its double immortal on the evening news. *Apocalypse Now* or total mobilization . . .

A Story of Doubles

Figure 1. Hanns Heinz Ewers, *Der Student von Prag. Romantisches Drama in vier Bildern. In Szene gesetzt vom Verfasser* (Deutsche Bioscop GmbH, 1913). The double (Paul Wegener) separates the Student (Paul Wegener) and his beloved (Grete Berger).

Ever since—to the understandable dismay of *Lebensphilosophie*[24]—film cameras have chopped apart, with a shutter and a Maltese Cross, bodies in front of viewfinders in order to shoot twenty-four pictures per second. Lacan's *corps morcelé* has become a matter of positive fact. The disarticulated body takes the place of the whole persons that Classical and Romantic poetry both celebrated and produced. The great arch of hysteria, for example—a physiological form of total mobilization—was not created only by Charcot's staff and hand, which (as is well known) he considerably passed over the nether regions and ovaries of his female patients.[25] The great psychiatrist was more modern than that, and he said as much, too. At the Salpêtrière, it became possible to record hysteria for the first time in the history of medicine because new machines and machine operators had transformed a run-down Parisian madhouse into a laboratory.[26] In 1883, Albert Londe—Charcot's engineer and the inventor of the Rolleiflex—had already built a camera with nine or twelve lenses; at the signal of a metronome, it took successive snapshots—that is, it made films avant la lettre. The objects of this chopping-apart were the hysterics at the Salpêtrière; the spectator of this spectacle of disarticulation was a young Sigmund Freud.[27] The hysterical arch must only have become grander and more majestic when cameras recorded it (or produced it in the first place . . .).

A total mobilization paved the way for psychoanalysis, but Freud paid no mind. The word "cinema" does not occur in his writings. Instead, Freud left it to his literary-historical adjutant to apply his theory to film. Cinema provides the point of departure for Rank's study of the double—which was published immediately after the first screening of the first German film by an auteur. Seeking "to uncover deeply buried and significant psychic material," Rank was not afraid to choose what he called a "random and banal subject": Hanns Heinz Ewers's *The Student of Prague*. He even speculated that

> cinematography, which in numerous ways reminds us of the dream-work, can also express certain psychological facts and relationships—that the writer often is unable to describe with verbal clarity—in such clear and conspicuous imagery that it facilitates our understanding of them.

With due precision, Rank recorded all the "shadowy, fleeting, but impressive scenes" that showed the titular student engaging in a sixty-minute duel with his mirror image and double. (After all, in 1914, videotape—that is, an optical means of rereading—had not been invented yet.)

Rank did so simply in order to unravel unconscious symbolism in a banal mass medium, as if Freud's manifest dream content and the entertainment industry occupied one and the same plane. Conversely, he viewed the latent content of dream and/or film—if only because Ewers commendably followed literary "models"[28] in his screenplay—as being constituted by discourse (and nothing but discourse). Of all things, it was a silent film that brought Rank to the Romantic poetry of doubles—and from there to mythology and/or psychoanalysis. Nothing, then, came of his promise to link dream-work and cinematic representation, Freud and Londe. The psychic apparatus obstructs all sense for technical apparatuses. Even when—at the end of his historical-methodological regression—Rank refers to the Fiji Islander who called his first look into a European mirror a gaze into the spirit world,[29] it does not occur to him that occult media have always required that technical ones precede them.

The psychoanalysis of film reverses filming. As if no technological thresholds existed, it verifies a literary work [*Dichtung*] that film has in fact replaced. Freud's primal scene—his year at the Salpêtrière—has been successfully repressed.

Therefore, it is only half true when Tzvetan Todorov, in *The Fantastic: A Structural Approach to a Literary Genre*, concludes that

> psychoanalysis, and the literature which is directly or indirectly inspired by it, deal with these matters in undisguised terms. The themes of fantastic literature have be-

come, literally, the very themes of the psychological investigations of the last fifty years. We have already seen several illustrations of this; here we need merely mention that the double was even in Freud's time the theme of a classic study (Otto Rank's *The Double*).[30]

Todorov is right when he has the Romantic doppelgänger meet its demise around 1900. However, it is impossible to believe that theory alone could have exercised such power. Only in the pincers of science and industry, psychoanalysis and film, did the empirical-transcendental doublet "man"—this substrate of Romantic fantasy—implode. All the shadows and mirrors of the subject were clinically verified by psychoanalysis, but cinema implemented them technically. Since then, literature that wishes to be Literature, writ large, must be *écriture*: writing without an author. And no one can derive [*herauslesen*] doubles—that is, possibilities of identification—from letters alone.

But because—as everyone knows—ghosts do not die, a new form of the fantastic has emerged, one wholly distinct from literature. Cinema and screenplay writers filled the positions left vacant by Romanticism. As the first theorist of film recognized, "every dream becomes real"[31] at the movies. What poetry promised but provided only in the imaginary space of reading experiences appeared in the Real [*im Realen*] on-screen. To be transported into an actual [*wirkliche*], visible world, correct reading—which Novalis had deemed absolutely necessary—was now superfluous. No longer did one need to be educated or drunk. Even, and especially, people who were illiterate could see the student of Prague, his beloved, and his mistress—all the "shadowy, fleeting scenes" to which Rank refers—for what they already were: doubles, the celluloid ghosts of actors' bodies.

All that was necessary was for the ingenious Méliès to take the stage and complete the documentary approaches of Londe or the Lumière brothers with his bag of tricks, so that cinematic doubles of the first order were joined by cinematic doubles whose potency had been squared. With mirrors and multiple exposures, it was easy to show the actor playing the student twice at the same time. When the film's protagonist practices fencing in front of a mirror, his reflection promptly steps out of the frame. Whether this "uniqueness of cinematography"—as Rank affirms—"visibly portrays psychological images"[32] remains an open question. It is clear, however, that it films filming itself. Cinematic doubles demonstrate what happens to people who step into the firing line of technical media. Their mechanized likeness wanders into data banks that store bodies.

The printed program for *The Student of Prague* had already presented "the double figure of the hero" as "a form of possible expression that only the cinema can show—in such perfection as the stage will never achieve."[33] On the stage, the single yet doubled student would have degenerated into two actors—and on the pages of a book, his presence would have remained an empty assertion. Conversely, as the "film problem of all film problems," as Willy Haas put it,[34] the doppelgänger effect set the course for early cinema. Ewers's *Student*, Paul Lindau's *The Other* [*Der Andere*], Gerhart Hauptmann's *Phantom*, Paul Wegener's *Golem*, Robert Wiene's *Caligari*—to say nothing of innumerable versions of *Jekyll and Hyde*—all present variations of the "special effect of all special effects" [*Filmtrick aller Filmtricks*], as one might put it more simply, and accurately.

The reason is perfectly clear: special effects—whether in film, love, or war—represent strategies of power. Only in the clichés of Germanist scholarship do Expressionist films critique the Wilhelmine bourgeoisie. In actual fact, they rehearse—which is to say, they drill—a new dispositive of power: "How to do things without words."

Lindau's *The Other* features a man who, because of a neurophysiological split in his personality, is both a public prosecutor and a criminal, the hunter and the hunted. With all the arguments of psychiatry and the whole arsenal of criminology, a civil servant who is simply not up-to-date with technology has it drilled into him that his juridical conception of personhood (to say nothing of other ideas) is over and done, now that even mute, corporeal traces can be captured. The film concerns powers that the protagonist himself represents—ones to which he fatefully belongs.[35]

It is only logical, then, when the magical power of Rabbi Löw in Wegener's *Golem* culminates in his screening a film-within-a-film to Kaiser Rudolf. (Kaiser Wilhelm, the media maniac of 1914, surely appreciated this.) The fact that the rabbi can build a motorized automaton by the name of "Golem" hardly allegorizes (as film historians contend) "the risk of a dictatorship established for a fixed duration by the ruling class, which turns against its initiators themselves."[36] Even if one leaves the "greatest cineaste of all time" (Syberberg) aside, Golems pose a danger: they are stupid doubles of a human being that has no longer existed ever since media—according to McLuhan, "extensions of the body"—started to replace even central nervous systems.

When a film begins in a projection hall—darkened as if for an air raid—(whose prototype, in the history of art, can only have been Wagner's *Festspielhaus*[37]), the substitute central nervous system extends to the audience. Whether viewers belong to the ruling classes like Rudolf, Wilhelm, or von Papen, or whether they belong to a ruled class like everybody else, everyone's retina fastens on the screen. "The spectator," Edgar Morin has written, "reacts before the screen as before an external retina telelinked to his brain."[38]

Film is total power, even—and especially (as in the case of Rabbi Löw and his magic tricks)—when it exposes itself. For as long as doublings remained literary, as in the case of the book-within-a-book in *Wilhelm Meister*, they could be read as reflections, as an invitation to so-called critique. In contrast, technical media and aversion strategies celebrate victory precisely by exhibiting themselves. After all, how can one "get behind" a prosthesis of the central nervous system—behind what used to be called "the soul"?

A few twentieth-century authors have understood as much. A form of the fantastic extends from Gustav Meyrink's *Golem* up to Thomas Pynchon's *Gravity's Rainbow* that has nothing to do with Hoffmann or Chamisso and everything to do with the movies. Literature of the central nervous system competes directly with other media—for this reason, perhaps, it has always already been destined for filming. Making present instead of narrating, simulating instead of authenticating: such is the motto. Meyrink's *Golem*, which appeared in 1915, begins with an unnamed speaker and a mounting physiological presence. The speaker "no longer possesses an organ" with which he might ask the question, "who is 'I' now?" Therefore, reflexive questions are replaced by the flow of pure neurological data—that is, by what always already has been a film on the retina [*Netzhautfilm*].

Bit 1: "The moonlight is shining on the foot of my bed, lying there like a large, bright, flat stone." This large, flat stone in the first sentence of the novel promptly loses its comparative function and shifts from being a literary metaphor into the Real of neurophysiology. Bit 2: "And the image of the stone that looked like a lump of fat grew in my mind to monstrous dimensions [*ins Ungeheuerliche*]." In keeping with the logic of camera "travel," the monstrous close-up fills the half-sleeper's whole optical nervous system. Bit 3: "I am walking along a dried-up river-bed, picking up smooth pebbles." This space—still the foot of the bed, yet already a riverbed, too—turns into time: the close-up transforms into a flashback. Bit 4: "All the stones that ever played a role in my life push up out of the earth around me."[39]

And so on and so forth in the first chapter—until an array of cinematic special effects has made a spot of moonlight in Life A into the old-city ghetto of Prague in Life B. The "cinematographic illusion of consciousness" that Bergson discusses in his contemporaneous theory[40] transfers the caesura between biographies and epochs into the perfect continuum of a retinal film: through the hole in his identity (which does not exist) the anonymous "I" of the framing narrative plunges into his double—"Pernath"—who, a lifetime ago, experienced the events of the framed narrative. The doubling of the doppelgänger motif proves that the old-city ghetto in Prague is a film. Just as the anonymous "I" has toppled into Pernath, Pernath in turn topples into a Golem, who—in explicitly photographic terms—is called Pernath's "negative."[41] The much-maligned mysticism of the novel, then, is simply a matter of media-technical precision. With Meyrink, literature, for the first time, made it evident how the brain's physiological processes correspond to film sequences. The soul is not real—celluloid is.

Dream-work and cinematic representation are more closely linked than Otto Rank dared imagine in 1914. No psychoanalytic theory of the double can adequately conceive Meyrink's endless flights of doppelgängers or Schreber's "fleeting-improvised-men."[42] The science of the epoch alone is responsible—the same science that lay the foundations which made film possible in the first place. Without the experimental psychology of parties such as Helmholtz and Wundt, there would have been no Edison and no brothers Lumière; without the physiological measurements of the retina and the optical nervous system, there would have been no moviegoers. For this reason, the first adequate theory of film was offered by the head of the Harvard Psychological Laboratory. In 1916, Hugo Münsterberg thought out what Meyrink had described in 1915. This occurred, simply enough, because the great experimental psychologist found the name for, and practice of, a new science: psychotechnics.[43]

Psychotechnics—this coupling of physiological and technical experimentation, of psychological and ergonomic data—made the theory of film possible in the first place (to say nothing of work on the assembly line and combat training). Münsterberg effortlessly demonstrated, for the first time in the history of art, that feature films are capable of implementing the neurological flow of data itself. Whereas the traditional arts process orders of the Symbolic or orders of objects, film broadcasts to viewers the process of perception itself—and with a precision, moreover, that is otherwise accessible only under experimental conditions (i.e., unavailable to consciousness or language). Münsterberg assigned to each camera technique an unconscious psychic mechanism: the close-up corresponds to selecting an item for attention, the flashback to involuntary memory, special effects to daydreaming, and so on.[44]

For all that, mathematical equations can be solved equally well from the left and from the right. The name "psychotechnics" already declares that the film theory offered by experimental psychology also provides a theory of the mind/soul in terms of technical media. As in Meyrink's *Golem*, involuntary memory equals "flashback," attention equals "close-up," and so on. Unconscious mechanisms—which previously were accessible only through experiments on human subjects—took leave of people and henceforth populated film studios as doubles of a soul that had died. One golem is a tripod or a set of muscles, another the celluloid or the retina, a third a flashback or memory...

Münsterberg, moreover—after leaving Freiburg im Breisgau for Harvard—took a decisive step, too. He made sure to visit, in New York, the film studios whose products he theorized. There lies the difference between Münsterberg and Rank, between the expert knowledge of an engineer and the standpoint of a consumer.

Over the course of time, Freud—that self-declared prophet—has come to enjoy all the glory due to other discourses. Today Hugo Münsterberg is mentioned only in Freud biographies—where, moreover, his name is incorrectly given as "Werner"—as one of many audience members during the tour of the United States that psychoanalysis made in 1908.[45] The truth about technical media [*Medientechnik*] has been thoroughly repressed ever since Münsterberg made a final, fateful step. In 1916, Münsterberg declared himself a strategist in the First World War; as a result, he suffered scientific excommunication.[46] No evidence can be secured without destroying evidence as well. And without repressing founding fathers, no film companies can be founded per the instructions of a General Staff. Now that all theories can be implemented, there no longer are any. Such is the uncanniness of our times.

7 Media and Drugs in Pynchon's Second World War

For David Wellbery

In 1983, during the "German Autumn," an announcement of the German News Agency (DPA) circulated in the press:

> The Chairman of the CSU and Bavarian Prime Minister Strauss has obtained, by his own indications, "fairly reliable information," according to which the German Democratic Republic has, for years, been fitting subterranean facilities that date from the Third Reich for the installation of nuclear missiles. These "natural fortresses" are, in part, up to three- or four hundred meters beneath a rock layer, so that the nuclear weapons are secure, Strauss announced at an international symposium organized by the Hanns Seidel Foundation.[1]

What the DPA withheld was that the "natural fortresses," especially the ones near Nordhausen in the Harz Mountains, had already housed missiles once—and even produced them on a massive scale. For this reason, the SS-20s in their rock bunkers or the Pershings traveling on our national highways[2] simply describe the trajectory—the rainbow—of an eccentric homecoming.

1. War

"Gravity's rainbow" names the flight trajectory of the V-2 rockets that were launched from sites in Holland or Lower Saxony during the last six months of the war—from 8 September 1944 until 28 March 1945[3]—and flew over the German-Allied fronts at metropolises like London and Antwerp. *Gravity's Rainbow* also refers to Thomas Pynchon's attempt to read the signs of the times as a novel. These signs, notwithstanding all postwar dreams,[4] were written by the last world war, the "mother"[5] of the technologies that have made us what we are and the "mother" of our postmodern condition—which "threatens the idea of cause and effect itself" (56).

The V-2, which Wernher von Braun and the Army Research Center [*Heeres-*

Media and Drugs in Pynchon's Second World War

versuchsanstalt] at Peenemünde transformed from the plaything of engineers into a "wonder weapon" ready for mass production and deployment, was the first rocket in the history of war to use liquid fuel. In Pynchon's abyssal fiction, it even anticipates, at the end of the fighting—and roughly following Braun's blueprints—the manned space travel of today. For this reason, it stands at the focus [*Brennpunkt*] of a novel that reads the signs of our own times. Conversely, and at the farthest horizon of both the novel and of theaters of war, parallel American weapon development occurs as well, in Hiroshima and Nagasaki (480, 505, 539). One need only replace the V-2's standard charge—a ton of burning amatol (96, 312) ignited, at Hitler's personal recommendation,[6] just before ground contact—with a payload of uranium or plutonium in order to reach the state of things in 1985. Whereas, according to a secret directive of 15 October 1942, the German Army High Command only saw in "atomic decay and chain reactions" "a possible propellant for rockets,"[7] Enrico Fermi and John von Neumann were already at work on a fitting payload that (as progress has meanwhile shown) was much too good for their own *Enola Gay*s (518) and bombers.

Accordingly, the alliance between Germany and the United States [*Deutschamerikanische Freundschaft*] occurring as a technology transfer provides the theme of Pynchon's novel. What started on the sandy beaches at Peenemünde and grew into mass production in the bunkers of Nordhausen (which IG Farben built and the *Reich* took over)[8]—where, moreover, the first *Düsenjäger* were manufactured (304)—has continued in Huntsville (558f.) and Baikonur (705f.). The sum of all innovations that the Second World War occasioned, from magnetic tape (522) to color film, radar (388f.), VHF (325f.), and computers (259f.), produced a postwar period whose simple secret is the marketing of wonder weapons. As a result, the future can already be foreseen.

To be sure, in the Second World War there were still people who believed that they were dying for their countries [*Vaterländer*]. However, the precise details provided by Pynchon—a former Boeing engineer—make it clear that the "enterprise of systematic death" (76) "serves as spectacle, as diversion from the real movements of the War" (105). "The real crises," namely, "[are] crises of allocation and priority, not among firms—it [is] only staged to look that way—but among different technologies, Plastics, Electronics, Aircraft," and so on (521).

If World War II was a theater of war in the literal sense, and if its sea of corpses was a simulacrum—a screen behind which various technologies fought

85

for their future (and our own)—then everything played out in the media. And media, from drama up to computers, simply convey information. Competition and disputes about priority between technologies have always already amounted to struggle for information about them. As one character working in industrial espionage puts it melancholically: "life was simple before the first war"; then, "dope and women" were the only matters of interest. But ever since 1939, "the world's gone insane, with information come to be the only real medium of exchange"; now even industrial espionage is in the process of switching over from agents (i.e., human beings) to "Information machines" (258ff.).

When conditions of totalizing semiotechnics prevail, the only real question involves the media they implement. And if as Pynchon puts it, "the more you dwell in the past and in the future, the thicker your bandwidth, the more solid your persona" (509), then media research [*Medienwissenschaften*] would do well to recall the military history of the objects it studies. It could be that the narrativity—that is, the entertainment—that media seem to offer is only a screen for semiotechnical operations [*Effizienzen*]. Media such as literature, film, and sound recording are all at war. That is the reason why *Gravity's Rainbow* pursues their systematic combination.

2. Literature

In that mythical prehistory—that is, when drugs or women were still of interest—war may have been a song sung by a soldier: a verbal, narrative matter. But ever since universal conscription declared that "no one may be absent from the field" [*im Felde niemand fehlen darf*], which Goethe recognized right away, there have been no more listeners for narratives: everyone is concerned.[9] The Wars of Liberation between 1806 and 1815—which made the people of Central Europe into the underlings of nation-states (i.e., set militia armies [*Volksheeren*] free)—needed a new medium. This medium was literature as writing and command [*Schrift und Kommando*]. The new and absolute enemy just had to be named and his destruction ordered. And that is exactly what dramas like Kleist's *Hermannsschlacht*—that commanding position [*Feldherrnhügel*] in the propaganda war—did.[10]

As is well known, writers' fortunes [*Schriftstellerglück*] of this kind did not last. When the hilltop command posts disappeared in battles of matériel during the First World War, literature had to descend to a perspective of grunts on the line of battle (as Paul Fussell's brilliant study has shown apropos of English texts[11]). Absolute enmity, now that it had been taken over by machines, no longer needed narratives, justifications, or plans. Facing obscure orders and invisible foes, literature became a matter of *Combat as Inner Experience,* as Ernst Jünger aptly put it. And that means—quite simply—film. At the outer limit of the book medium, where explosions refute all language,[12] its technological replacement appeared. Whenever Lieutenant Jünger—instead of continuing to write Expressionistic studies of experience—stumbled upon reality [*ein Reales*] in morning fog and barbed wire, he encountered a cinematically hallucinated doppelgänger.[13] For this same reason, novels written from the perspective of frontline soldiers—as Erich Maria Remarque also demonstrated—lend themselves to filming.

But if producing different ways to die [*Todesarten*] and simulating relationships of friend and foe only serve to mask competing technologies—which, for their part, are not based on experience or narration, but rather on blueprints, statistics, and secret commands—then the viewpoints of combatants on the front are obsolete. *Gravity's Rainbow*, which secures evidence from the second, technological world war, employs different narrative techniques from the inception.

In lieu of a war as inner experience, Pynchon presents a stochastic scattering of figures and scenes, of fronts and discourses, of Allied and German positions. Only when two chance distributions meet by chance does the perspective of a protagonist—a plot—emerge. To wit, the Poisson distribution, in which the V-2s rain down on London, matches, point for point, the private statistics that an American lieutenant named Tyrone Slothrop has compiled about his chance erotic encounters. Just as the rockets, which travel at a speed faster than sound, confuse cause and effect, audible threats and visible explosions (23),[14] Slothrop's erections provide the index (in the double sense of Charles Sanders Peirce and all prophets), which already marks the next site where a strike will occur. The V-2s follow his arousal just as the sound of flight comes after the explosion. In other words, Slothrop's love—or "imagination"—has the structure of a bomb.[15] That is reason enough for the Allies to employ the lieutenant as an experimental subject in the most technical capacity. Slothrop is smuggled into the collapsing *Reich* to track down that ultimate, one-of-a-kind, mythical rocket that is transporting his German double into space and/or death.

Except that Slothrop escapes the "operational paranoia" (25) of the Secret Service to the same extent that it grabs hold of him. This shift occurs through the medium of writing. The lieutenant is descended from Puritan paper manufacturers, that is, from people who "converted" America's "diminishing green reaches ... at a clip into paper—toilet paper, banknote stock, newsprint—a medium or ground for shit, money, and the Word" (28). This realm of the Symbolic, to adopt the language of Lacan, catches up with him while reviewing looted V-2 documents. Reading and paranoia coincide. All the traces that Slothrop learns to decode in the fortress of Europe, that is, point to the fact that the military-industrial complex has always already transcended all wartime fronts—that is, it has conditioned both the sexual reflexes of American GIs and the innovations produced by German missile engineers.

Poring over the dossiers that have long since governed so-called experiences or life stories, Slothrop learns that already when he was an infant—along parallel lines running between IG Farben and Rockefeller's Standard Oil (a connection that, moreover, is 100 percent accurate historically[16])—he was the subject of behaviorist tests conducted by the same "Professor Jamf" who, with his synthetic polymers, has also made manned space travel possible. Belatedly, as always, it emerges then that the detective and his doppelgänger meet up in the cockpit of a V-2. Likewise, it turns out that the meeting of two iconic patterns—the historically real map of missile strikes and the novel's map of erotic exploits in London—is anything but a matter of chance. When documents are thoroughly studied, coincidences always point to a conspiracy.

This sinister conclusion, however, does not—as readers trained for guilelessness might presume—rest on anything immanent to the fiction. Rather, it involves the historical precision of what the text itself calls "data retrieval" (582). Slothrop's paranoia, which is internal to the novel, repeats—and step by step, at that—the critical-paranoid method that novelists might make their own on the model of Salvador Dalí. Despite the fact that dossiers are open to writers and protagonists in reverse chronology, this does not mean they are fictitious. *Gravity's Rainbow* has been honored hundreds of times as the epitome of postmodern literature, yet scholars do not say a word about the extent and thoroughness of the research that it incorporates. To a degree comparable only to historical novels such as *Salammbô* or *The Temptation of St. Anthony*,[17] the book builds on documentary sources, even if—and for the first time—circuit diagrams, differential equations, business contracts, and organizational charts are also incorporated. (Admittedly, such matters are easy for literary critics to overlook.)

Gravity's Rainbow retrieves data from a world war whose secret files have become accessible to the extent that the goals they articulated have become reality [*ins Reale eingezogen*]—to the extent, that is, that they no longer need to be kept secret. For this very reason, paranoia, which according to Freud or Charles William Morris is, like all psychoses, just the confusion of words for things,[18] of designata and denotata,[19] is knowledge [*Erkenntnis*] itself. When the Symbolic of signs, numbers, and letters determines the course of so-called realities, securing evidence [*Spurensicherung*] becomes the first duty of the paranoiac.

Consequently, the novel's critical-paranoid method extends to its readers. They change from the consumers of a narrative into hackers of a system. Slothrop—notwithstanding all his puritanical love for the Word (207)—does not, by any means, decode all the war secrets that the novel has encrypted. It would be impossible for him to decipher that the fictive Major Marvy, who is responsible for transferring V-2 technology to the United States, just represents a cryptogram for the historically accurate name "Staver."[20] Or that Pointsman, the chief behaviorist of the British secret services in the novel, has this name to coincide, in the multinational conspiracy, with his onomastic double in Germany—a certain engineer named "Weichensteller," whose "responsibility" at Peenemünde involved, of all things, the "re-entry" (453) of V-2s in British airspace.

In *Gravity's Rainbow*, fictive names and narrative structures mask a level of information that also connects with other novels that are just as paranoid (cf. 587f.)—ones that, for practical reasons, it is better not to speak of openly. And with that, the novel is absolutely up-to-date. When technologies assume dominance over science and aesthetics, only information counts. After all, many of the roots of semiotics itself lie in those behaviorist semiotechnics that Pynchon analyzes in terms of wartime strategy.

When one analyzes and recombines data that are just as secret as they are scattered, however, this entails two problems: the closing and self-application of the system. It is not just because Slothrop's data retrieval takes place in 1945—that is, long before the relevant archives were opened—that he is "dancing on a ground of terror, contradiction, absurdity." First of all, it would have been easy for the military-industrial complex to have "brought programmers by the truckload to come in and make sure all the information fed out was harmless" (582)—as harmless as a novel, for example. Secondly, Tyrone Slothrop's paranoia makes him conclude that his desire is only his own (216)—even though it really was, always and already (and as Lacan would put it), the desire of the Other, that is, of a scientific investigator [*Versuchsleiter*]. Surpassing experiments on historical models (Watson and Baby Albert), Jamf has had the "elegant" idea, which is called this because it is "binary," of conditioning Baby

89

Tyrone not in terms of unquantifiable data such as fear, but rather in terms of the straightforward and unambiguous fact of erection (84ff.). With due consequence, then, Slothrop beholds, in dreams, a "very old dictionary of technical German," which translates "JAMF" (the proper name of the scientist) as "I," the index of speech in English (287; cf., 623).

The ego, then—to speak in other words, which are still Pynchon's—is only "a branch office in each of our brains, his corporate emblem is a white albatross, each local rep has a cover known as the Ego, and their mission in this world is Bad Shit" (712f.; cf. 285f.). There ends a quotation that could just as well have been written by Foucault, which, moreover, is also where all paranoia ends. For nothing and no one remains of an involuntary private eye who has cracked the *alibi* (etymologically, the "elsewhere") of his own I. Under the conditions of total remote control, the heroes of novels can no longer be narrated. In a never-ending series of exchanged clothing and metamorphoses, Lieutenant Slothrop loses his uniform, his proper name, and his literacy; he dissolves into episodes, comic strips, myths, and finally, album covers (742). In this way—and only in this way—does he escape the trap that the medium of writing, which is itself part of the military-industrial complex, sets for readers as such. If paranoia is said to consist of the foreboding reading of a single, coherent scheme that can be narrated (703), "there is still also anti-paranoia, where nothing is connected to anything" (434).

And if the historical genre "novel" was defined by the fact that possibilities of ramification of its Markov chains lessened in direct proportion to the course steered by the hero, until, finally, a structure or solution was achieved, then the antiparanoia of *Gravity's Rainbow* produces precisely the opposite: an increase of information and thereby (following Shannon) of entropy. In its progressive mixing of standing figures, organization, and fronts, the novel purposefully repeats the second principle of thermodynamics. The law that entropy always increases gives the arrow of time its direction and therefore—according to an apposite example provided by Sir Arthur Stanley Eddington—can determine whether films are running forward or backward in physical time.[21]

3. Film

In this technological and temporal sense, *Gravity's Rainbow* is cinema. Not because the novel lends itself to film adaptations, as in the case of Remarque's book, nor because it hallucinates invisible enemies, as in the case of Jünger's work, but rather because it sets its own progressive dissolution against the negentropy of military-industrial complexes. Already the present tense that is sustained throughout its many parts—in contrast to the classical past tense of novelistic narration—induces forgetfulness, which does not allow linear chains of cause and effect to emerge in the first place. "Each hit is independent of all the others. Bombs are not dogs. No link. No memory. No conditioning." Therefore, there can also be no question, "which places would be safest." And thanks to such training, "a whole *generation* has turned out like this"—one whose experience of the postwar consists of "nothing but 'events,' newly created one moment to the next" (56).

Accordingly, only the "Monte Carlo Fallacy" (56) might induce one to assume that a missile strike, a film image, or a novelistic event—the value N, as if it had a memory—could be determined by the series of 1 to $N-1$. To be sure, for the principal behaviorist in the text, rocket fire over London signals the fact that "the reality is not reversible." It could only end if "rockets" were to "dismantle" and "the entire film [ran] backward: faired skin back to sheet steel back to pigs to white incandescence to ore, to Earth" (139). However—and as (of all people) Walther Rathenau, the inventor of German war economies and, therefore, of Soviet Five Year Plans, says when his spirit is summoned—"all talk of cause and effect is secular history, and secular history is a diversionary tactic" (167), that is, "conspiracy" (164). As is well known, secular history once resided in the medium of the book. In contrast, technical media enable—besides the diversionary tactic called "entertainment"—the modification of precisely those parameters which they, and only they, control; that is, they enable changes of physical temporality. Just as a rocket strike reverses the sequence of explosion and noise, thus do the many fictive films in *Gravity's Rainbow* work with the trick that in the dialect of real-world electronics engineers is known as "Time Axis Manipulation."

The last film by Gerhardt von Göll—who stands in for contemporaries such as G. W. Pabst, Fritz Lang, and Ernst Lubitsch (112)—is called *New Dope*. "24 hours a day," it demonstrates how this drug makes one "incapable of ever telling anybody what it's like, or worse, where to get any."

> It is the dope that finds *you*, apparently. Part of a reverse world whose agents run around with guns which are like vacuum cleaners operating in the direction of life—pull the trigger and bullets are sucked back out of the recently dead into the barrel, and the Great Irreversible is actually reversed as the corpse comes to life to the accompaniment of a backwards gunshot (745).

Only, this kind of special effect does not remain limited to the Imaginary of hallucinations and trips to the movies. When the novel describes the British bombing of a V-2 launch site, "vehicles" are transformed "back to the hollow design envelopes of their earliest specs" (560). Thereby, it hints at the most sinister of its many paranoiac insights: the fact, namely, that Germany's industrial facilities—following the theory of ruins conceived by their chief, Albert Speer[22]—were already built with the destruction wrought by the Royal Air Force in mind; indeed, it is precisely as ruins that they will fulfill their role in multinational conspiracy, after the war (520f.).

Similar reversals of time—if not in such a calculated fashion—are also performed by Göll's first work: a fake documentary film made in keeping with all the rules of the Allies' "Black Game."[23] Britons made up to look like Hereros play one of Major General Kammler's motorized rocket batteries. Once completed, the film is artificially aged and damaged—that is, noise, which defines technical media by constituting their background (cf. 94), is added. Then, as the pseudo-evidence of a made-up V-2 location, it is deployed in order to trigger German rumors about "negroes" in the Waffen-SS (113ff.). "'It is my mission,'" von Göll declares, "with the profound humility that only a German movie director can summon, 'to sow in the Zone seeds of reality'" (388, cf. 275). Verily, in 1929, Lang's *Die Frau im Mond* (*The Woman in the Moon*) had sowed the seeds for countdown (753) and the V-2 project as a whole.[24]

Nor is this reversal of cause and effect—of programming and documentation—the end. The spiral expands. After the fact, it comes to light that von Göll's Hereros in the Waffen-SS were not the effect of propagandistic simulation, but rather its magical cause. Because they already exist, von Göll's forgery would have had to run backwards—just as countdowns do. And so, once more, the novel poses the $64,000 question: What is the relationship between programming and narrativity in media?

War and Cinema, by Paul Virilio, seeks to demonstrate that world wars and film technology are not just contemporaneous, but also stand in a relation of strict solidarity. Warfare that bets, in military, technological, and propagandistic terms, on speed and information cannot operate without temporal abbreviations, expansions, and reversals—without "time axis manipulation," that is. What would be impossible in the medium of writing or literature, notwithstanding Ilse Aichinger's "Mirror Story" [*Spiegelgeschichte*], has been the program of film from the very beginning (which, in turn, involves revolvers, among other things[25]). To be sure, literature has been able to manipulate those times that make believe [*vorspiegeln*] that a path of education [*Bildungsweg*] or "combat as inner experience" is occurring. But to be able to work with physical time itself—where paths of education and struggles for life and death actually take place—it is necessary to employ technical media. Missile technology needs film technology and vice versa. V-2s—to the disbelief of the newly created technical wing of the British Secret Service[26]—only hit their targets in London at all because of an ingenious innovation. The parameters measured for rockets concerned neither their paths, as had always been the case for armies, nor speed, as more recently had held for tanks. Rather, they involved acceleration—the only information available to the missiles themselves, which made speed calculable through simple integration, and then, through double integration, did the same for flight paths (301f.). A pendulum and two RC circuits in a row, and Virilio's "dromology" is already up and running, a matter that is just as easy as it is easy to overlook (which is precisely what British experts did).

Pynchon writes:

> There has been this strange connection between the German mind and the rapid flashing of successive stills to counterfeit movement for at least two centuries—since Leibniz, in the process of inventing calculus, used the same approach to break up the trajectories of cannonballs through the air. (407; cf. 567)

For all that, the technological medium that actually implements movement as infinitesimal calculus is called "film." Ever since Étienne-Jules Marey's photographic gun,[27] all cinematic illusions of continuous movement have been matters of single integration, like the speed of the V-2—dependent variables of time axis manipulation that is all that counts for optimizing weapons of destruction. As was already the case for the mechanical antecedents of cinema in 1885, the high-performance Ascania cameras used in 1941 were not developed for moviegoers, but rather for slow-motion study of the V-2 in flight (407). Needless to say, the matter hardly precludes the possibility of expanding such technology "past images on film, to human lives" (407).

One of the many narratives whose entropy constitutes *Gravity's Rainbow* puts narratability itself into question by means of technology. This component of the novel focuses on a Peenemünde engineer who is tricked by time axis manipulation. The simulacrum in his film (or life) is the man's twelve-year-old daughter, who, incidentally, owes her conception to the semiotechnics of film in the first place. To wit, one of the rape scenes in von Göll's late-Expressionist cinema—which was edited before coming to a climax in theatrical release, but performed to the bitter end in the studio (as well as in Joseph Goebbels's private archive)—led to the impregnation of the actress herself, and to the impregnation of countless wives and girlfriends when moviegoers returned home. Under high-tech conditions, children are simply the doubles of their doubles on the screen: boys are cannon fodder, and girls are pinups.

Fast forward to thirteen years later: film-bred cannon fodder is sent off for the Blitzkrieg, and pinups are needed. Being a Pynchon character, the rocket engineer has, of course, long since forgotten about his daughter and what she looks like. But from 1939 on, she has been showing up for every wartime summer vacation—a "bonus" provided by the Peenemünde Army Research Center. Once the pinup daughter has seduced the engineer, too, it is revealed that year after year, her visits have been staged by a series of doubles without original. *Konzentrationslager* Dora at Nordhausen—which also mass produces V-2s—has simply been sending inmates on furloughs. First it was a twelve-year-old, then a thirteen-year-old, and so on—up to the end of the war. As Pynchon puts it:

> The only continuity has been her name, and Zwölfkinder, and Pökler's love—love something like the persistence of vision, for They have used it to create for him the moving image of a daughter, flashing him only these summertime frames of her, leaving it to him to build the illusion of a single child. (422)

Moviegoers as such, then, are victims of semiotechnics, which makes believe [*vorspiegelt*] that coherent life conditions [*Lebenszusammenhänge*] exist, when, in fact, there are only snapshots and flash photography [*Blitzlichter*]. The feature film began—at least in Germany—with doppelgängers who filmed filming itself to propagate the medium.[28] For both Pynchon and Virilio,[29] the process culminates in the unnumbered Japanese whom the Bomb records "as a fine vapor-deposit of fat-cracklings wrinkled into the fused rubble" of their city Hiroshima (588).

The time of the exposure? Sixty-seven nanoseconds—*Blitzkrieg* in the most literal sense.

However, a war that fuses with representation proves unrepresentable. *Gravity's Rainbow* unites all the impossibilities of depicting technological warfare in the figure of Slothrop's German antipode. On one side stands the GI whom only coincidence and marching orders have set in motion, charged with tracking down the V-2. On the other side stands the boss who not only commands the production and launches of this wonder weapon, but also steers the sex life of his engineers with special effects that are true to life. Providing a portrait of the head of operations at Peenemünde would renew the wartime cliché of the evil German. That Pynchon avoids doing so and, instead, depicts the riddling relationship between fact and fiction has left his interpreters with nothing to say. But herein lies the novel's greatness.

Historically, as is well known, the HA Peenemünde was commanded by General Dornberger of the Weaponry Office [*Heereswaffenamt*]. Already in 1932, when still a captain and the adjutant of Professor Becker, Dornberger had discovered the young Wernher von Braun. The organization chart stayed this way from Kummersdorf to Peenemünde, until the methodically mounting entropies of the Hitler state (427) had made the SS into a state-within-a-state. In 1944, after the Weaponry Office had fulfilled its technical duty and the Wehrmacht entered its death throes, command over Peenemünde, Nordhausen, and the army corps that had been assigned to it "by special decree"[30] (the only instance in the history of German armies) fell to *Obergruppenführer* Dr. Kammler, who came from the Main Economic and Administrative Department [*Wirtschafts- Verwaltungshauptamt*] of the SS.[31] Hans Kammler, who was born in 1901, and Thomas Pynchon, born in 1937, share the strange trait of having destroyed all personal photos.[32] Kammler makes his way through the novel just as invisibly.

Pynchon's fictive head of rocket science erases the markers of his own identity because he is not a character at all, but rather the product of a double exposure. From 1932 on, this director of operations is called "Major Weissmann"—that is, he is an officer of the Wehrmacht and (like Dornberger) "a brand-new military type, part salesman, part scientist" (401). Up to and including his exchanges with subordinate engineers, which feign scientific interest only to camouflage the economic pressures of war (416ff.), Pynchon's Weissmann follows a single source: the involuntary openness of Dornberger's memoirs.[33] It is only logical that the name "Dornberger" not appear in this painstakingly exact novel—as if fact and fiction were two sides of the same sheet of paper.

Without any reason being provided, the same Weissmann later holds the SS rank of *Gruppenführer* (654) at Peenemünde. Finally, in 1944, he assumes the "SS code name" of "Blicero"—a periphrastic designation for death itself (322). As Blicero, Weissmann breaks with all the formalities that members of the German General Staff observe: he becomes a bellowing animal and chases the final rocket batteries over the bombed-out autobahns of the *Reich*. Dornberger, Braun, and their dismayed ghostwriters report the same thing of Kammler, who believed he could decide the war by himself.[34] As if all the entropies of the Hitler state had been made flesh.

The merging of Dornberger and Kammler, of Weissmann and Blicero, of

Wehrmacht and the SS, and of order and entropy, forms the eccentric core of the novel, the site of its nonrepresentability. It remains a riddle whether Blicero is dead or not (cf. 667). The same held for the real-life Kammler, and for many years after the war.[35] Weissmann/Blicero's deeds and delirious fits exist only as accounts of accounts made by witnesses who, for their part, were under the influence of the drug Oneirine (463, 669ff.). Oneirine, which was synthesized by none other than the fictive Professor Laszlo Jamf (348) of course, possesses the "property of time-modulation," a "peculiar" feature that "was one of the first to be discovered by investigators" (389; cf. 702f.). That is what makes it possible for Blicero—the double exposure of 1932 and 1944, of both Dornberger and Kammler—to exist in the first place. It is also why his madness can inaugurate, somewhere in the ruins of the *Reich*, the manned space flight that will, in fact, occur only twenty years later. Finally, this is why Pynchon's Second World War can end with the intercontinental weaponry of the next world war. On the last page of the novel, Blicero's manned V-2, which was launched in Lower Saxony in 1945, lands in Hollywood in 1973—the year of the novel's publication. Its base delayed-action fuse aims for the very movie theater where Pynchon and his readers are sitting. "For us, old fans who've always been at the movies," there is finally "a film we have not learned to see"—one we have dreamed of, however, ever since the inventions of Muybridge and Marey: the merger of film and war (760ff.).

But Oneirine also has other, less sensational properties, too. In contrast to the structuralist properties of *Cannabis indica* (cf. 347), the hallucinations that Oneirine induces "show a definite narrative continuity, as clearly as, say, the average *Reader's Digest* article." In other words, the visions are "ordinary" and "conventional"—that is, American (703). That would be Pynchon's contribution to the debate on narrativity in the media, as well as his explanation why every medium, including the novel, is a drug (and vice versa).

According to Gustav Stresemann, people pray "not only for their daily bread ... but also for their daily illusion" (452). After the collapse of the illusions afforded first by theology and then by the philosophy of history, business enterprises like the real-life IG Farben or Jamf's fictive Psychochemie AG (250) have done everything in their power to provide a positive—that is, psychopharmaceutical—solution to "the basic problem" of "getting other people to die for you" (701). Already in 1904, when "the American Food and Drug people took the cocaine out of Coca-Cola," we were given "an alcoholic and death-oriented generation of Yanks ideally equipped to fight WW II" (452). And so—in the words of the great Oneirine expert von Göll—all that remains to hope for is the ultimate merger between film and war. Even if Slothrop (who protests: "this ain't the fuckin' movies") may still rightfully fear that people get shot even though "they [aren't] supposed to," von Göll knows better. In the movie director's eyes, we are not in a film "yet." "Maybe not quite yet. You'd better enjoy it while you can. Someday, when the film is fast enough, the equipment pocket-size and burdenless and selling at people's prices, the lights and booms no longer necessary, *then* . . . *then* . . ." (527).

All the same—and already in 1973—*Gravity's Rainbow* organized a TV game show for readers: "A Moment of Fun with Takeshi and Ichizo, the Komical Kamikazes" (805). Whoever, like "Marine Captain Esberg of Pasadena," guesses that this spectacle is just "another World War II situation comedy" wins the grand prize: a free, one-way flight to where filming actually occurs. There the winner can experience "torrential tropical downpours," "make the acquaintance of the Kamikaze *Zero*," take charge of the plane, fly—and crash (691).

The narrative continuity of Oneirine hallucinations and/or films, then, haunts the novel that has made them its theme. Plotlines and dialogues seem as if they had been written under the influence of the drug (cf. 704ff). As a consequence, *Gravity's Rainbow* is, among other things, a *Reader's Digest* article, too: ordinary, conventional, and American. "There ought to be a punch line to it, but there isn't" (738). The riddling question—whether and how the technologies of the world wars programmed our so-called postwar era—remains unanswered. The novel stays just that: a novel. And its hero, Tyrone Slothrop, remains "a feeb." Failing to capture Weissmann-Blicero's manned space rocket, he finds that his lot is simple "mediocrity."

And that occurs "not only in his life," but also—as the narrator puts it explicitly and with due bitterness—"in his chronicler's too" (738).

4. Records

Writing stores symbolic matters, and feature film imaginary ones. The medium of stupidity, however, is constituted by the countless songs in the novel. Record grooves capture the vibrations of real bodies, whose stupidity—as is well known—knows no end. Accordingly, the ravages of wars, drugs, and media on the body keep on playing, as music. "Tape my head and mike my brain, stick that needle in my vein," begins a song in *Gravity's Rainbow*. Time and again, the novel comes to a halt because fictive rumbas, beguines, foxtrots, blues improvisations, and so on—which are all accompanied by exact instructions for performance yet distant from war games—bend events (i.e., conspiracies) into round dances, the eternal return of strophe and chorus. At the very end, as a new world war is starting high above California, the novel presents a consolatory song for a "crippl'd Zone" that reaches far beyond postwar Germany. At the end of the song and the book: "Now everybody—."

8
Heinrich von Ofterdingen as Data Feed

> And we know that messages are not simply reported; they also affect what people do and do not do.
> —Karl Knies, *Der Telegraph als Verkehrsmittel* (1857)

The story of *Heinrich von Ofterdingen* is simple. A youth of twenty years journeys with his mother through Germany. At the destination, he falls in love with a girl, but death steals her from him. On the final pages, an old man comforts the young man. Nothing more dramatic than this occurs, and the most momentous event is left out: on paper, the girl's death appears only in the hero's dream.

This plot is so meager that interpreters have overlooked it time and again. They only grow interested when the novel passes over to making theoretical statements. In this light, it becomes possible to read the plot as the illustration of theoretical discourse—to embed it in a history of the mind [*Geistesgeschichte*] transcending events. "Philosophy of history," "poetics," "conceptions of nature," or "representation of the Middle Ages in Romanticism" are the conventional rubrics under which the course of narration appears—and then disappears. It is as if the events recounted were merely a pretext for formulating theories. As if, in other words, speech amounted to nothing more than what it says.

To be sure, Hardenberg's novel consists of innumerable conversations, possibly more so than any other work. The text says so itself: "idle conscience, in a smooth world that offers no resistance, turns into a gripping exchange—a Fable that tells of all things [*zur alleserzählenden Fabel*]" (332).[1] On the one hand, *Heinrich von Ofterdingen* presents a minimal quantity of actions and obstacles (leaving aside the book's one catastrophe, which is elided). On the other hand, it offers a maximal quantity of words that are exchanged. Because it features neither a lady stealing by night into the hero's chamber, nor a rival giving cause for a duel or suicide, Fable (ultimately, a character in the work) can ascend to a position of uncontested power and—as occurs in the inset fairy tale [*Märchen*] narrated by Klingsohr—allegorize the flow of information itself.[2] Everything must be said, because there is nothing to say.

The innocence of speech even has a home in the text. Chapter five, in the first part of the novel, begins at a village inn. A simple observation is offered: "a large number of people, some of them travelers, others simply guests for a drink, sat in the room and conversed about anything and everything" (239). This place, then, is an earthly paradise—and one that stands far closer than poetic dreams of Atlantis or philosophical speculations about the Golden Age. Accordingly, it lies far below interpreters' threshold of perception. "People" (239) simply talk. Their names and what they say are not provided, much less recorded. This, everyone knows, is how it goes every day.

In a foreword which itself has vanished, Foucault described this everydayness:

> The great *oeuvre* of the history of the world is indelibly accompanied by the absence of an *oeuvre*, which renews itself at every instant, but which runs unaltered in its inevitable void the length of history: and from before history, as it is already there in the primitive decision, and after it again, as it will triumph in the last word uttered by history. The plenitude of history is only possible in the space, both empty and peopled at the same time, of all the words without language that appear to anyone who lends an ear, as a dull sound from beneath history, the obstinate murmur of a language talking *to itself*—without any speaking subject and without an interlocutor, wrapped up in itself, with a lump in its throat, collapsing before it ever reaches any formulation and returning without a fuss to the silence that it never shook off. The charred root of meaning.[3]

These words apply to literature every bit as much as they do to history. The work they perform is also measured and hemmed in by a murmur that cancels them out. No flow of information can occur without white noise, because channels of communication emit it themselves—as the chance distribution of interference. Whether they are drunken or not, the tavern murmurs constitute the unerasable background from which Hardenberg's novel extracts its characters' profiles and their words in the first place. Indeed, it is for this very reason that there are characters and words of a literary nature at all.

Right after the anonymous murmuring—which, in contrast to the systematic return of all other information [*Romaninformationen*], never receives further mention—the text passes over to a listener named Ofterdingen and to a narrator who, like all characters in novels (who occur verbally), not only has any number of things to say, but possesses knowledge as well. "The old man's discourse pleased Heinrich uncommonly, and he was inclined to hear yet more from him" (243). Interpreters overlook statements of this kind, too, because they possess no theoretical content and merely report the flow of information. Yet Hardenberg's novel absolutely abounds in them. Because this white noise provides the zero value of literature, the transmission of knowledge (for example, when the Miner speaks) marks a discursive event and should be analyzed as such. The fact that speaking does not amount only to what it says constitutes its reality or history. For instead of simply reflecting so-called reality or history, every stream of information switches between dispositives of power. The fact that a budding poet like Ofterdingen listens "uncommonly" gladly to old miners provides information about the information networks of 1800.

That is to say, there can be no fiction—and certainly no Romanticism—if one uses labels that seek to measure the zero grades of discursive effectiveness [*Schwundstufen von Wirksamkeit*]. As the data feed that it is, *Heinrich von Ofterdingen*, the most Romantic of all the novels that Romanticism produced, displays the absoluteness [*Unhintergehbarkeit*] of an event. Its seeming poverty of plot [*Handlung*] simply offers space and a site for other, more forceful action [*Handeln*]: the action of speech itself. Fable, who tells of everything [*die alleserzählende Fabel*], is not a fairy tale or a myth. Without German poetry [*Dichtung*] of the kind that the Age of Goethe both produced and inaugurated in the first place, the *Bildungsstaat* of the revolutionary nineteenth century would not have existed.

To be sure, information networks must be reconstructed as such in order to demonstrate as much. Neither the unity of authorial intention nor the unity of the artistic work proves decisive for discourse analysis. If, according to Claude Shannon's theorem, information networks fundamentally connect a source, a transmitter, a channel, a receiver, and a destination,[4] then messages consisting of words (discourses, that is) must be recorded [*angeschrieben*] as a network that always—and as a matter of necessity—incorporates numerous other books, documents, archives, libraries, and institutions.

The task, then, involves reading along the lines that Novalis drew in his first, fragmentary novel. Here, an anonymous master instructs the *Novices of Sais* how to search for "crystals or flowers" which he, in turn, arranges in series and columns—which he archives, that is. One morning, a pupil appears before him. Previously, the youth had "always looked sad." Now, however, he intones a "lofty, joyous song"—that is, he has grown up to be a poet. The novice has come to give his teacher "an unprepossessing little stone of a strange shape."

"The teacher took it in hand and kissed it for a long time; then, he looked upon us with wetted eyes and placed this stone on an empty space between other stones, precisely where the rows touched each other, like rays" (81). It is impossible to state more clearly that poetry does not consist of radiant substance [*in strahlender Substantialität*] or of aesthetic appearance [*Schein*], but rather is defined by its place value [*Stellenwert*].⁵ Even before they appear, all the stones, that is, units of information, count as part of a network plan or graph connecting them with each other. *Heinrich von Ofterdingen*—with a degree of precision that lies far ahead of conventional social histories of German literature—treats this same matter.

1.

At the beginning of the network plan—how could it be otherwise?—lies white noise. With its unheard-of opening words, the novel indicates the background against which it can become a novel in the first place. "His parents were already lying asleep, the clock on the wall struck its uniform beat, the wind rushed at the knocking windows; the room was intermittently lighted by the glow of the moon" (195). A twenty-year-old "youth," lying "restlessly upon his bed" in the same chamber, hears nothing of his parents' intercourse—verbal or otherwise. The information that reaches him consists solely of inhuman and stochastically distributed acoustic and optical data, which the text registers, but the hero does not. This is precisely how literature begins. To get started, it touches (on) other streams of information that—as rattling, ticking, whispering, and blinking—escape verbalization [*Sprachlichkeit*], because only gramophones can record actual acoustics and only films can record actual optics.

Around 1800, no medium besides words existed for serial data—that is, data in the succession of time. This is why the sleepless youth is predestined to become a poet, both in the opening scene and in general. Ofterdingen takes in all the sounds and faces that no word could possibly store as such; he does so in order, through the act of selection that he performs, to become estranged from his own presence [*aus seiner Gegenwart herauszufallen*]. Inasmuch as the Age of Goethe could only process serial data in and as language—which, Ofterdingen affirms, "man [*der Mensch*] commands" (287)—then he has no other choice. "The youth lay restlessly upon his bed, and thought of the Stranger and his tales" (195).

Therefore, it is a foreign person—about whom Ofterdingen "thinks" and, as the following makes clear, to whom he directs quiet conversations with himself—who makes him forget, first his parents, and second the white noise that surrounds him. From the outset, the novel determines that words, or more precisely, "tales" [*Erzählungen*], provide the source constituting its entire information network. The Stranger—their broadcaster—takes care of the initial selection. This also connects Ofterdingen, the aspiring poet, to the circuit of transmission called "literature." Both chance sounds and mere familial conversations fade away. A narrative takes their place, which individualizes its speaker as much as it does its auditor. According to Ofterdingen, no one has "ever" seen "a person similar" to the Stranger. Moreover—and even though everyone "has heard the same thing [*das Nämliche*]"—no one has ever "been gripped by his discourse [*Reden*]" as much as Ofterdingen himself (195). Therefore, the Stranger and his listener—the transmitter and the receiver—are separated by a marked proximity, which (as is the case everywhere in the novel) involves the object of discourse(s). This object both is and is not a word; it is like the Symbol, which Goethe defined as "the thing, without being it, yet being it after all [*die Sache, ohne die Sache zu sein, und doch die Sache*]."[6] Its name is "the Blue Flower" (195).

Of all the words or things in the novel, this flower stands apart, for it functions both as a name and as an intuition [*Anschauung*]—as signifier and signified in one.[7] Although the Blue Flower is only "given" in the Stranger's narratives, that is, it does not occur in sensory presence, it rouses a "passion" or "yearning" to "behold" it. At the same time, as a word that is transmitted [*weitergegebenes Wort*], it can also quiet this same passion: "Often, I am so delightfully well, and only when the Flower is not fully present [*wenn ich die Blume nicht recht gegenwärtig habe*] does a deep, heartfelt urging [*Treiben*] befall me" (195). Here, all at once, language proves capable of transporting an optical—that is, a sensory—flow of data.

In the quiet repetition of narratives that have already been heard, their referent really does become "present" [*gegenwärtig*]. Only when this wonder does not occur does Ofterdingen experience, in addition to the "deep, heartfelt urging," a fear that he "might be mad" (195). Here one may gauge just how hallucinatory the flower's appearance is—the extent to which, where this particular signifier is concerned, language trespasses its own borders. Ofterdingen's state, the ambiguity of deepest interiority and madness in one, describes the altogether poetic capacity of the soul [*das schlechthin poetische Seelenvermögen*] that Novalis—and all the aesthetics of the Age of Goethe[8]—called "imagination" [*Einbildungskraft*].

A fragment from 1798 offers the following definition:

> Imagination is the wondrous sense that can replace all our other senses—and which stands so beautifully subject to our will [*der so sehr schön in unsrer Willkür steht*]. Even if the outer senses seem to be governed by mechanical laws—the imagination is obviously not tied to the presence and touch of external stimuli. (II, 650)

In this precise sense the merchants tell Ofterdingen about poetry [*Dichtung*], affirming that "this art is a truly wondrous matter." The crafts of "painters and musicians [*Tonkünstler*]" pursue only the "artificial imitation of nature" for "eye" and "ear." In contrast,

> of the art of poetry [*Dichtkunst*], there is nothing to be encountered externally. It also creates nothing with tools or hands; the eye and the ear perceive nothing of it: simply hearing the words is not the particular effect [*die eigentliche Wirkung*] of this secret art. Everything is internal, and just as those artists fill the outer senses with pleasant sensations, thus does the poet fill the inner shrine of the soul [*Heiligthum des Gemüths*] with new, wonderful and pleasing thoughts. He knows how to rouse those secret powers in us at will, and he gives, through words, an unknown, majestic world to be heard. As if from deep caverns, there emerged ancient and future times, countless human beings, wondrous regions, and the strangest occurrences within us—which tear us from the present with which we are familiar. (209ff.)

Lying on his lonesome bed, Ofterdingen—who can "dream and think [*dichten und denken*] of nothing" but the Blue Flower (195)—obeys this definition of poetry to the letter. Imagination replaces all his senses. That is, it provides, instead of signifiers that are heard, what the signifiers signify. Because of the simple fact that "merely hearing the words" does not constitute "the actual effect" of the "secret art" of poetry, the Stranger's tales do not exist as something present to the senses. Memory [*Erinnerung*] and memory alone—to employ Hegel's terminology—"has preserved them." For this same reason, the words, to quote *Phenomenology of Spirit* once more, transform into a "picture gallery."[9] In his "dreaming and thinking," Ofterdingen performs the elementary act that defined poetry in the Age of Goethe—and philosophy, too. Here lies the foundation of their historical alliance.

The matter has simply escaped readers' attention: the people of "thinkers and poets" is a people of readers. What novels and systems of philosophical aesthetics formulate as wonders or enigmas can be explained in very simple—that is, technical—terms. "If one reads properly," Novalis wrote elsewhere, "then there unfolds within us [*in unserm Innern*] a real, visible world following the words" (III, 377). Accordingly, the wondrous sense that can substitute for all our senses is called "literacy." The unknown, majestic Word that poets make us hear through words opens a *fantastique de bibliothèque*—which, as Foucault has observed, represents the fundamental literary invention of the nineteenth century.[10] For the first time in the history of a culture of writing [*Schriftkultur*], letters no longer needed to be laboriously deciphered, or even read in a muted voice. Silent and automated reading[11] transported them immediately onto the "ground of subjective interiority" [*Boden der Innerlichkeit im Subjecte*],[12] which as a matter of course consisted of hallucinated signifieds. That is why the novel—once again, in keeping with all the aesthetic systems of the day—need not acknowledge that poetry had long existed in book form, too. As the presence of the Stranger makes plain, it finds expression as a disembodied [*unsinnliche*] and absent voice that only appears in recollection [*nur noch erinnerte Stimme*].

To be sure, schools also teach silent reading today. However, no student still believes that for this reason he or she is hallucinating the meaning of what stands printed. Now wonders of this kind occur when one watches films and video clips. Ever since writing lost its monopoly on serial data processing, it has appeared for what it is: meaningless marks in black and white on paper.

Precisely this fact was inadmissible in the Age of Goethe. To recruit new initiates, the alphabet learned to make a new promise. As Ofterdingen puts it shortly before falling asleep:

> I heard tell, once, of olden times, how the animals and trees and rocks spoke with mankind. It seems to me as if, at any moment, they would begin again, and as if I could see in them what they want to tell me. There must yet be many words I do not know: if I did know more, I would understand everything much better. (195)

In the phantasm of an originary language or writing [*Ursprache oder Urschrift*],[13] then—even though this vision is guaranteed only by discourse(s)—signifiers and signified coincide in such a way that the latter themselves speak. This is reason enough for one, as the hearer or reader of natural language, to learn "many more words" oneself. Anyone who wished to become a poet around 1800 had to desire his own literacy first. And because all desire is erotic, the originary language had to beckon with a reward that promised the Impossible[14]: a recording of the relationship between the sexes.

No letter, no word, and no book says what women are. That is the reason why Ofterdingen drifts off. The riddle posed by his parents lying there and sleeping—which remains unanswered—is "solved" in a dream. The inaugural dream in the novel does not simply conduct all the tales told by the Stranger out of the words that they are into a real and visible world; it does not simply turn off—because, after all, "the slightest sound is not to be heard" (196)—the unrecordable sources of incidental noise. Rather, at its radiant ending, the dream, inasmuch as it embodies imagination, presents the very meaning [*gibt die Bedeutung selber zu sehen*] around which all the Stranger's words and all his auditor's dreaming and thinking have been circling:

> He saw nothing but the Blue Flower, and he gazed upon it for a long while with ineffable tenderness. Finally, he sought to approach it, when, all of a sudden, it began to move and change: the leaves became more luminous and nestled on the growing stalk; the Flower inclined toward him, and the petals unfolded a blue collar in which a delicate face was floating. (197)

In 1916, Hugo Münsterberg—who invented both the word for, and the practice of, "psychotechnics" [*Psychotechnik*]—published the first scientific theory of the feature film. His study sought to demonstrate that narrative cinema is able to simulate, implement, and thereby render superfluous all the unconscious processes of the mind. The logical consequence was that the medium of literature—should it represent something more than, and be something different from, printer's ink—had been surpassed.

> No theater could ever try to match such wonders, but for the camera they are not difficult. [...] Rich artistic effects have been secured, and while on the stage every fairy play is clumsy and hardly able to create an illusion, in the film we really see the man transformed into a beast and the flower into a girl.[15]

The same fairy-tale wonders that have been simple matters of technology ever since Georges Méliès engineered his cinematic special effects were screened as literature and psychology around 1800. In the imagination of a dreamer enamored of words, the Blue Flower turns into a woman. The sleeper's hallucinatory vision receives an answer, and his "sight" [*Sehen*] sees a "face/sight" [*Gesicht*].[16] Such was media technology around 1800. "If one reads properly, then there unfolds, within us, a real, visible world following the words." In this *fantastique de bibliothèque*, words are not only capable of referring to women; they can *mean* them, too. Here speechless beings like a flower really speak with human beings—after all, the plant has become a girl. Ofterdingen will not be able to do otherwise than love the incarnate meaning of the signifier "flower" his whole life long. As soon as he encounters an empirically extant vision of a girl [*Mädchengesicht*] (277), his own transformation—into a poet—is complete.

The transformation of words into flowers,[17] and of flowers into words, sustains all poetry in the Age of Goethe. Hardenberg's novel attests as much in the tale of Atlantis, which the merchants tell the budding poet Ofterdingen. The tale features a king who, "from youth onward, had read the works of the poets with rapturous delight," "devoted himself, with great zeal and at great expense, to collecting them from all languages, and always esteemed intercourse with singers above all else" (214). The poor monarch seems to lose his only daughter to death—a girl who "had grown up surrounded by hymns [*unter Gesängen*]" and whose "entire soul had become a tender song." When, at court,

> she harkened to the competing songs of the inspired [*begeisterte*] singers with deep attention [*mit tiefem Lauschen*], one took her for the visible soul of the majestic art which had invoked those magical incantations, and ceased to marvel at the delights and melodies of the poets. (214)

It is no wonder, then, that her loss also robs a lover of letters like her father of the object of his desire. Without a woman to sign for it—as the "visible soul" of all songs and magical incantations—the medium of literature falls back into disconsolate literalness. Accordingly, the King "thinks" [*gedenkt*] to himself:

> what good does all this majesty, my high birth, do me now? Now I am more miserable than other men. Nothing can replace my daughter. Without her, the hymns are nothing but empty words and illusion [*Blendwerk*]. She was the magic that gave them life and joy, power and form. (223)

So that words would not be what they are—empty, that is—the poetry of the Age of Goethe underlaid them with a transcendental signified that transformed literacy into Desire itself. The transcendental signified could not be "replaced" by anything because—inasmuch as it involves the birth of a woman "out of the imagination"—"it can *replace* all the senses" (to say nothing of signifiers, which are defined by replaceability in the first place).

The idol of "Woman" forms the condition of possibility for Classical-Romantic poetry to the same extent that actual women remain silent. In the empirical sphere, women have nothing to do other than "listen" to actual singers—that is, men—"with deep attention." That is, women are consumers, a function of poetic discourse that is just as necessary as it is derivative. In transcendental terms, on the other hand—and as the visible soul of all words, which would otherwise be empty—the idol of "Woman" provides poets with an originary language whose depth is silence [*Stummheit*].[18] That is why Mathilde, the Blue Flower incarnate, appears to her lover in a dream and "speaks a wondrous, secret word into his mouth, which penetrate[s] his whole being." On waking,

Ofterdingen "would have given his very life still to know that word" (279). By the same token, "every future word will represent the effort to repeat that word, which is present within, yet unfixed."[19] In the first and final instance, poetry in the Age of Goethe means translating elementary, feminine speech—which never occurs—into articulated language.

As much is affirmed by another father in the novel when he discourses on his daughter and future son-in-law. Klingsohr—possibly an allegory for Goethe—speaks about Mathilde, the allegory "of Love," and Ofterdingen, the allegory "of Poetry":

> Just consider Love. Nowhere is the necessity of Poetry for the continued existence of mankind as clear as it is here. Love is mute, only Poetry can speak for it. Or Love is itself nothing but the highest Poetry of Nature. (287)[20]

With such technical precision does the mature, Classical Goethe impart a business secret to his Romantic heir. Even if the real Mathilde exchanges word after word with her lover, she remains mute all the same. "Since" women's "mere speech is already song" (276) mustering "scarcely audible words" (270), their discourse does not find its way into writing.[21] In order to be able to store the pure interiority of the transcendental signified, "natural poetry" [*Naturpoesie*] must first become "poetry" as a "strict art" (282). And that is a man's business.

2.

Secondly—and in empirical terms—literature in the Age of Goethe meant combining all the discourses that occur in disseminated form into unified poetic works. Klingsohr promises Ofterdingen "to read" with him "the strangest writings [*die merkwürdigsten Schriften*]" and to acquaint the aspiring poet "with all estates, all trades, all circumstances, and all demands of human society" (282). Even as a strict art, then, poetry involves translation. Discourses from the most varied times, places, and domains must be sampled, rewritten, and brought into a single channel. Put in terms of information technology, poetry means demultiplexing. Or as Novalis writes, "in the end, all poetry is translation" (IV, 237). That is precisely what Ofterdingen learns on his journey—that is, in the interval between meeting the Stranger, who initiates him, and encountering the Classical figure [*dem Klassiker*], who affords institutional recognition to his poet-heir [*Nachwuchsdichter*].

Ofterdingen's journey takes place without adventures or reversals of fortune. Everything happens so that he may be nothing more than an ear. "Heinrich listened very attentively to the new tales" (230), "heard their story, which was interrupted by many tears" (236), "paid attention to the conversation" (263), and so on, time and again. This occurs for good reason, for besides Mathilde—who must, after all, allegorize "Love" itself—no character in the novel suffers from muteness. All can speak of their station. Merchants speak of economic matters, knights of war, miners of geology and paleontology, Arab women of the Orient, and historians of history or literary history (cf. 265). All forms of contemporary knowledge, then, are provided in representative breadth and—as scholarly works put it so well—"in consideration of new findings in the cultural sciences." Therefore, even before pure poetic audition begins, an initial selection of discourses has already been made. The various forms of knowledge concern everyday lives, practices, and aptitudes, and they yield encyclopedic comprehensiveness. Delimited fields of knowledge emerge from the great murmur of daily routines—as if Ofterdingen were sitting in a library. Consequently, he can record all these discourses without effort (cf. 250).

And indeed, Ofterdingen's journey of education through oral narratives repeats Hardenberg's journey of education through all the books of the epoch. The one voyage provides the allegory of the other, and its orality represents a *fantastique de bibliothèque* that—as in the case of the merchants—simply emerges from translating quoted texts about the poet Arion back into speech and omitting the proper names of the protagonist and author (211–13).[22] With that, the novel guarantees the success of contemporary programs for instilling literacy. Ofterdingen is able to hear—that is, effortlessly absorb—everything that Hardenberg has read. When knowledge reaches the poet, it has already been distilled into meanings, that is, signifieds.

"Those calm, unknown people, whose world is their soul [*Gemüth*], whose action is contemplation, whose life is the quiet cultivation of their inner forces" (266) cannot be culturalized in any other way:

> Great and varied occurrences would disturb them. A simple life is their lot, and only through tales and writings need they become familiar with the rich contents and the infinite phenomena of the world. Only rarely in the course of their lives may an event pull them, for a while, into its rushing confusion—to teach them, through a little experience, more precisely of the circumstances and characters of those involved. For that, their perceptive sensibility is already busy with ... matters near at hand, which present a rejuvenated world, and they make no step without experienc-

ing the most surprising discoveries in themselves about the nature and meanings of the same. They are the poets. (267)

It makes no difference whether written works are at issue, as in Hardenberg's case, or oral accounts, as for Ofterdingen. For poets, the so-called world funnels into a news feed. The latter, however, functions along technologically precise lines—that is, in an altogether unromantic way. Time and again, the novel indicates the sources, broadcasters, channels, and receivers of messages that reach the poet; nor does it forget what goes missing en route (cf. 210f.). But above all, data feeds are a matter of economy—to receive pure signifieds or "meanings," it is unnecessary to convey "the innumerable phenomena of the world." Information, according to Shannon's theorem, is the reciprocal of redundancy.

Hardenberg's novel sets up and enacts a principle of complexity reduction in precisely this sense: "rejuvenated representation" [*verjüngte Darstellung*]. All factual forms of knowledge of the epoch that reach Ofterdingen's ears are, by definition, miniatures—that is, depictions of similitude—that preserve relations but not dimensions. This is already taken care of inasmuch as elements of the messages get lost as they make their way to the poet. Precise phrasings are eroded until only pure meaning remains. Moreover, the narrative inlays in *Ofterdingen* assure the same: a tale within a tale (for example, when the merchants quote the myth of Arion or when Klingsohr's tale repeats Ofterdingen's family romance[23]) must necessarily have a lesser size than its frame and depict "the broadest stories, drawn together into tiny, shining minutes [*Minuten*]" (325).

Third, and finally, all the psychic conditions that the Age of Goethe called "poetic" perform miniaturization. Dreaming, Ofterdingen's father sees "the earth only as a golden bowl with the most intricate engraving" (202). In Ofterdingen's own vision, the heavenly realm shrinks to a "distant, small, wondrous majesty," while the earth lies "before him," "like an old, dear dwelling place" (321f.). Finally, in "childhood,"

> we see the full richness of infinite life, the tremendous forces of later time, the majesty of the end of the world, and the golden future of all things still closely woven into each other [*noch innig in einander verschlungen*]—and yet delicately rejuvenated in the clearest and most distinct way. (329)

Poetic imagination and complexity reduction coincide then. For Goethe, "true poetry" provided a "bird's-eye-view" of the earth in general.[24] For Johann Christian Reil, the great psychiatrist, the mind itself was defined by the fact that it processes all matter that is given to it in keeping with its organization, and seeks at all times to introduce unity into the manifold. It winds together, in self-consciousness, the immeasurable thread of time into a knot, reproduces extinct centuries, and combines the elements of space extending into the Infinite—mountain ranges, rivers, forests, and the stars scattered in the firmament—into the miniature painting of imagination [*einer Vorstellung*].[25]

Of course, such abbreviations of space and time as occur throughout *Ofterdingen* are not materially performed by "the mind." The mind is neither a film nor any other technological medium. The sole medium it has at its disposal is called language. According to Ofterdingen, however, language is "really a tiny world of signs and sounds" (287)—that is, the space of all possible miniatures that constitute poetry. For this reason, the rejuvenated depictions omitting the redundancy of the Real represent metaphors of reading itself. Just as the young Hardenberg read the sciences of his day only to be able to excerpt from book after book, the news feed reaching Ofterdingen's ear consists wholly and exclusively of prefabricated excerpts.

The greatest innovation in communications technology that occurred in the Age of Goethe was to combine storage and deletion. Hardenberg's philosophical preceptor, Johann Gottlieb Fichte, said so time and again. Only so long as the "art of printing" still stood in its infancy could the "sciences" (i.e., universities) consider it their task to "set down the whole of book-learning once again." In the days when the early modern Republic of Letters was first instituted, storage simply involved repetition, either through written commentaries or oral lectures. However, ever since, first, the universal "spread of the book trade" and, second, the rise of literacy (a matter Fichte forgets)—since, that is, "there has existed no branch of learning about which there is not an excess of books"[26]—ROM (read-only memory) has become obsolete. "What the author has said, we cannot tell our reader once more; for the former has already said it, and our reader can learn it from him in all respects [*in alle Wege*]."[27] Storage technology, therefore, must be refitted for RAM (random-access memory), which not only provides data in ROM format, but can also erase it and replace it with new information.

Accordingly, Fichte calls for an entirely new discourse of interpretation:

> We must uncover what the author himself *is*, inside—which is perhaps hidden to his own eyes, and through which all that is said becomes what it becomes to him. We must draw the spirit [*Geist*] out of the letter.[28]

But if interpretations delete letters—that is, reduce them to "mind" or "spirit"—then they are identical to the miniature paintings that Reil extolled and Novalis implemented. As scientific as they may seem, inasmuch as they are discourses, they turn into works, art, and, more specifically, poetry.

> After all, one does not study to prepare for exams one's whole life long—to reproduce, verbally, what has been learned. Instead, one studies in order to apply knowledge to circumstances in life as they arise—and thereby to transform it into *works*; it is not just a matter of repetition, but of making something else out of, and with, it: and so, here too, the final purpose is in no way knowledge, but rather the art of using knowledge.[29]

No one followed Fichte more loyally than Novalis, and no one is truer to him than the hero of his novel. To be sure, literature—as a matter of definition and at all times—is a data stream.

All the same, in the Europe of old, there had been times when encyclopedic breadth and the literal reproduction of data made literature great and worthy of praise in the first place. Baroque novels—for example, Lohenstein's *Arminius*—restated the whole of book learning.[30] That is, they operated without the deletions that prove constitutive for *Ofterdingen* and, for this reason, are depicted internally. Thus, Klingsohr's tale tells of a scribe, who clearly stands for the erstwhile Republic of Letters; this figure, however, must submit all his encyclopedically exact records to the censorship of a woman whose very name stands for the new philosophy. "Sophia's" magic dish "with limpid water" has the power to "erase" most of what the Scribe has written (294). Only very different pages—which "little Fable" (i.e., Poetry) inscribes with the "quill of the Scribe"—pass the censor's office "fully shining and unscathed [*völlig glänzend und unversehrt*]" (295f.). After all, if one follows Fichte, these pages already are *works*, that is, reductions of complexity requiring no further reduction.

In an ingenious study, Heinrich Bosse has shown that the magical vessel in *Ofterdingen* and the wholly analogous magical water in E.T.A. Hoffmann's *Golden Pot*[31] are not mere symbols. Of course, the Age of Goethe had no writing materials that admitted thorough deletion. However, its schools invented a new surface of inscription that equipped even children with random-access memory: the slate [*Schiefertafel*]. Here, as everyone knows, chalk marks are made only to be corrected—that is, erased.

> The [old] writing exercises with pen and ink invariably formed—and endlessly, at that—the side-by-side arrangement of a tableau. Exercises with chalk and slate, on the other hand, opened the play of presence and absence—or, in less ludic terms, nothing other than the technology of spiritualization [*Technik der Vergeistigung*].[32] In writing and reading, they edged out mimicry—both the refined Old-European

art of *imitatio* as well as repetitive drills of spelling and copying. Instead, they encouraged students to do things themselves—a matter that was supposed to unfold in a framework of simulation that had already been shaped.[33]

The new storage technology of the Age of Goethe, then, gave form to schools and poetry in equal measure. Ofterdingen's journey through Germany does not occur out of necessity. He could just as well have stayed in the classroom and taken in all the subjects of contemporary learning, spiritualizing them and transforming them into poetry. Hegel's dictum holds that pupils learn "the history of the world's development [*die Geschichte der Bildung der Welt*]" only "as if traced in silhouette." That, however, is precisely the point of complexity reduction, whether poetic or "pedagogical."[34] Accordingly, between empirical and speculative approaches "for achieving knowledge of human history [*um zur Wissenschaft der menschlichen Geschichte zu gelangen*]," Ofterdingen chooses the second option: he considers each matter "in its living, manifold context"; on this basis, he can "easily compare it with all the others, like figures on a board" (208).

Through miniaturization, then, varying forms of knowledge—on the model of *The Novices in Sais* or, alternately, that of curricula invented around 1800—form constellations that permit the mathematical combination of their elements. This is how Hardenberg proceeds in *Allgemeines Brouillon*, and it is how his poetic protagonist makes his way as well.[35] Comparisons and tabulations transfer individual forms of knowledge into place values within a system called "philosophy" in *Allgemeines Brouillon* and "poetry" in *Ofterdingen*. It is also exactly how Klingsohr summarizes the stations of Ofterdingen's journey of education. "The narrative of your journey," he tells Heinrich,

> afforded me pleasant entertainment yesterday evening. Indeed, I remarked that the spirit of the poetic art is your friend and companion [*freundlicher Begleiter*]. Your fellow travelers have, unnoticed, become its voices. In proximity to the Poet, Poetry bursts forth everywhere. The land of poetry, the Romantic Orient, has greeted you with its sweet melancholy; War has addressed you in its wild majesty; and Nature and History have approached you in the form of a miner and a hermit. (283)

After the fact, individual voices heard in the course of events become pure instances of discourse, whose speech, while particular to a discipline, is already standardized, that is, poetical. Ofterdingen—as he has often done already (cf. 238) and the novel will continue to do as a whole—only needs to pass on what he hears as a coherent "tale" in order to produce real literature, which simply involves demultiplexing separate chains of transmission or channels of knowledge.

Heinrich von Ofterdingen as Data Feed

In so doing, Ofterdingen brings back what, according to the novel, represents the original condition of all discourses: primordial unity. "In the most ancient times, in the lands of the Greek empire today," poets are said "to have been prophets and priests, lawmakers and doctors, all at once" (211). All four branches of university study (assuming that prophets were the precursors of philosophers) spoke from the same mouth then. And inasmuch as the mouth, in general, is "simply a mobile and answering ear" (211), Ofterdingen receives the task of renewing such unity by translating all discourses into the one true Poetry. According to the novel, however, this is obstructed by a historical circumstance: the monopoly of the church on knowledge. "It is bad enough," the merchants say, "that the sciences have come into the hands of an estate so distant from worldly life, and that princes are advised by such unsociable and truly inexperienced men." Therefore, they urge Ofterdingen "not to become a cleric" (207), but rather to take up the new profession—entirely unknown to monks or chaplains (208)—of Poet.

Hardenberg's novel, it could scarcely be said more clearly, does not take place in the Middle Ages where it is set, but in the present day of its discourse. The program concerns the creation of a historically new estate of civil servants that will prove more effective politically than their theological counterparts and unify the disciplines of all four fields of university study. This same reform of discourse took place in the Age of Goethe. "The separation of the order in schools [*Schulregiment*] from that of churches"[36]—which occurred in Prussia from 1794 on—replaced the same chaplains from whom Ofterdingen is obliged to receive his first instruction (204) with *Gymnasium* professors salaried by the state, and it replaced the Bible as the foundational text for alphabetization with poetic primers.[37] Only in this discursive space was the project undertaken by Schlegel and Novalis to found a new, poetic mythology—or Bible—not sacrilegious. In it alone can Ofterdingen transform "the world and history," by way of poetry, "into holy writ" (334) and declare that "the Bible and doctrine of Fable [*Bibel und Fabellehre*]" represent "Constellations of Heavenly Revolution [*SternBilder Eines Umlaufs*]" (333). Around 1800, the all-telling Fable [*die alleserzählende Fabel*] replaced the Word made flesh, which was there in the Beginning.

"Fable," moreover, does not mean "fable." Instead, it means a discursive institution otherwise known as "German class" [*Deutschunterricht*]. Just as the novel translates and combines all forms of knowledge into poetry, so did the new field of instruction proceed. Teaching German, according to Friedrich Schleiermacher, is not just to be viewed as language learning; rather—because the mother tongue is the immediate organ of Understanding and the general organ of Fantasy—everything that can occur at schools for the free, formal cultivation of the Spirit flows into this instruction—all in preparation for Philosophy.[38]

Imagination or "Fantasy" offers the altogether wondrous sense for replacing all senses, and German the wondrous discipline [*Fach*] for replacing all disciplines [*Fächer*]. Only when slates eliminated rhetorical imitation and German essays took the place of orality in schools could novels like *Ofterdingen* come into being. Every *Abitur* essay "documents" (once again, in Schleiermacher's words) the "education [*Bildung*] of Understanding and Fantasy"[39]—just as Ofterdingen's miniature depictions do for scientific discourses. The latter all represent poetically exalted allegories for a new method of testing that, around 1800, replaced rhetorical oral cultures with the written interpretive essay. To pass his *Abitur* as a poet, then, Ofterdingen must also find a German teacher. This is exactly what happens when Klingsohr "approaches, leading a lovely girl by the hand, to open stupid lips through the tones [*Laute*] of the mother tongue and the touch of a sweet, delicate mouth" (268). The mother tongue, as Klingsohr teaches his pupil in their shared readings (282), forms the immediate organ of the Understanding—and the kiss, as the schoolboy dream of distant girls' lips, the general organ of Fantasy.

3.

Germany's higher school system—from the time it underwent fundamental reform in the Age of Goethe until 1908, when even Prussia finally admitted female students—was based on the exclusion of women. To be sure, girls could receive private tutoring, as Mathilde does from her father (282f.), and they could also attend schools for young ladies [*höhere Töchterschulen*], as Ottilie does in Goethe's *Elective Affinities*. However, since *Gymnasia* had the sole purpose of producing students (through the newly created *Abitur*), and universities bureaucrats (through the newly created *Staatsexamen*), women were left out of the system culminating in poets and thinkers, on the one hand, and civil servants, on the other (a development foreseen for the unwritten second half of Hardenberg's novel[40]). Because "Woman" meant "the Poetry of Nature," and because "the Poetry of Nature" was mute, her passing an *Abitur* or *Staatsexamen* was the very definition of the Impossible.

So that Ofterdingen may complete his course of education, then, Mathilde must return to the silence of blue flowers, which is where she came from in the first place. But this is also why Hardenberg's novel—even though it avoids all "great and manifold events [*große und vielfache Begebenheiten*]"—includes a "rushing whirl [*raschen Wirbel*]" (267) after all. Indeed, the words should be taken literally: Mathilde drowns in the waters of the Danube. Only when she dies are all the conditions of discourse fulfilled that lead Ofterdingen to the university and thereby make him a member of adult, male society [*Männergemeinschaft*]: "Terrible fear robbed him of consciousness. His heart [*das Herz*] was no longer beating. [...] His quiet mind [*Gemüth*] had disappeared" (278). In other words, Ofterdingen has lost his love, his soul, and his proper name all at once. As "the Pilgrim"—which he is called from this point on (319)—he represents a purely discursive instance: "the Student."

As a purely discursive instance, Ofterdingen has to rediscover what constitutes his sole and essential medium: language. At the beginning of the second half of the novel, he wanders alone in desolate mountains. He hears, at first, only "a strong wind" whose "muffled, varied voices were lost as soon as they came" (319, cf. 349f.). For a second time, then, Ofterdingen encounters sources of noise that openly scorn words and books. This only occurs, however, so that the noise can be filtered away and, as fading background static, make "language and voice live again" within him (22). Both in the novel and in what is commonly called "life," such signal selection harkens to the name of love.

> Love is a logical inversion. Because—in the field of sexuality, sexualized bodies, and the sexes and sexual desire—many things occur without functioning, nothing happens in love. On the other hand, this means that there are no disturbances, either. The factors of "noise"—sexuality and its unfulfilled desire—are filtered out. [...] The totality and *pleroma* of love is maintained only through the paradox that want is wanting [*daß der Mangel mangelt*]—that is, sexuality. Here, poetry stands in as the language of love.[41]

Therefore, against the background of the wind and other noises that "dully drone," there appears a human and articulated "voice"—which the hallucinating "Pilgrim" recognizes as belonging to Mathilde (321). Just as soon, the voice secures an optical frame as well, and turns into the vision of that "distant, small, wondrous majesty" in which a miniaturized Mathilde appears as the Heavenly Mother (322). Through this fusion of Virgin and Mother, love and religion, and eroticism and maternity,[42] a woman [*Frau*] in the Real becomes the wife [*Frau*] of all men in the Imaginary. And because "it seemed as if she wanted to speak with him"—even "though nothing was to be heard" (322)—Ofterdingen finally learns what pure instances of discourse really are.

In the first poem that Ofterdingen succeeds at composing, the new author hails "Mathilde" as the "Mother and Beloved of God" (324). Because the signifier "Mathilde" no longer possesses reference (Mathilde is dead) or materiality (she is inaudible)—but at the same time and for this very reason means Meaning in general—it makes poetic [*poetisiert*] all past discourses (which parties like Ofterdingen have the task of uniting). This is also why "everything seems much more familiar and more prophetic than before" (322). Doubly absented or neutralized—first in the discourse of a father and then in the dream where she drowns—a real woman becomes the *Alma mater* of all forms of knowledge.

Accordingly, the place where Ofterdingen hallucinates the presence of absent parties is introduced as "a cloister, altogether wondrous, like an entrance to Paradise" (340). The site excludes real women as a matter of course. And if—as Hegel proudly put it—"our schools and universities are our churches,"[43] then its name is certain as well. Around 1800, the monopoly on knowledge formerly held by monasteries and churches shifted to the new, state-run educational system. Thus, it is no wonder that—in the shadow of Mathilde, who has been transfigured into an *Alma mater*—there appears a wise old man, whose conversations with Ofterdingen may serve as an ending even in a novel that remained a fragment. Sylvester—of whom it is not said for nothing that he, "as a father, sits alone, eternally tearful, at the [mother's] grave" (327)—represents Ofterdingen's philosophy teacher at university (just as Klingsohr taught him German in high school). Sylvester's expertise encompasses medicine (325), history (326), theology (332), the natural sciences (334), and above all, philosophy (330–33). All disciplines—with the exception of jurisprudence—are represented when Sylvester initiates Ofterdingen into university learning. With that, both Poetry and the Poet achieve discursive legitimation.

Already in his exchange with Klingsohr, Ofterdingen had learned that poetry, despite the universality into which it can$_1$ translate all discourses, has a limit:

> If there exists a proper sphere for the individual poet, within which he must remain . . . , there is also, for all human faculties, a certain limit to what can be represented. . . . Mature experience first teaches one to avoid such irregularity of objects and to leave the detection of the Simplest and the Highest to Wisdom of the World. (285ff.)

Also, and especially, the poet Klingsohr accords philosophy (and not theology, as Dante and the Middle Ages in general had done) a superiority that, when Sylvester appears, finally enters the novel itself. His doctrine—that "the cosmos [*das Weltall*] dissolves into infinite worlds which are always encompassed by greater worlds," and that "all senses, ultimately, are a single sense" (331)—formulates precisely the "Simplest and the Highest," which, according to Klingsohr, only philosophers (and not poets) can pronounce.

In passing from Klingsohr to Sylvester, from "expectation" [*Erwartung*] to "fulfillment" [*Erfüllung*]—as the two parts of the novel are called—Ofterdingen performs the final steps necessary for education in the Age of Goethe. Graduating from a preparatory school, where there were still women and the highest point was occupied by German poetry alone, he arrives at the peak of contemporary discourses. In the European university system of old, philosophy had simply provided a propaedeutic course of study for medicine, theology, and jurisprudence (the three discourses of absolute power). But around 1800, in the new *Bildungsstaat*, it achieved the rank and title of supreme knowledge. The careers of Fichte and Hegel attest as much.

> Inasmuch as preparation for all university study passed to *Gymnasia*, the discipline of philosophy—previously the "preschool" of general knowledge for the three older fields of study—achieved an autonomous position at the beginning of the nineteenth century. In addition to the matter of cultivating scientific research, it was charged with the particular task of preparing [students] for the teaching profession.[44]

The formation of German teachers (in so-called reality) or poets (in so-called fiction) must occur by way of a discipline that surpasses even German. Not for nothing, in Schleiermacher's words, does "everything flow" into German lessons at the *Gymnasium*—everything, that is, "that can occur in schools for the free, formal education of the spirit, [which is] all preparation for philosophy." Klingsohr and Sylvester, Ofterdingen's two spiritual fathers,[45] are linked by this same information network. The one teaches the poetic, the other the philosophical, unity of all forms of learning.

Even though, as Sylvester tells the novel's protagonist, "Fable" is the "universal instrument [*Gesamtwerkzeug*]" of his "present world" (331), the wise old man considers Philosophy "the science of sciences" (III, 666). Here—at the highest level of reflection—resides his power. For all that, Sylvester is not unaware of the historical novelty of his position, nor does he leave room for others to doubt. He remarks "how far learning [*Wissenschaft*], which until now was called the Doctrine of Virtue or Ethics [*Tugend- oder Sittenlehre*]," has stood "from the pure form" of ethics as conceived by transcendental philosophy (332). In making this declaration, Sylvester is simply quoting Hardenberg, whose fragments and excerpts, in turn, quote "Fichte's moral doctrine [*Moral*]," which offers "the most correct views" (III, 685).

Hardenberg's novel is just that precise in securing its philosophical legitimation. A New Year that pronounces the words of Fichte [*ein Sylvester mit Fichte- Worten im Mund*] has nothing fictitious about it: the wise man's ascent to the highest instance of discourse in the text simply repeats the career path of philosophers in the disciplinary history of learning around 1800. Ever since a stranger told him about the Blue Flower, Ofterdingen has inhabited a realm of *Dichten und Denken*. Thus, when Ofterdingen and Sylvester meet at the end of the novel, the coupling of two discursive formations transforms into positive fact. The poetry that the Age of Goethe inaugurated cannot exist without the support of German Idealism.

From the inception, that is, starting with the works of Gustave Lanson and Georg Lukács, literary sociology has held a strangely vague conception of society. According to *The Theory of the Novel*, Hardenberg's work runs—indeed, it heightens—the "danger" of "lyrical, mood-dominated romanticizing of the structures of social reality," which "cannot, given the fact that reality at the present stage of development lacks pre-stabilized harmony, relate to the essential life of . . . interiority."[46] For all that, however, *Heinrich von Ofterdingen* does not set interiority in general against social formations that are (supposedly) hard at work dismantling the *deutsche Misere*. In actual fact, so-called interiority involves a student—that is, a man. By the same token, so-called social formations are actually the powers of discourse that institute the Social in the first place and steer it (for example, by way of readers' reactions to poetry). For this reason, Hardenberg's novel does not need to recruit readers—much less seduce them—with a world that is "beautiful and harmonious but closed within itself and unrelated to anything outside."[47] Rather, the work simply needs to regulate its relationship with that instance of discursive power which, around 1800—and only then—took over literature. Of course, the reception of literature has always occurred via channels that determine in advance what qualifies as a text, an author, a work, a letter, and so on. But only in the Age of Goethe was this task performed by a philosophy that invented the new field of investigation that, ever since, has been called "literary interpretation."

Once Sylvester has taught him the philosophy of history, nature, and proper conduct [*des guten Handelns*], Ofterdingen can finally articulate what he has been doing unconsciously all along—ever since his dreaming and thinking came to revolve around a blue flower. Whereas at first he "didn't even have a clue" about "poets and singers" or "their peculiar [*sonderbar*] art" (208), he now brings forth a concept of poetry that (as Hegel would say) promptly submerges in the Concept itself. "O splendid Father," Ofterdingen apostrophizes the philosopher,

> with what joy does the light fill me that issues from your words. And so the true Spirit of Fable is a friendly disguise for the Spirit of Virtue, and the actual purpose of the Art of Poetry, which is subordinate to it, is the Activity of a most exalted and authentic Existence [*Regsamkeit des höchsten, eigenthümlichsten Daseyns*]. (332)

The logic of the signifieds—as determined by the discursive space around 1800—has been achieved. In a "friendly" way (i.e., one that moves the general public), Poetry, boiled down to its core spirit or concept, disguises Spirit [*Geist*] or the Concept itself; in other words, it disguises Philosophy (per the latter's self-definition). At the same time, because it explicitly stands "subordinate" to Philosophy, the "Art of Poetry" has both a mistress and an address into which its discourse—this unified articulation of all the information channels of the epoch—can truly flow [*münden*]. Poetry has achieved legitimation, as well as a storage unit into which Hardenberg's *Ofterdingen* merges just as completely as Goethe's *Faust* into Hegel's *Phenomenology*.[48] That is why, to this very day, interpreters of the novel, instead of analyzing the poetic-philosophical networking of discourses, have affirmed, time and again, Ofterdingen's subordinate position—that is, subsumed the text under idealistic theorems.

Such is the power of authoritative discourses [*Herrendiskursen*]: whoever is able "freely" [*in "freyer Gewalt"*] to define what authority and a "master" possessed of "freedom" [*mit "freyer Gewalt"*] are makes what he says inescapable. Sylvester need not worry about leaving "the subordinate Art of Poetry"—this "disguise" of conscience—to the youthful Ofterdingen. As a philosopher, he has already claimed Conscience as such.

> Precisely this all-encompassing freedom, mastery, or dominion is the essence, the motor of conscience. In it is revealed the holy quality, the immediate creativity of personality; and every action of the master is, at the same time, the pronouncement of a lofty, simple, uncomplicated world—the Word of God. (331f.)

That is plain speech. Once again, Sylvester simply incarnates Fichte. A master of discourses says that God's dominion over discourse is over and done, now that Philosophy has assumed a position at the crown of all other disciplines. Formerly, perhaps, in the times of the "Bible," there existed "immediate exchange with Heaven"; maybe later, during the Middle Ages, "the Holy Ghost" spoke "to us" indirectly, "through the understanding of prudent and favorably-disposed men" (198). Around 1800, however, "the word of God" became identical with authorship. According to Sylvester, it appears as a "holy quality"—as the "immediate creativity of personality." According to Fichte, the "proof of the wrongfulness of reprinting books [*Unrechtmäßigkeit des Büchernachdrucks*]" can only be demonstrated when one recognizes that "everyone" has "his own course of thought [*Ideengang*]," "his own particular way of forming notions and connecting them with each other."

Around 1800 there arose for the first time a conception of copyright and authorship that understood books as the "exclusive property of [their] first master."[49] Whoever—like Sylvester or Fichte—is magisterial enough to trump masters of this kind through reflection and provide reasons legitimating their own dominion truly speaks the Word of God.

A novel like *Heinrich von Ofterdingen*, which cycles through the discursive space of its epoch from beginning to end—from unrecordable noise up to the system of universal storage called "Philosophy"—and moreover does so for each and every word or author, does not depict "actions" [*Handlungen*]. Instead, it *acts*.

9

World-Breath:
On Wagner's Media Technology

For Erika

In the nineteenth century, Germany also produced giant concerns in the arts. Only one of them, however, has survived without subventions or interventions from the state: Wagner's Bayreuth.

Unlike all programs of aesthetic education, and unlike the redemption promised by the Eternal Feminine, music drama has remained current. With his sense for public relations, Wagner knew full well that music drama would have been possible and successful in America, too—otherwise, it would not have occurred to him to emigrate from Bayreuth to Hollywood avant la lettre. What this means is that music drama represents the first mass medium in the modern sense of the word. It is simultaneous with our senses because of the technology it employs. The arts (to employ an old word for an old institution) entertain only symbolic relations with the sensory fields they take for granted. In contrast, media relate to the materiality with—and on—which they operate in the Real itself. Photo plates register chemical traces of light, and phonographs record mechanical traces of sound.[1] This distinction between arts and media was clear to Wagner. In "The Artwork of the Future"—an unambiguous title—he observed ironically that poetry offers readers the catalog of an art gallery, but not the actual paintings.[2] In order to fill this technological gap, he invented the first art apparatus capable of reproducing sensory data as such.

At one fell swoop, reflection and imagination, education and literacy—all the celebrated psychic faculties that Classical and Romantic poetry necessarily had to presuppose for its pages to affect people—were rendered obsolete.[3] In the revolutionary darkness of the *Festspielhaus*[4]—from which the darkness of all our cinemas derives—the medium called "music drama" began to play upon, and with, the nerves of the audience.

The Ring of the Nibelung stands for power, not money.[5] The sole power that does not founder when twilight descends on the gods at the end of the tetralogy is technical in nature. The great engineer Alberich—the inventor of a magic cap [*Tarnkappe*] that makes him as invisible as the conductor in Bayreuth's orchestra pit—survives: invisible, yet unvanquished. That is why he, and not his divine antagonist Wotan (who, after all, merely improvises Wagner's corporate politics of establishing family dynasties), provides the allegory of the *Festspielhaus*. Neither Wotan nor Wagner could ever keep their descendants—even if they had been programmed for unthinking loyalty—from bringing forth traitors like Siegfried or Wieland. The invisible power exercised by Alberich with his whip[6]—and conductors with their batons—over Nibelungs, musicians, and listeners remains physiologically inscribed in bodies and nerves. It is easy to demonstrate the innovations of Alberich, aka Wagner. Simply comparing the medium of music drama with traditional drama and opera is enough. That said, there is no need to distinguish between these three genres of art according to form, content, and meaning—that is, with the conventional means of philosophical reflection. One need only view them as media, which means with the same stupidity that characterizes Wagner's heroes, especially Siegfried.

In light of this stupidity, classical drama offered little more than the exchange of verbal information between people who could—it goes without saying—talk and listen. They knew each other by name or, if they had not yet met, at least by sight. When dramatic reasons called for the perfect transparency of verbal and optical data streams to grow cloudy, two—and only two—forms of interference could occur: misleading words (especially names), on the one hand, and masks, on the other. Even so, however, the meaning of words spoken and heard did not vanish in the noise of the Real. And even the distorting power of masks did not reach far enough to change their bearers into the invisible voice that Wagner's Alberich becomes beneath his magic cap. There was no place in drama for the acoustic field as such, with all its senseless noises and disembodied voices.

To be sure, opera proceeded as an acoustic data stream. However, its parameters were not all defined in this way. Exchanges, as formulated by recitative sequences, were more or less rudimentary, and they followed the model of drama: speech and/or sight informed characters about their respective positions in the play. On the other hand, when characters sang arias and thereby entered the acoustic field, it occurred to express so-called affects—which, for their part, had little repercussion upon the dramatic interaction itself. Only in exceptional instances did sounds (e.g., signals or cries) also convey information on an interpersonal level. Opera, then, was based on a separation between verbal and acoustic data—recitatives and arias. In the final analysis, the division may simply have repeated the division of labor between libretto and score, lyricist and composer.

Wagner's technical program can only be reconstructed in contrast to the traditions of drama and opera. Two art genres with different sensory fields could not simply be stuck together, and so—in order to achieve the structure of modern mass media in a fitting way—music drama had to intervene in the very materiality of data streams. In contrast to what occurs in drama, the interaction of characters had to be motivated by acoustic events, and unlike what happens in opera, acoustic events (whether vocal or instrumental) had to be motivated by dramatic interaction. These are two reasons why Wagner's texts are not simply opera libretti—and why his scores include so many stage directions.[7]

None of the traditional data streams—neither meaningful words and looks nor psychological affects—could guarantee the reciprocal motivation of different sensory fields. One, and only one, phenomenon can appear simultaneously in text and score, in drama and music. Everyone—except for Wagner scholars—knows what it is: breathing.

Siegfried, Act III: the hero has entered the circle of flames; there, in the middle, he finds a body lying on the ground, in full armor. Siegfried does not know if the body is dead or merely sleeping. Nor does he know if it is a man or a woman. Two oppositions fundamental to every culture—life and death, male and female—must be made clear again. The dramatic scene, one of the most beautiful that Wagner ever composed, begins as a "primal scene" in every sense of the word.

A single hint—one *bit* of information—passes through the field of complete uncertainty. Siegfried's ears hear that the body is breathing. Accordingly, as he approaches, the hero hymns the "rising breath" as a sign of life. "He draws his sword, and with gentle caution cuts the chain mail on both sides of the armor," thereby freeing the breathing from its "confining breastplate." In so doing, he also discovers signs of femininity—the breasts beneath the plate. Therefore, Brünnhilde's breathing itself, a sign of life and eroticism, becomes an erotically desired object. Siegfried enthuses about "this breath's blissfully warm fragrance." He does so for good reason, too, for all he has said and done until this point has failed to rouse the sleeper. Brünnhilde will only return to "perceiving earth and heaven" when Siegfried—even "should" he "expire in death"—"sucks in life from the sweetest of lips."[8]

Awakening, for Wagner, always means singing. The materiality of data streams in his music drama depends on the vital intensity of the diaphragm, lungs, throat, and mouth; this is also why singing represents the final, and most important, transformation of breathing. With the same breath that Siegfried's kiss has given to—or possibly taken from—her, the reawakened Brünnhilde begins to salute the Sun, Light, and Earth, the three media that sustain physical life. The radiant song gains in verbal, meaningful, and psychological depth as it unfolds. Awakened by Siegfried's desire or breath, Brünnhilde begins telling her lover that on the one hand, she is his departed mother and therefore protected by the incest taboo; on the other hand, she says that she is (now) a living woman, with whom he can sleep.

However, because the aria has originated in breathing itself, it remains operative on the level of physiology first theorized by Wagner's contemporaries—Alexander J. Ellis in England, Hermann von Helmholtz in Germany, and Ernst Wilhelm Brücke in Vienna.[9] And so, for the first time in literary history, meaningful, articulated speech meets with explicit rejection. When Brünnhilde declares that her love is both eternal and chaste, Siegfried responds:

Wie Wunder tönt	*Like wonder sounds*
was wonnig du singst,—	*what blissfully you sing,—*
doch dunkel dünkt mich der Sinn	*yet the meaning strikes me as dark.*
Deines Auges Leuchten	*Your eyes' sparkle*
seh ich licht;	*I see clearly;*
deines Atems Wehen	*the waft of your breath*
fühl ich warm	*I feel warmly,*
deiner Stimme Singen	*your voice's singing*
hör ich süß:—	*I hear sweetly:—*
doch was du singend mir sagst	*but what, singing, you tell me,*
staunend versteh ich's nicht.	*amazed, I understand it not.*
Nicht kann ich das Ferne	*I cannot, what is remote,*
sinnig erfassen,	*grasp by sense,*
wenn alle Sinne	*when all my senses*
dich nur sehen und fühlen!—	*see and feel only you!—*[10]

The burning presence of desire instead of eternal or Platonic love, *sound* (in the precise sense of Jimi Hendrix) in lieu of verbal meaning, and physically stimulated senses in the place of a psychological mother imago: Siegfried's response offers the definition of the music drama itself. His words speak only of the media that Bayreuth presents: optics, acoustics, lights, and heaving breaths. In traditional art, a response of this kind would have created a scandal. Dramatic plots would never have developed had it not been for the strange fact that our understanding finds meaning in words (and fails to hear their breathing). The bel canto of Italian operas, even though it verges on incomprehensibility, was never meant to reveal the physiological roots of song in respiration; if anything, it concealed them behind melodic figures and the performer's virtuosity. And so, no opera has ever permitted itself a finale like Wagner's *Siegfried*—that is, to put the physiology of love on stage.

Such "respiratory eroticism"[11] (in Lacanian terms) does not, by any means, constitute an exception in Wagner's music dramas. The same primal scene occurs time and again: one figure of drama listens in on another's breathing. It occurs at the beginning of *The Valkyrie*, between Sieglinde and the unconscious Siegmund; when the love affair is over, it happens between Siegmund and the unconscious Sieglinde.[12] It happens when Tristan lies dying—first as the servant Kurwenal "bends over him in grief" and, "full of care, listens to his breathing," and then, finally, when Isolde laments before Tristan's corpse that she no longer can hear "a breath's fleeting flutter."[13] Over and over, the other's breathing turns into the diagnostic sign of life or death—being able to sing or falling mute. Conversely, one's own breathing provides the necessary condition for acts that are musical and dramatic at once. In Act I of *Siegfried*, the hero's unarticulated cries and refrain—"blow, bellows, blow the flames!"—accompany the kindling of a fire that amounts to an industrial smelting furnace.[14] In Act II, the same breath activates Siegfried's horn and reed pipe.[15] Finally, in Act III, it swells into a fully articulated love song. Wagner's music dramas motivate and generate the music itself—whether it is vocal, like the love song, or instrumental, like the horn and reed pipe—out of the plot.

All the same, most critics agree in their ignorance of, or scorn for, Wagner's so-called libretti. Perhaps they have eyes only for the printed word, and no ears for the breathing, murmuring [*Rauschen*], and storming that Wagner discovered in his poetry. Maybe they have also been blinded by the grandiloquent philosophy with which Wagner framed his altogether simple texts. In any event, the facts of physiology and media technology seem too dumb or unconscious for them.

And yet, in mass media, the Unconscious turns into the main event [*Sache selber*]. The messaging channels of the traditional arts were to be switched consciously—and by the same token, interrupted consciously as well. Those who held a speech or understood discourse could also choose not to do so. Those who made or responded to eye contact could also shut their eyes. *Sound*, on the other hand, pierces the armor called Ego, for of the sensory organs the ears are the hardest to close. This is why, in *Twilight of the Gods*, Alberich succeeds in making his sleeping son Hagen "hear"—and even manages to dictate commands to him in his "sleep."[16] The all-pervading power of *sound* carries Wagner's artistic imperialism. The plots of his music dramas reveal that the composer knew its might every bit as well as did Alberich, his media technician.

Time and again, critics have remarked that the whole of *The Flying Dutchman* builds on an optical hallucination: Senta's "dreamily" fascinated gaze at the portrait of the Dutchman on the wall leads him to materialize.[17] No one, however, seems to have realized that in *Lohengrin*—that is, with the onset of Wagner's maturity—optical hallucinations are replaced by acoustic ones. Their content is nothing more, and nothing less, than the all-pervasive power of acoustics. Elsa, Lohengrin's future bride, both says and sings as much:

Einsam in trüben Tagen	Alone in gloomy days
hab ich zu Gott gefleht	I begged to God;
des Herzens tiefstes Klagen	my heart's deepest lament
ergös ich im Gebet:—	I poured forth in prayer:—
da drang aus meinem Stöhnen	there, from my moans,
ein Laut so klagevoll	arose a sound so piteous
der zu gewalt'gem Tönen	that in mighty resonance
weit in die Lüfte schwoll:—	it swelled far into the air:—
ich hört ihn fernhin hallen,	I heard it echo far away
bis kaum mein Ohr er traf;	until it scarcely met my ear;
mein Aug' ist zugefallen	my eye fell closed
ich sank in süßen Schlaf.	and I sank in sweet sleep.
[...]	*[...]*
In lichter Waffen Scheine	In the shine of bright armor,
ein Ritter nahte da.	a knight then drew near.[18]

The knight is first hallucinated by Elsa's closed eyes. Accordingly—and just like Senta's Dutchman—he promptly appears on the stage. All the same, his presence, which after all coincides with the dramatic interaction as a whole, derives from an acoustic hallucination. Elsa's pleas, laments, and moans have ordered Lohengrin to appear from some four hundred miles away—the distance between her duchy, Brabant, and his holy mountain, Monsalvat. Such an achievement would be impossible were the medium not the message. Because Elsa omits the "contents" of her laments, pleas, and moans and mentions only the fact that these sounds occur, McLuhan's theory turns into reality. As when Siegfried listens to Brünnhilde or Kundry speaks (or rather, makes a "hoarse and broken" effort "to regain speech"[19]), the discourse shrinks down to the modalities of vocal physiology. Scarcely audible sounds, freed from the mouth and will of their speaker, swell into "mighty"—that is, absolute—"resonance," which then travels through space and time as *sound* "[echoing] far away."

Neither Elsa's medieval times nor Wagner's nineteenth century were able to implement this acoustic effect. Our ears are the first in history to know it by heart [*auswendig*]: night after night, the PA systems of rock music (amplifiers, delay lines, equalizers, and mixing boards) generate vocal noises, surround sounds, and reverberation effects.[20] In other words—the words of Jimi Hendrix—Wagner's Elsa was the first resident of "Electric Ladyland." What she describes, with accuracy beyond measure, as resonance, swelling, and reverberation has little to do with prayers or Christian belief. Rather it announces, before the fact, the theory of positive feedback and oscillators.

Under technical conditions as they were given, Wagner could not implement the feedback of sound. Instead, he composed it, which represents a further innovation. Fantasies like Elsa's can be traced back to German Romanticism—to Friedrich Schelling or Bettina Brentano.[21] To achieve realization, however, it was necessary for them to wait until Wagner. The orchestral background of Elsa's prayer—and even more, the prelude to *Lohengrin* as a whole—performs in actual fact what Elsa describes in the unending crescendo of her voice. Breathing and its constituent gradations (sighing, pleading, moaning), then, simply provide points for a second feedback—this time between vocal and instrumental effects—to arise. To make Elsa's barely audible laments into *sound* echoing far away, the orchestra, especially the brass instruments, must take them up.

Wagner's orchestra functions exactly as an amplifier does. This is also why his autobiography expresses such fascination, over and over again, with echoes, instances of feedback, fade effects, and acoustic illusions.[22] Moreover, it explains why Adorno—who remained loyal to European art and musical logic—foundered when he encountered Wagner. Amplifiers put philosophy out of commission. They break with traditional musical values such as thematic workmanship and polyphonic arrangement—data that are fundamentally written—and replace them with *sound*. With Wagner, music became a matter of pure dynamics and unadulterated acoustics.

One finds ready proof of this—both in the libretto and in the score—in *Tristan*. Everyone knows, of course, that Wagner's most modern music drama also derives from a medieval romance. This is the case, however, for a reason that is less well known. Gottfried von Strassburg wove acrostics and anagrams throughout the text of his own *Tristan*—that is, he was the first writer in the vernacular to emphasize writtenness itself. It is no accident that he bore the title of "master" [*Magister*], that is, "master writer" [*Schreibkundiger*]. Unlike his many knightly forebears, Gottfried no longer addressed a host of highborn earwitnesses. With the play of letters [*Buchstabenspiele*] that necessarily escaped hearing, he instituted a new audience composed of literate parties, that is, readers.[23]

Wagner's *Tristan* utterly revokes the communicative system that reigned from Gottfried, via Gutenberg, up to Goethe; that is, it revokes literature itself. In the courtly romance, Tristan and Isolde used their initials—"T" and "I"—as a secret code, which the equally literate author then scattered throughout his text as a whole. Wagner's music drama places a distinctive *sound* at the exact site of this alphabetic code. Act II opens with a whirring, ambiguous toning of the orchestra, which Isolde's maidservant Brangaene hears—only too correctly—as the horn signal of King Marke. Isolde, for her part, is brought by the "wildness of her wish" for Tristan to "hear [*vernehmen*] what" she "thinks [*wähnt*]." That is the very definition of an acoustic hallucination. Her maidservant tells her: "The clangor of horns sounds not so fair"; only "the spring's gently purling waters rush along so lovely."[24] The ambiguous tone of the orchestra, then, yields the work's theme on an interpersonal level. It gives rise to an acoustic hallucination that literally removes the unbeloved Marke and replaces him, through the natural sound of a spring, with the presence of Tristan, who is dear. And because text and score motivate each other over and over in Wagner's works, the textual oscillation between the sound of nature and the instruments of the orchestra—between random noise and a hunting signal—correspond to two equally illiterate horns playing C major and F major at once.[25] Such an effect counted as illicit for as long as music stood under the dominion of scores—and for as long as scores, in turn, stood under the dominion of writing. Wagner's new medium, *sound*, exploded six hundred years of literal and literary practice.

Everywhere in *Tristan*, from beginning to end, acoustic effects take the place of the symbolic (i.e., written) structure of drama and music. The substitution concerns both voices and instruments because breathing forms their common root. Writing to Mathilde Wesendonk—his own Isolde—Wagner explained that (and how) the dynamics of the prelude to *Tristan* elaborated, in composition, and materialized, in performance, "the Buddhist theory of the world's origin." At the very beginning, before the first sound, there reigns endless silence—"Nirvana," or "heavenly clarity." Then, at the cello solo, which is explicitly called "a breath [*Hauch*]," "the heavenly clarity darkens." Third and last, the "Tristan chord" steps in, and the sound of the orchestra "begins to swell" and "become denser"—"ultimately, the whole world in its impenetrable massiveness once again stands before me."[26] That is plain talk and, as far as massiveness is concerned, pure dynamism in terms of technical media. From Nirvana, over a primordial gasp of breath [*Atemhauch*], up to a fully composed world: the orchestral prelude to *Tristan* formed the first circuit of acoustic feedback.

The second circuit—this time, a vocal one—opens when the curtain rises. A "young sailor" is singing. In the first place, the actor is not visible; second, the song is a cappella—without orchestral accompaniment, that is. The youth sings of the "wind freshly blowing to home," which, in so doing, drives the ship and the sailor farther and farther away from his "Irish child." Therefore, with his next breath, the sailor asks of his faraway love: "Is it the blowing of your sighs that fills my sail?" Distantly echoing sighs (to use Elsa's language) should themselves create the distance that they in turn lament: such is the paradox of respiratory eroticism. And so, wind and breathing—natural sound and the human voice—become indistinguishable, even in the seaman's wordplay. For everything he says and sings simply exploits the nearly perfect homonym of the German words *Weh* [woe] and *wehen* [to blow]. His song ends with the dreamily sad verses: "*Wehe, wehe du Wind! Weh, ach wehe, mein Kind!* [Blow, blow, you wind; woe, oh woe, my child!]"[27]

Human voices as winds, and winds as human voices: only the linguistics of a Wagner or a Siegfried—in their disdain for meaning—allow such equations, which, moreover, are acoustic puns. Music drama thrives on them, however, because they alone can switch between voices and instruments, between lyrics and score. When the sailor, whose a cappella song does not occur by chance, transforms the sounds of nature into human voices, he is already anticipating and providing motivation for the scene that follows. Now nonhuman—that is, orchestral—sounds strike up again. The seaman's song wanders over into the strings to provide the background for an entrance that presents the actual (and only) "Irish child" of the piece: Isolde.

Now a woman has the floor [*das Wort*]. As one might expect, she turns all the seaman's words around. Isolde—who suffers from Tristan's distance and his lack of desire—straightforwardly desires that all human voices once again sink into noise, or in other words, Nirvana. Accordingly, she wishes that a magical power might return, which her mother once possessed and bequeathed to her:

Wohin, o Mutter,	Wherefore, O Mother,
vergabst du die Macht,	did you bestow the power
über Meer und Sturm zu gebieten?	to command sea and storm?
O zahme Kunst	O feeble art
der Zauberin	of the sorceress,
die nur Balsamtränke noch braut!	who now brews only balsam!
Erwache mir wieder,	Rouse in me again,
kühne Gewalt	intrepid force;
herauf aus dem Busen,	out from that bosom
wo du dich bargst!	where you hid yourself!
Hört meinen Willen,	Hear my will,
zagende Winde!	trembling winds!
Heran zu Kampf	Onward, to arms
und Wettergetös!	and elemental roar!
Zu tobender Stürme	To raging storms'
wütendem Wirbel!	Riotous whirl!
Treibt aus dem Schlaf	Drive out of sleep
dies träumende Meer,	this dreaming sea,
weckt aus dem Grund	rouse from the depth
seine grollende Gier!	its rumbling hunger!
Zeigt ihm die Beute,	Show it the spoils
die ich ihm biete!	that I offer!
Zerschlag es, dies trotzige Schiff,	Destroy this defiant craft,
des zerschellten Trümmer verschling's!	may it gulp the shattered debris!
Und was auf ihm lebt,	And all that lives upon it,
den wehenden Atem,	the wafting breath,
den laß ich euch Winden zum Lohn!	I leave you winds as your reward![28]

Until this point, then, Isolde's magic had been reduced to interiority [*eine Innerlichkeit*]—which, not coincidentally, constituted all the magic of Classical and Romantic poetry. When Wagner's music drama begins, however, a more archaic and more exterior magic returns. Isolde's command extends to two sites at once: first, to the winds and woodwinds, and second, to nature and its technological correlative. With every word that she sings, the dynamism of the orchestra swells. A single, human voice wishes to drown, in instrumental feedback, along with all the other voices on the ship. Therefore, it is absolutely telling that the orchestral fortissimo behind the female voice pauses for one, and only one, measure. Unaccompanied—and as if to recall the sailor's a cappella—Isolde sings the word "breath," the opposite, that is, of nonhuman sounds.[29] In such an unheard-of manner does music drama operate when it switches between textual and acoustic events.

World-Breath

Operas before Wagner had been limited in dynamic range. Sound effects simply were not permitted to drown out human voices and human language. But that is precisely what happens when Isolde gives "the breath" of all living beings on the ship "to the winds" as their "reward." Vocal physiology, that is, forms only a small part of acoustics in general. And so, Isolde's phantasmagorical wish offers a further definition of the music drama. Moreover, it is also why, in the final scene, her wish goes into fulfillment. Isolde's so-called *Liebestod* has no function other than this. Under the immeasurably exact title of "World-Breath" [*Weltatem*], it celebrates an acoustic power above and beyond all humanity.

Once again, the beginning is simple, gentle, and human. As Isolde remembers, she sings an old "air" that is called "so wonderful and quiet" because it stands, as a leitmotif, for her dead lover. The air ascends from the wind instruments, which Isolde herself follows—in keeping with the utter technical precision of the score—after a delay of exactly one eighth-note.[30] Given such feedback between orchestra and voice, the quiet soon ends. What now occurs is a literal crescendo: "growing." Within Isolde's auditory—or hallucinatory—field of perception, Tristan's corpse begins to live again, swelling and breathing:

Mild und leise,	*Soft and quiet,*
wie er lächelt,	*how he smiles;*
wie das Auge	*how his eye*
hold er öffnet—	*opens sweetly—*
seht ihr's, Freunde?	*do you see it, friends?*
Säht ihr's nicht?	*Don't you see?*
[. . .]	*[. . .]*
Wie den Lippen	*How from lips,*
wonnig mild,	*blissfully mild,*
süßer Atem	*sweet breath*
sanft entweht?	*softly draws?*
Freunde! Seht!	*Friends! See!*
Fühlt und seht ihr's nicht?'	*Do you not feel and see it?—*
Höre ich nur	*Do I alone hear*
diese Weise	*this air,*
die so wunder-	*so wondrous*
voll und leise	*and quiet,*
Wonne klagend,	*lamenting joy,*
alles sagend	*telling all,*
mild versöhnend	*mildly reconciling,*
aus ihm tönend	*sounding from him,*

in mich dringet	*it pierces me,*
auf sich schwinget,	*rises upward,*
hold erhallend	*sweetly resounding,*
um mich klinget?	*rings all around me?*[31]

A crescendo in both the text and the score makes it possible to bring a dead body—or (in musical terms) a body no longer capable of breathing and singing—back to life. Tristan's expired breath returns as an orchestral melody: the notes that sound from him penetrate listeners. However, it is Isolde who sings or hymns [*besingt*] all of this, from the crescendo to the sound effects. And so, her orchestrally amplified voice supplements the absent [*vermißte*] voice of her lover. In Wagner's works, the voice is so "un-individual" and the acoustics so ecstatic,[32] that to the ear of the woman singing, her own voice essentially seems to be the voice of the Other.[33]

When in *Twilight of the Gods* Siegfried loses his breath and his life, he celebrates the memory of Brünnhilde (which had gone missing) as a kind of (artificial) respiration—as if the "lovely wafting" of her "breath" were welcoming and reanimating him, the singer. It is as if his death were the exact counterpart to Brünnhilde's erstwhile reawakening.[34] Under such conditions, even the most hallucinatory and phantasmagorical claims come true, simply because they cannot be sung. "Friends! ... Do you not feel and see it?" is a rhetorical question. Like Jimi Hendrix's question—"Have you ever been to Electric Ladyland?"[35]—it provides its own answer by means of the sound effects that it triggers. In the orchestra, the dead Tristan experiences an acoustic erection. And because the "friends" whom Isolde addresses stand for the audience already programmed into music drama, the unthinkable actually becomes audible. Isolde and her listeners "drown," as she says (or predicts), in the "highest"—that is, unconscious—"pleasure" of a "surging gush [*wogenden Schwall*]," a "sounding echo [*tönenden Schall*]." Its name is "World-Breath,"[36] and its technology orchestral fortissimo.

The world premiere of *Tristan und Isolde*, on 10 June 1865 in Munich, was the beginning of modern mass media. Wagner was apprehensive—and for good reason—that the "last Act," even in an "entirely *good* performance," would either be "banned" or "drive people crazy."[37] Tristan's acoustic resur-erection, which forms the pillar of the orchestra's World-Breath, explodes all the possibilities of traditional art. Only media can implement what Isolde calls—as much in a technical sense as an erotic one—a "surging gush" and "sounding echo."

World-Breath

Wagner demonstrated uncharacteristic modesty when he said it was "merely one of his plans" to add invisible actors to the invisible orchestra he had already invented. He did so both in fact and in truth. Tristan displayed his acoustic—and therefore invisible—erection[38]; Alberich vanished under his magic cap; the young seaman was "audible from the heights, as if from the masthead"[39]; and the Rhine Maidens occupied "the depths of the valley, invisible," beneath Valhalla.[40] All these voices—and others, too—are inhabitants of a "total world of hearing [*vollkommene Hörwelt*]," as Nietzsche clearly recognized.[41]

Only when the total world of hearing is created with media-technical precision can it be coupled with a "world of seeing" and enter the technical era. A space of sound that, thanks to its feedbacks, no longer needs the old-fashioned visibility of actors' bodies enables parallel connections with the new (i.e., technical) visibility of film. Already when the *Ring* premiered in 1876, the Bayreuth *Festspielhaus* employed a *laterna magica* to create the hallucination of the nine Valkyries riding on horseback—that is, born aloft on orchestral sounds.[42] Finally, in 1890—five years before the feature film was introduced—Wagner's son-in-law suggested a "night-dark" room, in whose "background" moving "pictures" would "fly by" to the sound of his father-in-law's "sunken orchestra" and put all spectators into a state of "ecstasy."[43]

In the meanwhile, precisely this ecstasy has been produced in the films of Hollywood—and in stereo, too. At the time, however, only Wagner's technical innovations made it possible. Music drama functioned as a machine that worked on three levels—that is, in three data fields: first, that of verbal information; second, through the invisible orchestra of Bayreuth; and third, with scenic visuality, which involved "tracking shots" and spotlights avant la lettre. The text was fed into a singer's throat, the throat's output was fed into an amplifier called "orchestra," and the orchestral output was fed into a light show; finally, all of the above was fed into the nervous system of the audience. Ultimately—when everyone had gone crazy—every last trace of the alphabet had been erased. Data, instead of being encoded in the alphabet of books and scores, were amplified, stored, and reproduced through media. (For Wagner, even scores—as if they were already phonographs—had the sole function of timing discourses and sound effects.[44]) Music drama defeated all literature.

For this reason, "World-Breath," Isolde's final word, is no metaphor. Instead, it names the orchestra itself. Just like the *division*—a fighting unit constituted by the three weapons systems of infantry, cavalry, and artillery—the orchestra, as drill, power, and combination of instruments, was invented by the grand

Nineteenth Century.[45] Wagner knew as much and said so. His god Wotan—a god of armies and ecstasies, of initiation and death—signifies, both in etymological terms and as a character in the tetralogy, the rage of a superhuman and prophetic voice. In like fashion, the army of Wotan's nine Valkyrie daughters means nothing but "storm." All this power, all this sound and fury, originates with the goddess Erda—that is to say, once more, with the World-Breath. As Wotan tells her, the mother of his storm daughters or storm troops, "Wherever life [*Wesen*] exists, there blows your breath."[46]

The earth in its materiality—this givenness [*Vorgegebenheit*] that is unthinkable for Classical forms of art[47]—dominates music drama as a whole. It reigns as Breath rising from the depth of graves or mineshafts, which all contain the bottomless abyss of the body. Wagner equated such graves not just with the cave of prophecy at Delphi, but also with the orchestra pit at Bayreuth. There is no difference between technical and psychedelic vapors.

The name of "World-Breath" encompasses all of Wagner's innovations in just such a precise and technical manner. It offers proof of Wagner's thesis that "music" is the "breath" of "language."[48]

That, then, would be my inaugural lecture on Wagner. However, even lectures need not always conclude with the standard hermeneutic trick of quoting positive proof offered by the author's own words. Today, in 1985, Wagner's media technology merits a brief epilogue. And so I will end—in every sense of the word—with *Apocalypse Now*. In Coppola's film, when the U.S. Airborne Cavalry undertakes its notorious operations against villages suspected of being Viet Cong (missions that General Westmoreland christened "search and destroy"), light music—Muzak—accompanies them. Wagner's "Ride of the Valkyries"—the pleasant, old-fashioned light show from 1876—drones in the earphones of the military helicopters. A feedback loop between the music drama and war technology has transformed the Valkyries, Wotan's deadly daughters, into onboard gunners, their storm horses into helicopters, and Bayreuth into Hollywood.

In this way, the capitalist medium recalls its prehistory in the works of Wagner. General staffs and directors are more attentive than critics. But all the same, *Apocalypse Now*—the posthistorical incarnation of Wagner's riding Valkyries—has a prehistory of its own in two world wars.

From 1941 to 1944, Major Ernst Jünger—both an operations officer in, and the poet of, the Wehrmacht—resided in Paris at the Hotel Raphael, which housed a German government office in the occupied administrative zone

World-Breath

Figure 2

called "France." Whenever Royal Air Force bombers attacked the City of Light by night from bases in the south of England, Jünger would ascend to the rooftop terrace in order to enjoy the "great beauty" and "demonic power" of the multimedia "show" [*Schauspiel*]. For the same *radiance* [*Strahlungen*] that his wartime diary promises in his title was then to be seen—arranged by Field Marshal Harris, and performed by Avro Lancasters and Bristol Blenheims over Paris in flames. At such times, Jünger would hold "a glass of burgundy with strawberries swimming in it" in his hand.[49]

Recently, French critics have held this wineglass up as evidence of the drinker's nihilism and aestheticism. Interpreters are just that poorly informed. In fact, Jünger at his rooftop perch was quoting another world war and another writer. It is a matter of literary record that already in 1915 two Parisian residents stepped out to the balcony to enjoy the play of light between attacking German zeppelins and French defensive installations. Bomb warfare as a world premiere . . . One of the two was Robert, Marquis de Saint-Loup—a brilliant young officer on leave from the trenches that would be his grave. The other, less well-known party was a certain Proust. Because neither a world war nor an airborne attack could cloud his love for Wagner and Germany, the Marquis told the writer about the beauty of moments when the zeppelins "form constellations"—and the even more beautiful times when they crash and "form an apocalypse." Then, Saint-Loup's discerning Wagner-ears recognized, zeppelins become Valkyries and the sirens herald their wild ride.[50]

This test of Wagner's media technology could not have produced more empirical results.

10   The City Is a Medium

> Just as we are accustomed—if not required—to receive energy in various forms at home, we will find it altogether easy, there, also to receive or absorb those overaccelerated changes or oscillations from which our sensory organs—which pick them up and integrate them—make all that we know. I do not know if philosophers have ever dreamed of a company for the domestic distribution of sensory reality.
>
> — Paul Valéry

CAPITAL. The name says it all: capitals are named after the human body. Since the Greeks, the state has qualified as an organism, and the capital as its head. In turn, the capital is ruled by a chief, whose name means—yet again—head.

Historically, this equation has proved true enough. Lewis Mumford has demonstrated that what he calls the prehistoric "implosion" of villages and regions, from which cities emerged, did not occur out of economic necessity, but because of the arms monopoly of warlords. Plato, the legislator of an ideal city, limits its size to the reach of a voice giving laws or orders.

And for a long time—from the prehistoric foundations of cities, with which high culture or history began in the first place, up to the residences [*Residenzstädte*] of the Baroque—the military-administrative head remained architectonically recognizable: as a mountaintop fortress, acropolis, citadel, or palace. Only with the Industrial Revolution did the proliferation occur whose excrescences, in Mumford's eyes, disfigured the face of the city by going, in the name of pure technology, beyond the ecological necessities of communal life: thus, the Megalopolis was born.

Except that, when describing the course of an error, one often errs oneself. Perhaps holding fast to the self-evident centrality of the head only means that—as Foucault put it for "political thought and analysis"—"we still have not cut off the head of the king"[1] when it comes to the concept of the "capital." Were this so, then the monarchies to which Europe owes most of its capitals would have assured their survival, above and beyond architecture, in the head of theory itself. If "man," with his ecological needs, were only the miniature of such rulers, one might discern the possibility of decoding "head" and "technology" on the basis of technology, and not vice versa.

TECHNOLOGY. What to passers-by seems like growth run amok or entropy is technology—that is to say, information. Ever since it has become impossible to survey cities from a cathedral tower or a castle, and ever since walls and fortifications have ceased to contain them, cities have been traversed and connected by a network of innumerable networks, also (and especially) at their margins, points of tangency, and frayed edges. No matter whether these networks convey information or energy—that is, whether they are called "telephone," "radio," and "television," or "water supply," "electricity," and "highway"—they all are information (if only because every modern stream of energy needs a parallel control network). However, even before recorded time [*in unvordenklichen Zeiten*], when energy still needed physical carriers like Sinbad and information required messengers like the runner from Marathon, these networks did not *not* exist. They simply had not been built yet—or "implemented," as one says in technical jargon. The meager mule trail leading through the wild was replaced by train tracks and highways, which in turn were replaced by copper wires or fiber optic cables (which are no less ephemeral).

NETWORKS. For this reason, the reverse side of buildings—in the open space of the city—reveals their structures. They are networks, too.

To reconstruct the way out of a labyrinth (as the Greeks are supposed to have done when reading the ruined design of cities in Knossos, Phaistos, or Gournia), one does well not to note the walls that remain visibly connected; instead, one should pay attention to the opposite: the invisible connections between pathways and gates. Thereby—and in the mathematical sense of the word—a "tree" takes shape, whose forks make the difference between dead ends and exits plain.

Alternately, and like Claude Shannon in the official capacity of head mathematician at Bell Telephone Labs, one may construct a mechanical mouse whose muzzle rummages through the labyrinth in a process of trial and error. Such a mouse would be able to optimize city planning without the thread of Ariadne—Shannon, for his part, optimized something else, which was invisible: the telephone network of the United States.

GRAPHS. Only since 1770 has mathematics performed operations with networks of this kind. Topology and graph theory do not just describe modernity—they started it in the first place.

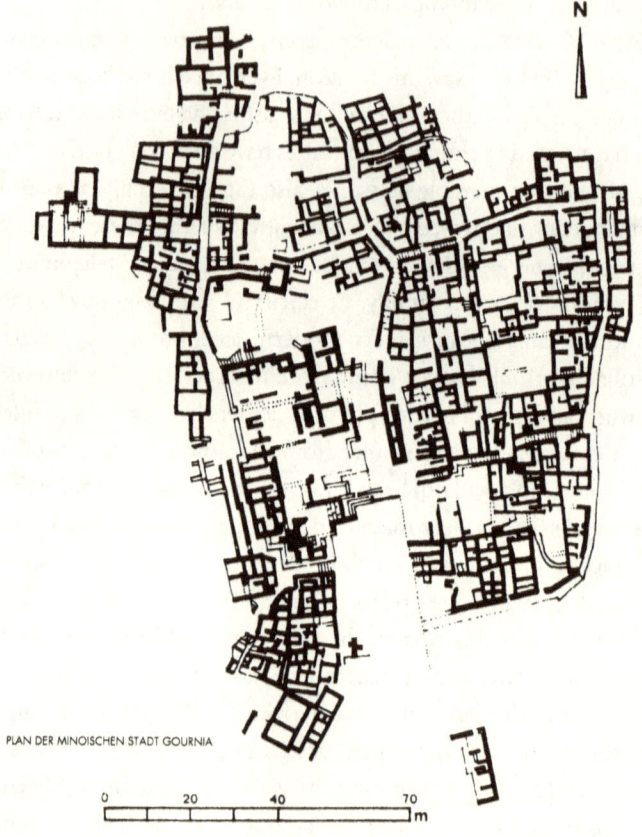

Figure 3. Map of the Minoan city of Gournia. From Dietmar Steiner, Georg Schöllhammer, Gregor Eichinger, and Christian Knechtl (eds.), *Geburt einer Hauptstadt am Horizont* (Vienna: BuchQuadrat, 1988).

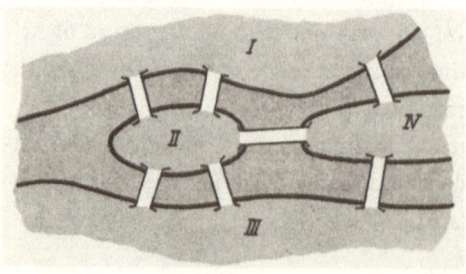

Figure 4. Map of Königsberg. From Dietmar Steiner, Georg Schöllhammer, Gregor Eichinger, and Christian Knechtl (eds.), *Geburt einer Hauptstadt am Horizont* (Vienna: BuchQuadrat, 1988).

The City Is a Medium

In a city that was still called Königsberg at the time, there were seven bridges leading over the river Pregel. A city is not just "the corollary of a street."[2] Rather, given its rivers, canals, and news channels, it is "the point at which all these paths meet."[3] This insight prompted Leonhard Euler—the mathematician summoned from medieval Basel to the new capital of St. Petersburg—to ask whether it would be possible, on one and the same journey through the city, to cross all seven bridges over the Pregel exactly once. By disregarding topographical data such as the length of streets and their twists and turns, Euler proved it could never be done. He might just as well have drawn the map of Königsberg on a piece of rubber one can stretch to any size. Graph theory, namely, recognizes only two abstract elements: vertices and edges. On this basis, it is possible to reconstruct all spatial structures: trees and stars, junctions and bridges, rings and hubs, regions, countries, and maps.

Place de l'Étoile, Ringstrasse, Anulare—these graphs have long since made their impressions felt. But maps of city traffic do not record streets and rail lines any more concretely than does geometry on a piece of rubber. "The space in which the modern city unfolds its structures is clearly an abstract space in which the individual constraints are of a topological order; seen from the point of view of the unfolding of these structures, the territory is simply the surface effect of its own topicality."[4]

What returned following the nineteenth century's passion for topography—that is, the passions of general staffs—amounted to the oldest maps in existence. On the *Tabula Peutingeriana*, which charts the St. Pölten of old as a relay station within the Roman postal system, the distances between north and south—presumably to facilitate the transport of the medium "map" itself across the country—have been squeezed together so tightly that hardly any traces of land, sea, and mountain ranges remain. Here an empire, and the Roman one no less, appears solely as a media landscape.

INTERSECTIONS. Streets between cities were the sole connection that the *Tabula Peutingeriana* recorded. The Roman postal system could disregard arteries of life such as aqueducts—to say nothing of what Hölderlin called the "shadowless streets" of the sea. Border cities, then, remained "vertices" along an "edge." Relay stations represented coordinate points where two lines met—whereas Rome, where all roads proverbially led, constituted the axis of all intersections. Because no other system traversed the network of roads, the graph could be represented on a single plane. But ever since technology has given rise to innumerable media channels, it has been impossible to continue to do so.

The City Is a Medium

"DIE SYSTEME VON FREUNDEN UND BEKANNTEN BILDEN EINEN HALBVERBAND, ABER KEINEN BAUM." [CHRISTOPHER ALEXANDER]

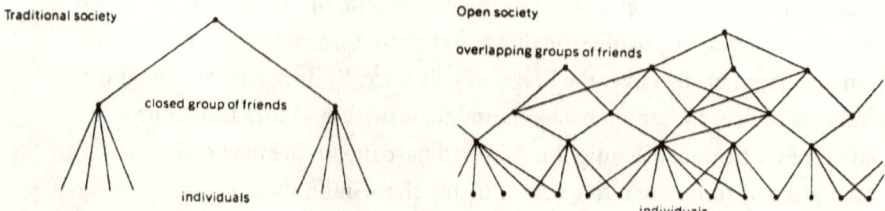

Figure 5. From Dietmar Steiner, Georg Schöllhammer, Gregor Eichinger, and Christian Knechtl (eds.), *Geburt einer Hauptstadt am Horizont* (Vienna: BuchQuadrat, 1988).

A well-known exercise from school demands that three houses be connected to three energy systems—gas, water, and electricity—without any of the connections crossing. The so-called GWE graph this requires is not flat; that is, it is impossible to "iron out" the various lines. A city is not a graph that might be flattened. Here networks overlap each other. Each traffic light, each subway interchange, and each post office—and all the bars and bordellos, too—tell as much. There are bridges that cross rivers other than the Pregel, and railway viaducts that extend over rivers other than the Traisen. To be sure, planners have tried to order networks in Chandigarh, Brasilia, and other newly founded cities on the model of the tree graph, which admits no intersections between branches and boughs and is therefore two-dimensional. But for all that, "a city is not a tree" so much as a "half-grid" whose points of overlap belong to the system itself.[5]

CAPITALS expand on the rule exponentially. These cities are not simply defined by the state with its *limes* or system of borders—its self-induced "resonance," that is. Rather, networks between cities overlap with other networks between cities. Tangled knots beneath the earth, on the surface, and in the air make a mockery of all streamlining efforts. Transfers and switches govern time in the capital. As Walter Benjamin noted, Jacques Offenbach's *Parisian Life* (1866) was the first theater piece to begin in a railway station. In Vienna, imperial Austria linked the point of intersection between its four rail stations opening onto Europe [*europäische Aufmarscheisenbahnen*] and their terminus stations with a metropolitan system that was, in turn, connected to narrow gauge rails extending to the surrounding area. The sheer frequency of crossings in capitals and metropolises assures the rule of Tyche, Fortuna, or simply Chance—which appeared to Valéry in a dream as the unceasing sound of the sea [*Meeresrauschen*] before he woke up and celebrated urban life as the precondition for all felicitous meetings. If one leaves aside, for a moment, the severed head of the king (Foucault), then the capital appears as the "daughter of the great number."[6]

MEDIA exist to calculate, store, and transmit numbers. A Greek city—presumably Miletus—produced our oldest media: coins and the vocalic alphabet.[7] In transforming from a city into a state, Rome adopted the Orient's most highly developed medium of transmission: the Persian postal system.[8]

And so our concepts for media, when they do not derive from the body (like the "heart" or "brain of a circuit"), learn from the city. Ever since Shannon implemented George Boole's switching algebra with a few telegraphic relays, media's simplest elements, in logical terms—which, at this stage, possessed no memory—have been known as "gates" or "ports." Sequential circuits, on the other hand, whose output is not a function of input gates but of their own prehistory, presume (and this, too, is an urban feature) a built-in memory. When John von Neumann—a mathematician in the Second World War, among other things—made this principle of sequential processing or calculation the basis of almost all of today's computer "architecture," he gave the apt name of "bus" to the parallel channels between the calculator units, gates, and memories; this designation renewed an order of urban traffic that had existed since the Biedermeier. Finally, ever since von Neumann accurately predicted that only computers themselves would be able to design the next, more refined generation to succeed them—because the tangled knot of necessary networks exceeds the capacities even of engineers—there have been computer programs called "routing." Network design, as was already the case for Shannon's mouse, occurs by motion onward [*Strassenbahnung*] (albeit with all the problems that attend lateral movement and multiple levels of activity).

Entire cities of silicon, silicon oxide, and gold wire have arisen, but the "houses" in them are measured in molecules, and their total surface area—even after being magnified a million times—barely exceeds square millimeters. Technical media have miniaturized the city to the precise extent that they have also made it expand toward the entropy of the megalopolis. Not only does the time-honored module "human-sized" seem to have been rendered obsolete, as well-known travails in parking garages and airports attest, but modularity in general seems obsolete. Graph theory takes this state of affairs into account. The more one "thinks" a capital like Paris, Valéry wrote, the more one knows that one is "thought" by it oneself. All the same, no system is self-governing—and that goes for cities and modules, too.

In a gray area without scales of measurement, it is advisable, then, to connect networks without attaching values to them; that is, to part with MUMFORD'S PARTING WORDS. "Through its concentration of physical and cultural power," Mumford writes at the end of his study,

> the city heightened the tempo of human intercourse and translated its products into forms that could be stored and reproduced. Through its monuments, written records and orderly habits of association, the city enlarged the scope of all human activities, extending them backwards and forwards in time. By means of its storage facilities (buildings, vaults, archives, monuments, tablets, books), the city became capable of transmitting a complex culture from generation to generation, for it marshaled together not only the physical means but the human agents needed to pass on and enlarge this heritage. That remains the greatest of the city's gifts. As compared with the complex human order of the city, our present ingenious electronic mechanisms for storing and transmitting information are crude and limited.[9]

It is clear that Mumford considers cities comparable to or compatible with computers—that is, he considers them media. However, the analogy and the points he enumerates are based only on functions of information storage and transmission. What is more, the perspective is limited to diachrony and, accordingly, suppresses (other) networks. And so, Mumford does not even posit the fundamental third function—data processing; doing so would pull the rug out from under his humanistic value judgments. It seems the historian of cities forgot his insight that part of the greatness of the Florence of old was that in building the Uffizi—the first office building—the city also set up an exemplary center for data processing.

MEDIA. Storage, transmission, and processing of information: such is the basic definition of media in general. Media include old-fashioned things like books, familiar ones like cities, and new ones like computers. However, von Neumann's computer architecture, for the first time in history (or as its end), implemented this definition technically. A microprocessor contains—not among other things, but exclusively—processing units, memory, and buses. The processor performs logical or arithmetical commands according to the specifications of the program memory; the buses transmit commands, addresses, and data as specified by the processor and its most recent command;

The City Is a Medium

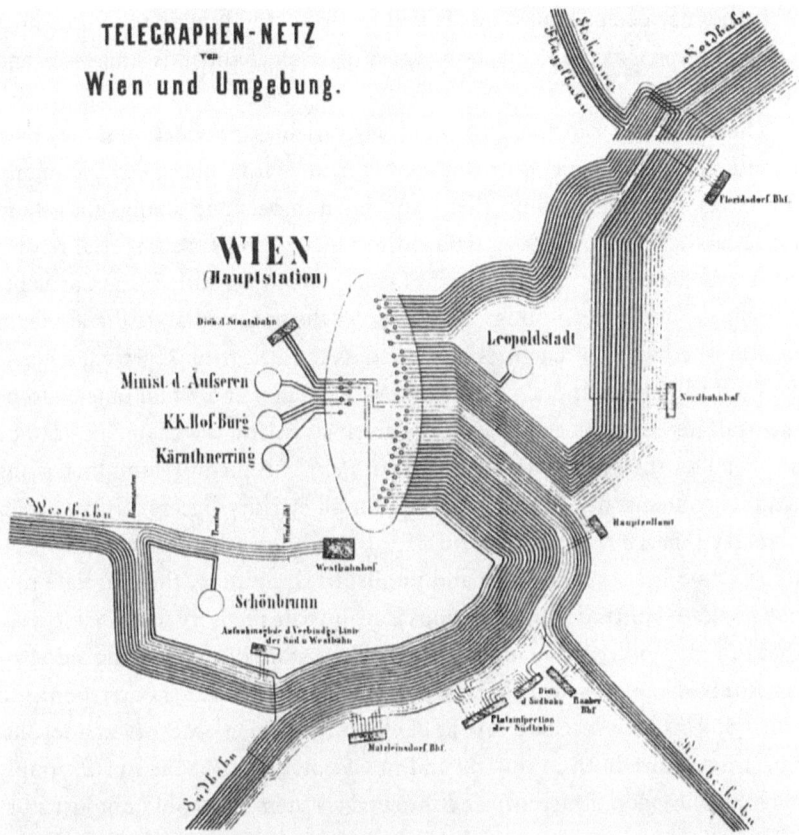

Figure 6. From Dietmar Steiner, Georg Schöllhammer, Gregor Eichinger, and Christian Knechtl (eds.), *Geburt einer Hauptstadt am Horizont* (Vienna: BuchQuadrat, 1988).

finally, the memory makes it possible to read or write commands or data at determinate addresses.

This network of processing, transmission, and storage—in other words, of commands, addresses, and data—suffices, according to Turing's famous proof of 1936, to calculate all that is calculable.[10] Thus, the development of technical media—from the digital communication medium of telegraphy, over the analog storage media of sound recording and film, up to the media for their transmission, radio and television—comes, in logical terms, full circle. As a matter of principle, all media can be transferred into Turing's discrete universal machine. This is reason enough also to understand the workings of the city in terms of general computer science [*Informatik*]. It is also reason enough why one may decipher past media and the historical functions of the entity known as "man" as the interaction [*Spiel*] between commands, addresses, and data.

Thereby, DATA can consist of arbitrary variables, provided that they have a defined format (analog or digital, bytes or words, etc.). Von Neumann machines can assign strings that stand for numbers and strings that stand for letters to the same site of data storage. Accordingly, an imperial reform edict issued on 12 January 1782 in the city St. Pölten permitted "the Carmelite cloister (with nineteen nuns), which serves the contemplative life alone, to be abolished and the facilities employed as the Pelegrini Regiment's house for educating boys and a garrison, the ornamental and ritual objects to be incorporated into the chapel, sold, or given away, and the chapel itself to be arranged for storage [*als Magazin einzurichten*]."[11] A memory unit that, along with its contents, had been established for all eternity, became direct-access memory [*Speicher mit wahlfreiem Zugriff*] and henceforth served the disciplined mobilization of troops and pupils. In computing, the correlate of a boy possessed of read/write functions is read/write memory for variable data, that is, random-access memory. Conversely, a store of unchanging values— read-only memory—for program commands and constants corresponds to ritual objects. The so-called late Enlightenment, which occurred as a revolution from above both in Austria and in German lands to the north, simply exchanged modes of memory and installed a system that could not just store information, but erase it as well. The eraser swept over the "individual" and passed on to the capital.

Ever since, one has been free—if not obliged—to forget that cities used to do just fine without a state. The matter of data formats, however, is more delicate than the business of data exchange. Where the city is concerned, the modules according to which it is built determine the formatting. The railway stations that, as Napoleon III observed, achieved the status of city gates in the mid-nineteenth century could not transform what they had replaced as easily as Joseph II had repurposed Austria's cloisters. City gates had long offered the site of input/output for a postal system whose carriages conveyed persons, goods, and messages—that is, addresses, data, and commands. The rails did not just take the traffic of persons and goods away from the postal system; rather, they assigned the information a new module or format: they mobilized officers in first class, noncommissioned officers in second class, and the battalions of infantries in third class.[12] Benjamin euphemistically remarked of "the historical signature of the railroad" that it represented "the first means of transport—and, until the big ocean liners, no doubt also the last—to form masses."[13]

City traffic—masses of automobiles—also needs to be formed or formatted. In 1935, Richard Euringer, the chair of the National Association of German Writers [*Nationalverband Deutscher Schriftsteller*], still hoped one might prevent all the "collisions, damages, injuries, and stoppages" that were caused by "freedom of self-movement"—auto-mobility, that is—through traffic regulation by means of *Führerprinzip*.[14] Engineers know better, however. Today's computer gate, notwithstanding all myths (or horror stories) about binary operations, allows for not two but three switching states: in addition to positive state, 1, and negative state, 0, it recognizes a state of high impedance, which isolates the data source from its output channel and thereby makes it possible, without collisions and after a short period of transition, to shift other data sources onto the same bus. The yellow light at every intersection does the same thing. In the endless changes between green, yellow, and red—or 1, 3-state, and 0—all streams of urban traffic (from pedestrians to public transportation) arrive in a digital format that, moreover, a computer somewhere in the city's CPU clocks. Only observers from air corridors or skyscrapers—like Claude Lévi-Strauss in the megalopolis known as "New York"—can still discern the analog (i.e., continuous) flow of traffic that was once called "traffic" but now, as part of the universal machine made up of streets, is better called "frequency."

ADDRESSES are data that permit other data to appear in the first place. To switch computer memory over to a data bus, first the address bus must have addressed a single storage location, and second the instruction bus must have addressed the memory as a whole. Media are only as good and as fast as their distribution keys. When books were still ancient, unending scrolls, passages could hardly be looked up. And even in handwritten medieval codices, page numbers helped little because different scribes had, each to a different extent, distributed the text on individual copies with varying intervals of space. It was only Gutenberg's printing press that guaranteed that "this page here is identical to a thousand others"[15]—that is, that a passage could be found by means of a table of contents or index that would be the same for all copies. Cities were no different. The police prefects of absolutism (like Gabriel Nicolas de la Reynie in Paris) first saw to it that the hand-painted guild signs on old houses received the same format and, ultimately, were replaced by a house number specified by location.[16]

Stephen Dedalus, James Joyce's fictive double, wrote on the flyleaf of his geography textbook (of all places):

> Stephen Dedalus
> Class of Elements
> Clongowes Wood College
> Sallins
> County Kildare
> Ireland
> Europe
> The World
> The Universe[17]

Somewhat more prosaically—but no less precisely—personal ads today provide telephone numbers and/or regional information based on codes for license plates. Whether or not anyone picks up the phone is of secondary importance. And for good reason. In the nineteenth century, it was enough—for legal purposes—for a registered letter from a governmental agency to land in a mailbox, even if its recipient was demonstrably never at home. "The nymphs are departed, have left no addresses,"[18] T. S. Eliot wrote. To be sure, Eliot was talking about nymphs and their companions—but even river deities are addresses. When the course of the Nadelbach, which came to be known as the Tragisa or Traisen, was rerouted, St. Pölten received its first historical inscription: Marcus Aurelius Julius, the Roman vice governor, consecrated an altar to Neptune, the Lord of All Waters.

In this altogether literal way, the address creates a channel. It separates mountain streams and waterways, people and subjects, cities and capitals. Indeed, under highly technical conditions, capitals hardly need to be built anymore—it is enough to give them an address. Paul Hindemith's 1931 play, *Wir bauen eine Stadt*, was not written for masons and architects, but for the midrange frequencies of the *Südwestdeutsche Rundfunk AG*. More specifically, the piece was written for his brother-in-law, the Frankfurt programming director Hans Flesch.[19]

Founding a capital today simply means that at freeway interchanges and railway stations, in timetables and computer networks, a new "hub" emerges, which centralizes the flow of energy and information. In the 1920s, some cities in Central Europe—if only so they could continue to dream—still resisted putting their names on road signs. Indeed, "it was often the case that agencies responsible for building roads had no knowledge of places outside their borders in the stricter sense and therefore did not indicate them on signs—sometimes deliberately failing to do so."[20] It was only after space had been strategically disclosed that the hub intervened in the animal kingdom and assigned rights-of-way. It is 3-state commands that on computer buses assign the priority of "masters" over "slaves." Napoleon first introduced driving on the right-hand side, which eliminated the chaos of traffic and paved the way for the columns of his autonomously operating divisions to march down national roads with rows of poplars to either side. Finally, it was the railroad that—to use computing terminology—installed bidirectional traffic and gave all modern media the model of separate lanes of travel. Ever since, encounters are really derailments and passers-by have stopped walking.

From 16 February on, a dividing line—which at first was more an ideal than a reality—banished French pedestrians, bicyclists, oxcarts, and so on from the poplar-lined national roads in order to facilitate the transportation of munitions supplies in one direction and corpses in the other. This is what saved the besieged city of Verdun from the "blood mill" [*Blutmühle*] of the German Kaiser. Heinz Guderian, the chief officer for armored vehicles [*Panzerchef*] in the First World War, then applied the enemy's innovation to the highways that he engineered, the next world war already in mind. "The counterattack—as a general rule in the art of war—never attacks the same with same. Rather, against artillery you have the tank, against the tank the helicopter, and so forth. Thus, the war machine possesses a factor of innovation that differs radically from the innovations of machines for production."[21]

COMMANDS—which Anglo-American computer inventors, with due pedagogical modesty, first called "instructions"—are orders. An equation without an algorithm calling for self-execution might, as in the past, have been left to the inventiveness of mathematicians, but data processing has taken care of geniuses and bosses [*das Genie oder den Chef*] like everything else.

In the final analysis, "to command" simply means "to address." This holds for the lowest level of digital computation, in so-called microcode, where the patent wars are fiercest. It also holds for the lowest level of everyday life in the city, as Althusser demonstrated: one is a citizen if a policeman's call ("Hey! You there!") makes one stop and turn around on the street.

Command centers, then, do not lie where authority plants the forest of its mightiest symbols. Instead, they dwell in much less conspicuous lines, which shoot at right angles—as occur in bridges traversing surfaces that cannot be flattened into planes.

The City Is a Medium

Even if the first Prussian ministries originated with a central—indeed, a privy—council [*Geheimer Rat*], the bureaucracies of Kafka and Austria knew better. Ever since the time of Kaiser Maximilian's reign, the offices of central administration did not, in any way, proceed from the noble organs of Roman-German imperial sovereignty. Instead, the technical back offices of Austria ascended to power by degrees, led by bourgeois jurists in the court chancery. Chancery courts for individual administrative regions [*Einzelländer*] followed, which connected cities and provinces to the hub in the capital.[22] Then as now, power meant occupying—at the right time—the channels of technical data processing. Centrality is a variable that depends on media functions, not the other way around.

On 9 April 1809, Emperor Franz II declared war on France. On the morrow, his patriotically charged armies crossed the river Inn. A missive to the Bavarian king, calling for him to break his treaty with Napoleon, went unheeded. Consequently, the martial forces of Austria marched themselves to deliver the information contained in the royal epistle and moved on Munich. King Max fled, and the French envoy had just enough time to dispatch a courier to Strasbourg, where Louis-Alexandre Berthier, Napoleon's chief of staff, was quartered.

Ever since the creation of fourteen autonomous revolutionary armies in 1794, all the border cities of France had been connected with the capital by means of optical telegraph—the first high-speed transmission medium in history. Thus, Berthier had no trouble sending telegrams to Napoleon; and Napoleon, from Paris, easily dispatched telegrams to his armies. In record time, two weeks, the French settled accounts in Munich. Promptly, the Bavarian king charged his academy of science with developing an improved—electrical—telegraph.[23]

Meanwhile, Napoleon's war machine marched on to Wagram, wiring Europe with optical telegraphs (just as, formerly, the Romans had done with the pony express of their postal system). Of all things it was church steeples—whose bells had, for centuries, provided the sole channel of communication between the authorities and the populace—that were repurposed. "On the northern side of the cathedral's spire" in St. Pölten, the occupation army set up

> a "telegraphic device" that was part of a military line of communication running from Vienna to Strasbourg. It consisted of military posts that, in intervals of one or two hours of travel time, had been installed on towers and hills; by means of three flags that were blue, red, and white, they transmitted signals whose meaning was known only to the "directors at either end of the line."[24]

And so, as an altogether functional tricolor flew over the cities of Austria, enemy reconnaissance divisions [Aufklärungsabteilungen] surveyed her territory. Since the time of the *Tabula Peutingeriana*, maps had more or less ignored these lands. Marshal Auguste de Marmont, on the other hand, sent out a vanguard of cavalry officers, who made sure to document everything cartographically—especially the mountains, valleys, and swamps surrounding St. Pölten; thereby, inviability itself was deciphered for new tactics of warfare.

Ever since, armies have been able to leave cities—indeed, capitals—by the wayside. Over mountain ranges, swamplands, and desert sands, the *Blitzkrieg* attacks the enemy from behind. The objective is not to "kettle in" cities, but to delimit spaces. The only requirement is that one have utterly precise maps—once a matter of utmost state secrecy and, in increasing measure after 1800, the monopoly held by general staffs in France, Prussia, and Austria.

It was not until 1942 that total war—now coming from the skies—again made an example of urban centers. Now, however, the module of destructibility was no longer human. For white phosphorous bombs, the unit was the city; for uranium bombs, a major city; and finally, for hydrogen bombs, megalopolis. The broad expanses of green space of German urban habitations offer little consolation, then—even if they stem from plans made to avert the next bomb in a world war.[25]

And so, the "invisible city," with which Mumford's urban world history concludes, does not consist solely of information technologies that operate weightlessly and at the speed of light. The computer commands for extinction are already online. "This is the last and worst bequest of the citadel (read 'Pentagon' or 'Kremlin') to the culture of cities."[26]

11

Rock Music: A Misuse of Military Equipment

> Out in a bloody rain to feed our fields
> Amid the Maenad roar of nitre's song
> And sulfur's cantus firmus.
> — Richard Wharfinger, *The Courier's Tragedy*

Nietzsche, who had read enough contemporary physiology to found a *Gay Science*, also discoursed, under this rubric, *On the Origin of Poetry*. "Lovers of what is fantastic," he observed, hold that poetry and, more specifically, lyric "counteracts rather than contributes to the clarity of communication" by "making speech rhythm"; that is, poetry appears to them as "a mockery of all useful expediency." For his part, Nietzsche countered with mockery of his own—with utilitarian and, more specifically, media-technical scorn. In *The Gay Science* he declares:

> In those ancient times that called poetry into being, one really did aim at utility, and a very great utility at that; back then, when one let rhythm penetrate speech—that rhythmic force that reorganizes all the atoms of a sentence, bids one to select one's words and gives thoughts a new colour and makes them darker, stranger, more distant: a *superstitious utility*, of course! Rhythm was supposed to make a human request impress the gods more deeply after it was noticed that humans remember a verse better than ordinary speech; one also thought one could make oneself audible over greater distances with the rhythmic tick-tock; the rhythmic prayer seemed to get closer to the ears of the gods. Above all, one wanted to take advantage of that elemental overpowering force that humans experience in themselves when listening to music: rhythm is a compulsion; it engenders an unconquerable desire to yield, to join in; not only the stride of the feet but also the soul itself gives in to the beat—probably also, one inferred, the soul of the gods! By means of rhythm, one thus tried to *compel* them and to exercise a power over them: one cast poetry around them like a magical snare.[1]

Ninety years after Nietzsche, Jim Morrison announced on his last record, "I wanna tell you about Texas Radio and the Big Beat." Morrison was the lead singer of a group called *The Doors*—a name that counts for something, for they, too, wanted to open the way to the Superman. Nietzsche and rock musicians speak of one and the same thing, because media technology provides their shared historical a priori. Ever since *The Gay Science*, the arts have cast aside their old title, which declared them to represent capacities of so-called man. They have become information technologies [*Nachrichtentechniken*]—and nothing but.

Precisely because, as one can readily discern, Nietzsche's analysis does not believe in the gods, both senders and receivers—that is, the immortals and their human counterparts—disappear, along with their messages, against the medium and its channel of transmission. It is hardly a coincidence that Nietzsche was the first philosopher to use machinery to write (i.e., the typewriter—*Schreibmaschine*).[2] Nietzsche described, apropos of Greek lyric, one method for storing and another for transmitting information; first, a mnemotechnics that makes verses more unforgettable than prose, and second, a channeling of discourse that conveys them over great distances. After all, mortals are so forgetful, and the gods so hard of hearing.

Storing information and transmitting information without having to employ such obscure instances as the human "spirit" or "soul": such is the very definition of media. To be sure, if one follows Nietzsche's analysis, human—or divine—ears, memories, and feet had to supplement the "rhythmic tick-tock" of the Greeks, for other apparatuses had not yet been invented. Bodies became interfaces of a circuit [*Schaltung*] linking them with their environment. As is well known, in the quantitative meter of antiquity the rhythmical tick-tock did not follow any meaning at all—whereas the meaning of words governs the qualitative accents of modern European lyric.[3] In order to assure its storage and transmission, ancient verse coupled the metrical foot to the feet of dancers, an altogether simple and physiological matter. For this same reason, the rhythm that once tied lyric and music together subsequently went missing. Storage and transmission facilities made of flesh and blood do not last.

With today's electric media, however, all of this has returned, possibly because gods cannot die. The separation between literature and music, which formed the basis for European culture—and consequently, the basis of our own literary study—has disappeared again, to make room for a matter that, to adopt Nietzsche's formulation freely, could be called "the end of poetry." "Texas Radio and the Big Beat" represents the "magical snare" that new gods cast on people or listeners.

Rock music—lyric as it exists today in actual fact—is equipped with all the attributes of a world power: inescapability, familiarity [*Auswendigkeit*], and omnipresence, from the tinkling of pianos in department stores to the subsonic forte of discos.

As is well known, the coup leading to global power occurred in two steps. First, the storage technology and then, later, the technology of transmission had to be taken from people—these interfaces of traditional literature—and installed in machines. Thomas Edison performed the first step in 1877 when he presented the prototype of his phonograph. The tinfoil cylinder and, ten years later, Emile Berliner's more up-to-date gramophone record (which achieved mass production at the expense of keeping consumers from making their own recordings, as the phonograph had allowed) took away people's memory for words and sounds. When, in 1897, Ernst von Wildenbruch—presumably the first writer in Germany to do so—composed the poem "Für die phonographische Aufnahme seiner Stimme" [For the Phonographic Recording of His Voice] and promptly declaimed it into the recording device [*Schalltrichter*], it was unclear why these verses still displayed rhyme and meter at all. After all, according to Nietzsche's analysis (which had just been published), they were simply techniques of a mnemotechnics (i.e., storage system) that a machine could perform much more thoroughly—and without concern for euphony, rhythm, or even meaning.

Under these conditions, writers—if they did not, like Wildenbruch and all lyricists after him, defect to gramophone recordings—had the sole option of optimizing the superannuated medium of writing technically, in the same way that Edison's phonograph had done for the medium of acoustics and his kinetoscope (the forerunner of all film projectors) had done for the medium of optics. And lo, the typewriter, which was developed at the same time, made it possible to record the Symbolic of writing—just like the Real of noises and the Imaginary of cinematic doublings in contemporary media.[4] Modern literature, that special effect by and for fetishists of writing [*Buchstabenfetischisten*], could begin. And so it did, with Mallarmé and Stefan George.

For all that, writing and books—in contrast to inscriptions—are not just storage media but also media of transmission. Only recently have we come to hold their mediocre frequency in contempt. The fact that the Torah or the Koran, those shrines of nomadic peoples, were transportable, enabled them to conquer the gods of Greece. Whatever dwells near its origin—say, a votive statue at its temple—has trouble leaving.

An imperium, as Harold Innis has shown in *Empire and Communications*, does not need to manipulate only time with storage arrangements such as statues, inscriptions, or even musical notation. Rather, it must also conquer space and place(s) with technologies of transmission. Just as Greece was formerly notorious for bad postal connections, the new storage systems of the Founding Age of Media [*Mediengründerzeit*] around 1890 lacked a technology of transmission that was up to the task. Only telegraphy and telephony were available to transmit signals with electrical—that is, unsurpassable—speed. The former transmitted the Symbolic through a writing system that had been internationally optimized for differentiality and economy: Morse Code (which, long before Saussure, made structuralism practicable). The latter, like Edison's phonograph, could convey the Real of stochastic noises (and not just the coded differences of language or music). Both technologies, however—telegraphy and telephony alike—could operate only over wires, which is to say, via *matter*.

Engineers in Budapest or entrepreneurs in London, who at the end of the 1890s provided recorded music to telephone subscribers (including ones in Queen Victoria's palace) at a special price, were utterly reliant on the state of wiring. Even after Maxwell's field equations had postulated electromagnetic waves—and thereby wireless transmission—in theory, a practical limit existed: all the available converters between physiology and the media landscape, acoustics and electrics, were low-frequency technologies. The free distribution of waves, in contrast, begins only above thirty thousand hertz. The carbon microphone invented in 1878—which far surpassed Bell's contemporaneous telephone receiver—could make a fly's footsteps audible (when, again in loose adaptation of Nietzsche's theory of metrics, the insect's feet traversed the membrane—an amplification of low frequencies whose enduring monument would be the stereophonic fly on Pink Floyd's *Ummagumma*). All the same, however, the technical precondition for radio—amplification and oscillation occurring at higher frequencies as well—was achieved only when Robert von Lieben and Lee de Forest engineered vacuum tubes. Long before our transistors and chips, in 1906, the tube created solved a problem that Thomas Pynchon, in *Gravity's Rainbow*, identified as elementary for the twentieth century: energy-free (i.e., perfect) guidance of energy (i.e., data) as far and as fast as one wishes.[5]

Since then, radio has been possible—and not just in principle, as was already the case since the time of Hertz, Marconi, and Braun, but practically and on a mass scale. On Christmas Eve 1909, from Brant Rock in Massachusetts, Reginald A. Fessenden is supposed to have entertained the first listening public—which admittedly consisted only of Marconi's radio operators on ships in

Rock Music

an area of some eighty kilometers—with a wireless speech delivered by a man, and a wireless verse recitation delivered by a woman (in this order, and with this distribution of roles). According to other sources, the holiday spirit came through an excerpt of Händel's *Messiah* on record—as if to prove that the content of a medium is always, and strictly following McLuhan, another medium: in the case of the typewriter, handwriting; for film, the novel; for the gramophone, the voice; and for entertainment radio, the record industry.

However, the mass production of tubes did not occur for Händel's *Messiah* or operatic arias, which in 1904 (that is, immediately after Caruso's acoustic immortalization upon the orders of Kaiser Wilhelm) Professor Adolf Slaby of the Technical University in Berlin transmitted from Potsdam to Charlottenburg.[6] When all industrialized countries—as Major William R. Blair of the U.S. Signal Corps wrote—"put huge amounts of money and energy into scientific radio research" and "pushed the development of more sensitive amplifiers through the use of vacuum tubes" as "the biggest improvement,"[7] they did so for a single reason: the First World War. Three new weapons systems, on land, in the skies, and on the sea, needed precisely the steering with neither energy nor matter that the German Chief of the General Staff, Alfred Graf von Schlieffen, had already described (or brought about [*herbeigeschrieben*]) in 1909 under the title *Krieg in der Gegenwart*:

> But as large as the fields of battle may be, so little will they offer the eye. Nothing is to be seen in the vast desolation. No Napoleon stands upon a hill surrounded by his brilliant retinue. Even with the best field glasses, he would not get to see much more. His white horse would be an easy target for countless batteries. The commander in chief finds himself further back in a house with a spacious office, where telegraphs, telephones, and signals apparatuses are to hand and from where fleets of motorized vehicles and motorcycles, equipped for the longest of journeys, await instruction. There, on a comfortable chair in front of a broad table, the modern Alexander has the entire field of battle laid out in a map before him; from there he telegraphs rousing [*zündende*] words; there he receives reports from the army and corps commanders, from the observation balloons and the dirigibles that observe the movement of the enemy along the whole line.[8]

Two years later, in 1911, Schlieffen realized his prophetic words with deeds: he created, in the capacity of highest weapons authority for Transport and Telecommunications [*oberste Waffenbehörde für das Nachrichten- und Verkehrswesen*]—that is, as if to prove that motorization and electrification of war now coincided—an autonomous office for General Inspection of Military Transport [*Generalinspektion des Militär-Verkehrswesens*]. The office employed all that contributed to the system to come, all the radio operators of Germany up to the radio concern *Telefunken* (which was founded especially for the army). Without wireless telephony, the new weapons systems of the First World War would have remained blind: like the U-boats at sea and the double-deckers in the air, the tanks of 1917 were also supposed to be steered by radio. Unfortunately, however, their antennas were so fundamentally foiled by the wire entanglements of no-man's-lands and trenches that it proved necessary to abandon two-way communications facilities [*Wechselsprechanlagen*] and use carrier pigeons.[9] The situation first changed through the intervention of a certain Heinz Guderian, chief radio operator [*Funkerhauptmann*] in World War I and colonel general in World War II. The latter-day consequences still entertain us around the clock today.

But even if the primitive tanks of 1917 escaped control, the troops of radio operators on both sides (or fronts) of the trenches grew and grew. The German Telegraph Troop [*Telegraphentruppe*], for example, counted 800 officers and 25,000 enlisted men on 2 August 1914, the day of mobilization. In November 1918, upon demobilization, 4,381 officers and 185,000 troops returned to the defeated *Reich*:[10]: growth matched by no other kind of weapon—and which ultimately produced the civilian radio of our everyday lives.

In the first place, the hundreds of thousands of radio operators in their foxholes wanted to be entertained. Broadcasters and receivers were available in untold abundance. And so, radio as the "misuse of military equipment" could begin. In the words of General von Wedel, the head of Wehrmacht propaganda during the following world war:

> That more recent facilities such as radio were foreshadowed stems from the fact that, from May until August 1917, Dr. Hans Bredow—an officer of the Communications Troop [*Nachrichtentruppe*] who later became the Secretary of State in the Ministry of the Imperial Post [and the creator of civilian radio in Germany]—broadcast, with a primitive tube radio, a program for an entire section of the front near Rethel in northern France from May until August 1917, during which records were played and newspaper articles were read. Listening, throughout the entire forward area, occurred by means of military radio devices. The overall success [in terms of propaganda] was all over, however, when a higher instance of command learned about it and prohibited the "misuse of military equipment"—and, with that, all further transmission of music and verbal communication [*Wortsendungen*]![11]

Second, even though a world war came to an end in November 1918, this did not mean that technical knowledge did the same. Even demobilized radio operators remained radio operators—especially given the massive "plundering of military property"[12] that occurred and exceeded the musical abuses of the preceding year. That was reason enough for the Spartacus League and Karl Liebknecht himself to assign the 190,000 radio operators coming from the field, with all their technological booty, to a Central Radio Commission [*Zentralfunkrat*], which in turn would serve the revolution that was being planned. The discourse of the ruling interests had a single word for such terror (or Central Radio Commissions) in November 1918: *Funkerspuk*—approximately, "radio haunting."[13] Hi-tech media do not belong in hands that have not received the blessings of the postal system [*Postregal*] or general staff. The alternatives were simple: Weimar Republic or Spooky Radio. For this reason, the exorcism of the haunted airwaves began with secret telephone calls between Friedrich Ebert and the Army Supreme Command, which effectively remained in power; it ended with the founding of civilian radio.

In October 1923, it had come that far. From the building of a record company in Berlin, over the medium-wave antennas of the former military transmitter Nauen, music was heard—as formerly had occurred in the mud of Flanders. One song was called "Have Pity" (presumably this referred to the sound quality), and another assured the listener that "My Heart Beats Only for You." At the end of this German radio premiere—so that listeners would not fail to hear the point of all the anti-*Funkerspuk*—the band of Infantry Regiment III/9 played "Deutschland, Deutschland über Alles."[14]

The Imperial Post, in keeping with a secret circular from its minister, did not just want, by means of radio, "1. To provide, to the broadest circles in the populace, good entertainment and instruction through wireless music, lectures, and the like"; nor did it wish only "2. To open a new, important source of revenue for the *Reich*." Instead, it wanted ultimately—or in the first place—to "steer a course that can prove significant for state security."[15] What civilian radio excluded, thanks to built-in technical handicaps, was *Funkerspuk*, or in other words, the misuse of military equipment. It was not for nothing that the Reichswehr gave civilian government the green light for radio only in 1923—the same year that Berlin witnessed the founding of the *Chiffriermaschinen Aktiengesellschaft*, whose products created a new level of secrecy for what was said on the radio[16] and whose anti- or decrypting machines (in the Second World War) led to the invention of the first computer. Mass communication, in other words, was first admitted when it was a matter of consuming or hearing everything, and no longer listening in on anything. "Reception"—possibly also as the guiding concept of literary sociology—is just a euphemism for systematically obstructed interception.

A Misuse of Military Equipment

Listen to Guglielmo Marconi, the founding hero of radio, a marchese and senator of Fascist Italy, speaking hours after his death in our world's new—phonographic—immortality, talking over the radio about radio:

> I confess that, forty-two years ago, when I achieved the first successful wireless transmission in Pontecchio, I already anticipated the possibility of transmitting electric waves over large distances, but in spite of that I could not hope for the great satisfaction I am enjoying today. For in those days a major shortcoming was ascribed to my invention: the possible interception of transmissions. This defect preoccupied me so much that, for many years, my principal research was focused on its elimination.
>
> Thirty years later, however, precisely this defect was exploited and turned into radio—into that medium of reception that now reaches more than 40 million listeners every day.[17]

In terms of technical standards, of course, Marconi's "principal research" sought a "wooden iron": a secret transmitter without any possibility of interception. Empty space as the medium of radio is difficult to fill [*besetzen*]. In terms of political standards, however, even wooden irons are viable: for example, mass radio without informational content—a device, that is, that has rendered the flaw of Marconi's equipment inoperative.

Accordingly, the truth can only reside in the medium itself, not in its messages—that is, it lies in the self-promotion of the delivery system [*Post*] or recording industry. What electrical information technology affords, in maximally exploiting all modules and parameters, is the self-referential business of rock music. Ultimately, "Texas Radio and the Big Beat" is not about just any received themes—neither love nor authorial biography. The truth of the songs coincides with the media that world power has afforded them. That said, the truth was soon to collapse back into the military-industrial complex that stands at the origin of radio. For even if the Rolling Stones are said to have composed the lyrics to *Beggar's Banquet* by putting together newspaper headlines chosen at random, "Sympathy for the Devil" says to what devil, radio ghost, or phantom army it owes its music as such:

> I rode a tank,
> held a gen'ral's rank,
> when the Blitzkrieg raged
> and the bodies stank.[18]

Already with the modest tools of interpretation, an origin of rock music—lyric poetry as it exists in actual fact today—follows from these lines: the *Blitzkrieg* raged from 1939 to 1941. Without its media-technical innovations, *sound* would

Rock Music

still be a mishmash of AM and steam-engine radio, which "Have Pity," the first broadcast over German radio, called by name. A nice symmetry holds: just as the misuse of military equipment that had been constructed for the positional warfare of 1917 led to medium-wave monophony, the misuse of military equipment that had been devised for tank divisions, bomber squadrons, and packs of U-boats led to rock music.

As is well known, the sonic space of Abbey Road and all the British studios that led American commercialism to musical electronics arose when musicians themselves took the helm of mixing boards and did away with the old separation between lyricists, composers, arrangers, and studio technicians. Tape machines for sound montage, hi-fi technology for liberating overtones, stereophony for simulated spaces, synthesizers and vocoders for songs beyond the human sphere [*jenseits der Menschen*], and finally, FM radio for signal quality reaching the masses: without them, all the innovations of the Beatles would have gone up in smoke.[19] Every single one of these technologies goes back to the Second World War. Fortunately, perhaps, this war still provides the basis for our sensory perception.

For this reason, one can also use more unconventional means of studying documents to name the *Blitzkrieg* general in Mick Jagger's tank. In 1934, the same *Funkerhauptmann* of 1914 who had recognized the weak spots of German telecommunications on the basis of the "Miracle of the Marne"[20] had Colonel Willy Gimmler of the Army Weapons Agency test whether bushes on the battlefield really intercepted VHF transmissions. The test result—all learned opinion notwithstanding—proved negative. And so, Guderian had every single tank in the Wehrmacht equipped with a VHF radio.[21] The carrier pigeons of 1917 could return to their roosts: when tanks were equipped with antennas, there began the motorized, remote-controlled autonomy possessed by the cars of today—with pop music, on the one hand, and traffic announcements, on the other.

Besides what occurred by means of radar, still higher frequencies—which would convey television signals in the postwar era—filled the air in battles over England. As is well known, transmission first began when the Wehrmacht and Luftwaffe broadened the basis for stereo by occupying Belgium and northern France. A coordinating radio transmitter on the right, at Antwerp, transmitted endless Morse dashes into the ether, while another on the left, at Calais, sent endless Morse dots—and precisely during the pauses of the signals coming from Belgium. The bomber pilots of the German Air Force wore headphones. On the basis of the volume in their right or left ear, they sensed whether they were straying from the remote-control flight path. It was the same ping-pong stereophony one hears today from the speakers of any well-furnished living room. When in 1940 the two Morse signals (the "dot" and the "dash") coincided in a sound outside of space—as if coming from the center of the brain—the pilot knew he was flying above London or Coventry and released the bombs. The Technical Department of the Secret Service needed half a year to develop receivers for the high-frequency transmissions of the enemy at all. Then, however, it was possible to intercept their signals, to disrupt them, and even to simulate them—until the bombs, instead of continuing to rain on city centers, fell on the gentle meadows or the vast wastelands of England.[22]

Defense against submarines worked the same way. Shortly after the beginning of the war, the Royal Air Force, Coastal Command, commissioned the record company Decca to develop a perfect storage medium: FFRR, or full frequency range reproduction. Brilliant overtones and heavy basses were recorded for the first time, but not for the ears of consumers. Air Force officers in training were to learn from these records how to distinguish between British and German submarines, on the basis of motor sounds.[23] Afterward—or during the "postwar dream"—FFRR, thanks to the same British enterprise, changed its name and turned into commercial high fidelity.

Hi-fi and stereo, then, both derive from localization technologies. Bomber pilots experienced where they themselves were, and pilots hunting submarines learned where the enemy was. Since then, the ears of consumers have also learned how to localize any guitar in the sonic field of two speakers, between bookshelves and radiators. Two guitars, bass, and drums are at work; the noise of a ship motor, the hissing of steam, and brass band music wander along the wall of the room from right to left and back, while a British voice, familiar to everyone, sings a truth of history:

> In the town where I was born
> Lived a man who sailed to sea
> And he told us of his life
> In the land of submarines
> So we sailed on to the sun
> Til we found a sea of green
> And we lived beneath the waves
> In our yellow submarine

Decoding songs like this is elementary. The sole possibility for the city in question is Liverpool, and the speaker can only be a man of the war generation. And

Rock Music

since civilian submarines simply did not exist, "Yellow Submarine," with all its military marches and sound effects, follows the recollections of a sailor in the Royal Navy. The song is postwar lyric in the literal sense.

However, in order to be able to mix and manipulate guitars and motor noise—the notes of music and the sounds of an environment—rock music needed a storage medium that the old-fashioned record could not provide. Cuts and fades, already integrated into film as a technical principle, created problems for the storage of sound. Therefore, medium-wave radio until 1940 could only broadcast what was stored on record or entered the microphone directly. The Beatles' Abbey Road Studio, however, was equipped with rather famous recording devices, the BTR (British Tape Recorder) series. In 1946, Berth Jones, together with other audio engineers from England and the United States, had paid a visit to Berlin. "[A]mongst the military equipment that had been captured," there was "a system of monitoring which the German command had used in an effort to break codes."[24] The BTR at Abbey Road—and therefore of "Yellow Submarine" as well—was simply its civilian reconstruction.

In his history of Wehrmacht propaganda, *Generalmajor* von Wedel, the head of operations, wrote about the acoustic intransmissibility of certain effects of the *Blitzkrieg* for the eye and ear—problems of *sound* in World War II:

> In the Tank Division, Air Force, and parts of the Navy, all possibilities for recording actual combat [*Möglichkeiten zu Originalkampfaufnahmen*] suffered from the fact that conditions for making records [*Schallplatten*], which require stable and level surfaces, could not be secured. In such cases, reportage provided after the fact was the only help at first. A fundamental change occurred when the tape recorder was invented and adapted for the purpose of making wartime news broadcasts. Only now did actual reports of fighting from the air, mobile combat vehicles, submarines, etc. become impressive accounts of experience.[25]

In much the same way, Francis Ford Coppola made it clear, in *Apocalypse Now*, that sound montages like "Machine Gun" by Jimi Hendrix were actual reportage of fighting from Vietnam. Indeed, that is why the heroic, foundational age of rock music lasted for precisely as long as that war. But because technologies of storage and transmission have, since then, almost achieved optimal status, the "impressive accounts of experience" from heroic ages of battle continue into all the presents and futures. Every discotheque—which, after all, further amplifies tape effects and couples them with the corresponding optics of stroboscopes or *Blitzlicht* in real time—brings the war back. More still, the technology, instead of reproducing only pasts, trains for a future, which might otherwise fail to be mastered on account of people's thresholds of perception. In order to be able to read and operate the displays in cockpits during Star Wars, reaction times in milliseconds are key. Not for nothing did President Reagan welcome the fans of Atari videogames as future bomber pilots.

It could be, then, that the epoch of media glamour is coming to an end. Computers—even if their user interfaces are growing more and more friendly to the senses [*sinnenfreudiger*]—are no longer devices on a human scale. Not even the magnetic tape that revolutionized all recording studios thanks to the Berlin visit in 1946 was limited to the manipulation of the senses [*Sinnlichkeiten*]. During the Second World War, it also stored enemy radio transmissions, in order to make them decipherable.[26] Although the Allies possessed no suitable reception device to counter it with, they did possess the only decipherment technology with a future. In 1936, on the basis of a typewriter stripped down to its bare essentials,[27] Alan Turing developed his Universal Discrete Machine, whose principle of design provided the basis for all thinkable computers. A few years later, during the war, the Secret Service turned it into the first electronic computer (which was still made with tubes). Success was not long in coming: from 1943 on, the computer by the name of "Colossus" read—in real time and in plain language [*Klartext*]—all the secret radio transmissions of the Wehrmacht, the same ones that, thanks to the *Chiffriermaschinen AG* (established the same year as civilian radio), had seemed so secure. Marconi was right: what is decisive—and decisive for war—is radio interception.[28]

For good reason, rock music, when it stood and stands at the height of technical perfection, does not limit itself to offering just anything from the musical reserves. In a play of strategy with their public, records can become secret relays that transmit, somewhere between the album cover and the last groove, a coded message. In the technologically perfect storage medium—that is, beyond the capacities of human minds—there is, as a matter of principle, never a lack of room for transforming consumers into potential paranoiacs. When fans of the Beatles puzzled whether a certain song by John Lennon, played backwards, contained the whispered message of Paul McCartney's death, the tape tricks of espionage during the world war simply returned.

However, deciphering has nothing to do with "Texas Radio and the Big Beat." Ever since Turing, it is a matter of bits and bytes. With the digitalization of all data streams, the media glamour has come to an end, if not for consumers then for technicians. And music made of binary codes—that is, a misuse of military equipment from World War $N+1$—has not yet come into view. (To say nothing of the corresponding literature.)

For now, then, it is simply a matter of misusing secret weapons from the Second World War to decode what media powers such as the radio put back into a state of innocence—or nonsense. That occurs, for example, through vocoders, which can distort voices in every individual frequency subband, and stand at the ready for any rock band of the better sort. In Laurie Anderson's lyrical performance art, they simulate everyday life in the United States itself. No one knows the prehistory of the vocoder principle, for the simple fact that it was developed in the context of secret weapons research. In 1942, the masters of war Roosevelt and Churchill commissioned Turing and his American colleague Claude Shannon, the founder of information theory, to build one—and only one—vocoder. Its purpose was simple and straightforward, namely, to maximize the security of transatlantic telephone lines. From that point on, the prime minister spoke into the telephone in London and had his voice distorted to the point of unrecognizability; immediately before it reached the president's ear in Washington, it became articulate again. Legend holds that Alan Turing had fun putting on a record with Churchill's wartime addresses, setting up his vocoder prototype, and showing guests how technical knowledge can lead even the most rhetorical speeches of politicians into absolutely information-free, utter, and complete white noise.[29]

Fittingly, "And the Gods Made Love" is the title of the first track on Jimi Hendrix's *Electric Ladyland*. But the masters of the world no longer have a voice or ears, as they did for Nietzsche. All one hears is tape hiss, jet noise, and gunshots. Shortwave—between the transmitters, which is to say intercepted from the military-industrial complex—sounds similar. Perhaps, under the conditions of a world war, love must come from white noise.

12 Signal-to-Noise Ratio

> If the place were not so distant,
> If words were known, and spoken,
> Then the God might be a gold ikon,
> Or a page in a paper book.
> But It comes as the Kirghiz light—
> There is no other way to know it.
> —Thomas Pynchon, "The Aqyn's Song"

Materialities of communication are a modern riddle, possibly modernity itself. It makes sense to inquire about them only after two things are clear. First, no sense exists—such as philosophy and hermeneutics have always sought between the lines—without physical carriers. Second, no materialities exist which themselves are information or, alternately, might create communication. When at the turn of the century, that hypothetical "Ether"—which Heinrich Rudolph Hertz and many of his contemporaries believed necessary to explain the distribution of wireless high-frequency signals (which would soon yield radio)—sank into the theoretical void, information channels without any materiality became part of the everyday itself. Electromagnetic waves as the modern outbidding [*Überbietung*] of all writing simply follow Maxwell's field equations and work even in a vacuum.

The information technologies of the last two centuries first made it possible to formulate (as Claude Shannon put it) a *mathematical theory of information*. As is well known, this theory not only disregards the fact that "frequently ... messages have *meaning*; that is they refer to or are correlated according to some system with certain physical or conceptual entities."[1] Rather, because systems of communication that would transmit a single message (e.g., the number π, a determinate sine wave, or the Ten Commandments) are now superfluous and can be replaced by two separate signal generators,[2] the messages themselves are as meaningless to information theory as their statistics are meaningful. The messenger of Marathon, whose life and course coincided with a single message, has forfeited his heroic glory.

That happened not long ago. For until the parallel development of railways and telegraphy, Europe's state postal systems—which functioned more or less

Signal-to-Noise Ratio

regularly after the Thirty Years' War[3]—transported people, letters or printed matter, and goods in the same carriage. In other words, because all three elements of the transport system were material beyond doubt, there was no need to distinguish further between addresses and persons,[4] commands and messages, or data and goods in terms of communication. Accordingly, philosophers could write of the "spirit of man" or the "sense of things" on the basis of actual material reality.

Modernity, in contrast, began with a process of differentiation that relieved the postal system of goods and persons and made them relatively mobile on tracks or national roads. As a matter of course, it placed officers in first class, noncommissioned officers in second, and troops in third; weapons were loaded onto freight cars.[5] All this occurred, however, to separate material entities from pure streams of command, which it brought up to the absolute speed of light or electricity. In North America, the new system was instituted during the Civil War—the "first 'technical' or 'total' war, which, unfortunately, has been studied far too little."[6] In Europe, the shift occurred through Field Marshal Helmuth von Moltke's two campaigns in 1866 and 1870. The flight path of the postcard—which, according to Derrida, is one with Destiny or History itself—no longer went straight from Socrates and Plato to Freud and beyond.[7] It abandoned the routes of literature and philosophy—that is, the path of the alphabet and its restricted possibilities of communication—in order to become a mathematical algorithm.

Shannon's famous formula reads:

$$H = -\sum_{i=1}^{n} p_i \log p_i$$

Here quantity H measures how much freedom of choice—that is, how much uncertainty—governs the output when an information network [*Nachrichtensystem*] selects a specific event out of a number of possible events with probabilities that are all known. If the system—for example, in the orthographically standardized sequence of q and u—worked with a single signal of material certainty, H would sink to its minimum level of 0.[8] According to Lacan, the sign of the sign is that, by definition, it can be replaced[9]; in contrast, all that is Real sticks in place.[10] Even measuring its travels through the space and time of an information channel would yield only physical data about energy or speed, but nothing concerning a code.

Therein lie the difficulties for materialism; for example, when Marx, contemplating the Second Industrial Revolution, affirmed the law of the conservation of energy. Messages are calculable, but not determinate. Also (and especially) if Shannon's formula for information, including the controversial sign that precedes it,[11] is identical with Boltzmann's formula for entropy, the possibility of information does not derive from physical necessity—that is, from a Laplace universe—but from chance. Only if system elements have the chance, here or there, to be open or closed, does the system produce information. That is why combinatorics came about on the basis of dice,[12] and computer technology through endlessly repeated grids.[13] In the elementary—that is, the binary—case, H achieves its maximum of 1 when p_1 and p_2, that is, the presence and absence of modern philosophemes, have the equal probability of 0.5. Both would reject a die whose six faces had unequal chances of occurring, even if a player, who bets on advantages for either side, might not.

The fact that the maximum of information means nothing other than highest improbability, however, makes it almost impossible to distinguish it from the maximum degree of interference. In contrast to the concept of logical depth, which IBM researchers have been working on recently, Shannon's index H serves "as a measure of the statistical characteristics of a source of information, not as a step towards finding the information value of any given waveform or function."[14] And so it happens that on the one hand, the highest information rate per time unit makes it advisable to use "all parts of the available frequency [in the channel]," while on the other, "one of the main characteristics of random noise is that its power spectrum is uniformly spread over the frequency band."[15] In other words, signals, whenever possible, mimic interferences. And because the thermal noise that all matter—and therefore also resistors or transistors—radiates when operating (according to another one of Boltzmann's formulas) is white noise of the same kind, information without matter and matter without information are coupled just like the two ways of reading a picture puzzle.

As strange as it sounds, applied engineering solves problems of this sort through what is called "idealization." One treats every signal, which after passing through a real channel is necessarily laden with noise, as if it had been generated by two different sources: a signal source and a noise source, which in the most straightforward case are simply added to each other. For all that, it is equally valid to assume that the signal already coded was coded once more by an enemy intelligence, and that this second coding is successful and enigmatic in proportion to the whiteness of the noise. According to Shannon's "Communication Theory of Secrecy Systems"—a paper that for good Pentagon reasons itself remained sealed for years—the only way out of this fundamental undecidability is offered by the experiential fact that encrypting systems are mostly selections from a number of chance events that, while large as possible, are ultimately finite, whereas noise can assume infinitely many values.[16] For this reason, numbers theory, which was formerly so purpose-free,[17] has today become a hunt for the highest possible prime numbers, which—as encryptions of military-industrial secret messages—necessarily appear as noise to an enemy who has not yet cracked them. Turing, the well-known computer theorist and unknown cryptographer of the World War, formulated that laws of nature can be replaced by code systems, matters of evidence by intercepted messages, and physical constants by daily keying elements—that is, the natural sciences as a whole can be replaced by cryptanalysis.[18] The difference between chaos and strategy has become just that slight.

It is this "return of the Chaos of old within the inside of bodies and beyond their reality" with which Valéry's technical Faust terrifies a Devil whose "entirely elementary science" is, as everyone knows, simply speech. Experimental interconnection of information and noise makes "discourse a side issue."[19] After all, the orders of a culture of writing, whether literary or philosophical, could only construct meaning out of elements that had meaning themselves. Sentences emerged from words, but words did not come from letters. In contrast:

> Let us consider the signifier quite simply in the irreducible materiality that structure entails, insofar as this materiality is its own, and let us conjure the signifier up in the form of a lottery. It will be clear then that the signifier is the only thing in the world that can underpin the coexistence—constituted by disorder (synchronically)—of elements among which the most indestructible order ever to be deployed subsists (diachronically).[20]

Shannon demonstrated just such a logic of the diachronic chaining of chaos all the more strikingly for purposefully shaping his writing experiment—in contrast to the ancient play of letters that occurs in Cabbala—in a way that does without semantics. His point of departure is our conventional alphabet, that is, not some twenty-six letters, but rather these same letters and a *space* (as one finds on typewriters).

Here, in a purely statistical sense, a finite quantity of signs is to approach or simulate a language; in this case, English. As a matter of course, zero-order approximation, with twenty-seven symbols that are equally probable and independent of each other, provides only noise or gobbledygook: "xfcml rxkhrjffjuj zlpwcfwkcyl . . ." First-order approximation, that is, given probabilities or frequencies of letters as they occur in texts written in English, begins to admit articulation: "ocro hli rgwr nmielsswis eu ll . . ." Second-order approximation, which as a Markov chain also considers diachrony (that is, the probability of transition between all possible pairs of letters in a language), readily yields short words such as "are" or "be." Approximation of the third order, involving triads of English letters, can already compete with the mad, with Surrealists, or (as Shannon observed[21]) with *Finnegan's Wake*: "in no ist lat whey cractict froure birs grocid pondenome of demonstures of the raptagin is regoactiona of cre." Finally, when Markov chains no longer draw their elements from letters, but from words, second-order approximation already produces the neatest self-references of orality, typography, and literature: "the head and in frontal attack on an English writer that the character of this point is therefore another method for the letters that the time of who ever told the problem for an unexpected."[22]

This frontal attack on English writers (or, alternately, devils) is led of course by noise, which Shannon's formula—as "another method for . . . letters"—introduced to written culture. Henceforth, letters received no better treatment than numbers (which exhibit unlimited manipulability); henceforth, signals and noises were defined only numerically. Communication (to use Shannon's language) is always "Communication in the Presence of Noise"[23]—and not just because real channels never do not emit noise, but because messages themselves can be generated as selections or filterings of noise.

Technical idealization, according to which the noise-laden output of networks counts as the function of two variables—of a signal input presumed to be noise-free and a separate source of noise—enables nothing more and nothing less than the specification of signal-to-noise ratios. In a first step, this interval indicates (on the basis of voltages, currents, or power) only the quotient of medium signal amplitude and the initial degree of interference. However, simply because electric networks, via their interfaces, are connected to human senses and these senses—according to Fechner's constitutional law of psychophysics—react to a geometric increase of stimulation as if it occurred only arithmetically, it is better to record the signal-to-noise ratio logarithmically. Accordingly, the unit *decibel* (named in technological—i.e., nearly unrecognizable—honor of the inventor of the telephone, Alexander Graham Bell) transforms a fraction into twenty times or (in the case of output) ten times its logarithm:

Signal-to-Noise Ratio

$$S_N^{\#} = 20 \text{ dB lg} \frac{U_{g \text{ eff}}}{U_{I \text{ eff}}}$$

Hereupon, spoken language—once, for the ears of philosophers, the auto-affection of consciousness itself—loses all interiority and becomes just as measurable as otherwise only the quality transmission of radio and television systems is.

A signal-to-noise ratio of 60 dB guarantees the seemingly noise-free communication that others would call "undistorted." One between 40 and 0 dB still affords understanding (albeit understanding that is not hermeneutical). Beginning at –6 dB, the hearer is left only with the general impression that language is "happening." And because our senses—as has been clear since psychophysical experiments, at the latest—are themselves information technology [*Nachrichtentechnik*] by nature, "the realm between the threshold of hearing and the threshold of sensation" (that is, between the minimum and maximum of acoustic perception) "practically" bridges "the entire realm for which air possibly can provide the transmission medium for sound: at the lower end, the threshold of hearing lies between 20 and 30 dB above the noise level, which is determined by the thermal noise of air molecules; and at a sound pressure level of 160 dB"—approximately 30 dB above the pain threshold—undesired, "non-linear effects of sound distribution in the air occur,"[24] as is the case with bad stereo systems. More poetically, and to speak with Rudolf Borchardt, if our ears were ten times more sensitive, we would hear matter roar—and presumably nothing else.

Poetry, however, Borchardt and Adorno notwithstanding,[25] is not supposed to admit noise. Ever since the Greeks invented an alphabet with vowels that also served the purpose of musical notation—which, that is, was lyric and therefore constituted the first "total analysis of the sound-forms of a language"[26]—its system of communication has rested on the interconnection [*Verschaltung*] of voice and writing. At the same time, however, the quantity of operations that was possible with these graphic-phonetic elements also limited the degree of literary complexity. To this extent, poetry formed an autopoietic system that produced its own elements as self-referential components—and for this same reason (and like any system of the kind) could not make further distinctions between elements and operations.[27] Necessarily, then, there was no possibility for analyzing the input and output elements of this Greek analytic system again, until the voices or graphic traits had vanished again into the *quanta* of noise that, in physical terms, they are. On the contrary, according to Jakobson's definition, the "poetic function" assured focus "on the message as such," an immediate "palpability of signs,"[28] and therefore maximized the signal-to-noise ratio.

"What is it, everywhere, / That Man is well [*Worauf kommt es überall an, / Daß der Mensch gesundet*]?" asked Goethe—poet and psychiatrist in one—in *West-Eastern Divan*. He answered his own question with the self-referential emphasis of rhyme and spondaic meter: "All hear the sound gladly / That rounds itself into a note [*Jeder höret gern den Schall an, / Der zum Ton sich rundet*]." In strict fashion, poetry excommunicated, in the name of the articulated communication that it is, its environment—inhuman sound or "primordial echo [*Erzklang*]."[29] Only madmen, like the anonymous "N.N." of 1831, whose verses represent the oldest poetry left behind in German asylums, had the audacity to choose, of all things, Goethe's poem "Audacity" [*Dreistigkeit*] as the motto for verses that hymned the very opposite: not articulated notes of speech but rather "Carnival's Good Friday-Easter-Cross-Wood-Hammer-Bell-Sound" [*Des Carnevals-Chartag-Ostern-Kreuz-Holz-Hammer-Glocken-Klang*].[30]

Of all the instruments, wood and hammers, metals and bells, have the highest quotient of noise. They function phatically—as a call to church or to a conflagration—and not poetically. For this reason, idiophones do not produce pure intervals, which Greek musical notation made storable and Pythagoras considered *Logos* itself. Mixtures of sound of innumerable frequencies—which moreover do not form integral relations[31]—cannot be recorded as sheet music. However, where the system of poetry and music stops, the mathematical "return of the Chaos of old" (as Valéry put it) begins. In the same Age of Goethe, which for solid poetic reasons had to excommunicate and lock up self-declared "sound-catchers" [*Klänge-Fänger*] like the anonymous N.N., a departmental prefect appointed by Napoleon, Baron Jean Baptiste Joseph Fourier, developed a method of calculation that paved the way not just for thermodynamics but also for all media of technological sound-catching, from Edison's cylinder phonograph up to the music computer.

Fourier analysis made it possible for the first time, through integration and series expansion, to evaluate periodic signals of finite energy—that is, all physical signals, whether their harmonics were integral multiples of a tonic note or not—as numbers. The equation,

$$S_c(f) = \int_{-\infty}^{\infty} s(t) \cdot {}^{-2j\pi ft} \, dt$$

transfers quadratically integrable functions of time, t, into functions of frequency, f, and in trigonometric conversion, provides the entire spectrum of partial sounds, S_c, according to magnitude and phase. A fundamental operation of poetry and music—repetition—is now thoroughly quantifiable, whether in the case of perceptible rhythms or in that of sounds which human ears hear as such only because they cannot break down their complexity into discrete elements. Above 60 hertz (or vibrations per second), our physiological capacity for distinction ends—if only because one's own vocal cords begin at this frequency.

With all its applications—from convoluting and correlating given signals up to the fundamental sampling theorem demonstrated by Nyquist and Shannon at Bell Labs—Fourier analysis changed the signal space just as much as, once upon a time, the vowel alphabet of the Greeks had done, this anonymous act that founded our culture. To be sure, in everyday life, the fundamental law of systems theory continues to hold that "communications systems cannot undermine communication" by reverting to, say, the frequency range of nervous impulses.[32] Only Thomas Pynchon's novels present mathematical-neurological characters such as, in *The Crying of Lot 49*, the drug-addled disc jockey Mucho Maas or, in *Gravity's Rainbow*, Private First Class Eddie Pensiero (89th U.S. Infantry Division): their perception has already learned, whether by "measuring" or "thinking," to oscillate [*einschwingen*] into feedback loops by way of technical Fourier analysis; that is, to circumvent their own limitations and separate elements of communication from their operations.[33] However, for the voices of people to be subject to spectral analysis—which after 1894 proved the superiority of female employees to male ones in telecommunications [*Fernsprechdienst*] even to the Reichstag deputy August Bebel,[34] and after 1977 made it possible for the U.S. Air Force to establish an optimal and infallible means of regulating personnel access[35]—the system of everyday communication has also changed in an institutional framework.

Therefore, under modern—that is to say, media-technical—conditions that mock all phenomenology, media have taken the place of the arts. A "new illiteracy," as Salomo Friedlaender called it long before McLuhan or Ong declared the end (in a celebratory tone) of the "Gutenberg Era," erected "antibabylonian towers." These "radio towers"[36] in cities and in brains have positivized the anonymous madman of 1831. All "guitars" and "bells"—about which "N.N." could only dream or write verses—achieve the honor they are due in the Real. Chuck Berry (and with him our own communication system, the Libertas disco in Dubrovnik) hymned an illiterate electric-guitar player, who—as if that were not yet enough—is called "Johnny A. B. C. Goode."

> There was a lonely country boy
> Named Johnny B Goode
> Who never ever learned to read and write so well
> But he could play the guitar like ringing the bell.

Entertainment electronics simply means feeding back all operative rooms of free play [*Spielräume*] in analog—and more recently, digital—signal processing into the ears and eyes: as a trick, gadget, or special effect.[37] As is well known, the founding hero of such effects was Wagner. In the form of *The Ring of the Nibelungen*, music abandoned its native realm of *logoi* or intervals in order to measure out all the possible spaces and transitions between sound and noise. The prelude to *The Rhine Gold*, because its Rhine is a pure river of signals [*reiner Signalfluß*], begins with an E-flat major chord at the lowest bass register, over which eight horns then lay an initial melodic motif. However, it is not melody but rather (and as if to test out the musical transmission bandwidth) a Fourier analysis of that E flat from the first to the eighth overtone. (Only the seventh, somewhere between C and D flat, cannot occur, because European instruments will not play it.)

And so, after the absolute beginning of Wagner's tetralogy has revoked, via music drama, Goethe's poetic filtering of "sound" into a "note," the absolute ending—Act III of *Twilight of the Gods*—can again leave overtones and again submerge into pure noise, that is, liquidate the signal-to-noise ratio.[38] Brünnhilde, who as the excommunicated Unconscious of a god can communicate with Wotan, the imperial author of her days, just as little as N.N. could communicate with Goethe, instead sings to him, as a finale, an "uninhibited lullaby"[39]:

Weiß ich nun, was dir frommt?	*Do I now know what avails you?*
Alles, Alles,	*All, all,*
Alles weiß ich,	*All do I know,*
Alles ward mir nun frei.	*All now is free to me.*
Auch deine Raben	*Even your ravens*
hör ich rauschen:	*I hear rushing:*
mit bang ersehnter Botschaft	*With anxiously desired embassy*
kehren die beiden nun heim.—	*Now they both homeward wing.—*
Ruhe, ruhe, du Gott!	*Rest, rest, you God!*[40]

Wotan's unconscious desire goes into fulfillment, then, as soon as a heroic soprano and a full orchestra implement it. What ends with the *fading* of a god in

Signal-to-Noise Ratio

Valhalla's sea of flames is European art itself. For the two ravens—dark messengers or angels of media technology—neither speak nor sing; in their flight, the transmission and emission of information—indeed, "message" and "noise"—collapse. *Twilight of the Gods* means the materiality of communication, as well as the communication of matter.

In the years between Fourier analysis and Wagner's tetralogy, the same thing motivated the Scottish botanist Robert Brown. To be sure, matter has been noisy since time immemorial, but Brown's chance discovery first transferred this stochastic message into a fitting concept. In 1872, the strange zigzag movements that pollens dissolved in water were performing under a microscope inspired him, like another Antonie van Leeuwenhoek, to believe he had discerned the hidden sex life of living matter for the first time. This sexualization of the realm of plants was in a sense appropriate for the Age of Goethe and its eponymous hero.[41] Unfortunately, however, Brown's further experiments revealed the same phenomenon occurring with dead pollens—indeed, with pulverized rocks. A spontaneous irregularity, the noise of matter, dissolved the fundamental concept of the Age of Goethe, just as Fourier had dismantled the articulated music of language [*Sprachton*]. But instead of providing an explanation that does not exist, Brown simply lent the phenomenon his name: "Brownian motion."[42]

It was only half a century later, when Maxwell and Boltzmann opposed an atomic-statistical model to the received physical theory of constant energy, that Brown's item of curiosity arrived at the touchstone of scientific truth. To the technologically equipped eye, the zigzags demonstrated nothing less than the infinite ping-pong that molecules play with each other above absolute temperature T. A Brownian particle experiences approximately 10^{20} collisions with other molecules per second, so that "the periods during which [it] moves without abrupt change in direction are too rare and too brief to be caught even by modern high-speed photography."[43] For this reason, Boltzmann's formula expressed the noise of matter simply as the statistical mean:

$$\frac{\Delta PN}{\Delta f} = 4kT$$

Telecommunications specialists [*Nachrichtentechniker*] may content themselves with medium-level noise on wave bands, but not modern mathematicians. Whereas classical analysis limited its realm to regular forms and constant functions, the twentieth century—very much to its "fear and horror"[44]—went over to formalizing irregularity. In 1920, Norbert Wiener formulated Brownian movement as a function that cannot be differentiated at any point, that is, as a function whose zigzags form innumerable angles without tangents. On this basis, he was able to assign a measure to thermal noise that not only includes average values, but also its actual paths.

After this mathematical formalization of the Chaos of old, it was no longer difficult to approach the materiality of music and language as well. Wiener's Linear Prediction Code (LPC) has become one of the foundational procedures enabling computers to simulate the random generators in our larynxes. On the basis of past but discretely sampled (and therefore storable) sonic events (x_{n-1} to x_{n-k}), linear prediction prophesies a probable future event:

$$x_n^* = -\sum_{k=1}^{k} a_k x_{n-k}$$

Needless to say, it thereby miscalculates the Real in its contingency, yet this very error (as the difference between x_n and x^*_n) determines the next valuation, in order to minimize it progressively and adapt the coefficient a_k to the signal as it actually occurs.

During the Age of Goethe—according to standing psychiatric definitions—the madness [*Wahnsinn*] or "idiocy" [*Blödsinn*] of patients like "N.N." consisted of "hearing a wild noise everywhere, but no intelligible tone, because they are not capable of extracting one of them from the multitude, of tracing it back to its cause, and thereby recognizing its meaning."[45] Wiener's Linear Prediction Code positivized this very Chaos. That is, his Fourier analysis can demonstrate mathematically that "the minimization of middle quadratic prediction error is equivalent to the determination of a digital filter that reduces the power density spectrum of the linguistic signal [at the input] as close as possible to zero"—or alternately, "transforms the spectrum of the prediction error into a white spectrum."[46] Whereas other filters (for example, in Shannon's writing experiment) also introduce, by way of transition probabilities, redundancy as the simulacrum of meaning, the Whitening Filter literally makes discourses "a side issue."

For this same reason, Shannon's mathematics of signals and Wiener's mathematics of noise return in structural psychoanalysis—which, after all, analyzes (or eliminates) discourses in the same way that Freud analyzed souls (or translated them into "psychic apparatuses"). In the first place, Lacan's concept of the Real refers to nothing but white noise. It celebrates "jam"—this keyword of information technicians—as modernity itself:

The quantity of information then began to be codified [i.e., by Shannon]. This doesn't mean that fundamental things happen between human beings. It concerns what goes down the wires, and what can be measured. Except, one then begins to wonder whether it does go, or whether it doesn't, when it deteriorates, when it is no longer communication. It is the first time that confusion as such—this tendency there is in communication to cease being a communication, that is to say, of no longer communicating anything at all—appears as a fundamental concept. That makes for one more symbol.[47]

Second, and as a matter of due consequence, Lacan's symbolic order—far from what philosophical interpretations hold—is a law of probability that builds on the noise of the Real; in other words, a Markov chain.[48] Psychoanalysts must intercept the improbabilities in (and out of) repetition compulsions just as cryptographers extract a secret message from what seems to be noise. Third, this media-technical [*nachrichtentechnische*] access to the Unconscious liquidates the Imaginary—which as a function of initial optical pattern recognition has already equated the philosophical concept of insight [*den Erkenntnisbegriff der Philosophie*] with misrecognition.[49] That is why it is only by means of psychoanalysis that a subject's chances can be tallied in terms of game theory—that is, calculated.[50]

What can be calculated by means of computerized mathematics is another subject, and a strategic one: self-guided weaponry. Wiener developed his new cybernetics not to analyze human or even biological communication. As he put it himself, "the deciding factor in this new step was the war."[51] On the eve of the Second World War—given the extremely accelerated air forces of the enemy—it was strictly a matter of optimizing Anglo-American artillery systems to compete. Because the actual flight path of bombers involves the complex interplay of commands, errors of navigation, air turbulence, turning circles, maneuvers evading artillery fire, and so on, it cannot—inasmuch as it is the chance movement of human beings—be predicted. And yet, prediction proves vital simply because artillery projectiles, whose speed exceeds their target's only in relative terms (unlike that of human targets), must intercept the bomber in its future position, and not at its present location. Therefore, to minimize the problem of incomplete information—this noise from a future[52]—Wiener implemented the Linear Prediction Code in an automated artillery system, which soon operated on a computer basis. The United States of America entered the Second World War armed in this capacity.

In less than two hundred years, mathematical information technology transformed signal-to-noise ratios into thoroughly manipulable variables. Along with the operational borders of the system known as everyday language, those of poetry and hermeneutics were exceeded, and media established whose address (all advertising to consumers notwithstanding) can no longer be called "human" with any certainty. Ever since its foundation [*Stiftung*] in Greece, poetry had the function of reducing the chaos of sound to recordable and therefore articulated tones, whereas hermeneutics—ever since it was instituted by Romanticism—secured this complexity reduction intellectually [*geisteswissenschaftlich*]: by assigning it to the address of a poetic subject called the "author." Interpretation purified an interior space of all noise, which in the beyond of events, in fits of delirium and wars, never ceased not to stop.

Ever since noise, through the interception of enemy signals, has not been evaluated by interpreting articulated discourses or sounds, the yoke of subjectivity has been lifted from our shoulders. For automated weapons systems are subjects themselves. An unoccupied space has emerged, where one might substitute the practice of interception for the theory of reception, and polemics for hermeneutics. Indeed, one might inaugurate *hermenautics*—a pilot's understanding of signals, whether they stem from gods, machines, or sources of noise.

13 The Artificial Intelligence of World War: Alan Turing

Alan Mathison Turing came into the world in Paddington in 1912, at a time when, thanks to the military might of war fleets and undersea telegraph cables, this world was still called "the Empire." His parents, like those of Rudyard Kipling, its poet, served Her Majesty as colonial officials in Kurnool, Visakhapatnam, Madras, and other cultural outposts in the imperial dominion of India. When Turing died in 1954—after he had given his country the gift of making computers possible—the Empire no longer existed. Power, *imperium*, had moved, in an enormous transfer of technology, across the Atlantic, along with all the computers, liquid rockets, and guided weaponry of wartime European research laboratories. Britannia had ruled the waves of the sea and the Third Reich (in direct competition) those of the ether,[1] but the *Pax Americana* began with calculators, artificial intelligence, and spy satellites. Turing had done his duty as an inventor. In 1951—during the McCarthy Era—the United States excluded homosexuals and other security risks from all sensitive government positions; the United Kingdom, so as not to be cut off from American intelligence, had to follow suit immediately. Put on ice, Turing committed suicide, at least according to official investigations. Like Snow White, he bit into an apple soaked in cyanide next to his bed.

Hans Magnus Enzensberger dedicated one of his beautiful and new poems of "human science fiction" to Turing. In allusion to this eccentric's wont to wear gas masks against hay fever, his habit of repairing his sole means of transportation—a bicycle—with twine, and his fondness for tinkering together cooking pots, high-frequency coils, and chessboards, the verses promise the inventor of the computer eternal life. "Especially on damp October days, in the environs of Cambridge," the poem reads, one can "see him, or his simulacrum, on mowed stubblefields, hiking in the fog cross country, unpredictably doubling back."[2]

But ever since artificial intelligence has existed, even the most intelligent poetry has transformed into myth and anecdote. It is more likely that Alan Turing's ghost haunts the computers of department stores than that it spooks about on the meadows of the city where he studied. Turing was an intellectual who wanted to break with the power of intellectuals—the priestly caste of modernity. Thanks to him, knowledge disappeared from human heads and moved into the small machines that (as technicians put it neatly) "implement" it. The computer hardware of today, with the monotony of hundreds of thousands of transistor cells on a silicon chip, is organized in far too labyrinthine a fashion even to be designed by engineers. And so circuitry development is now performed more and more by CAD, Computer Aided Design. Just as John von Neumann's "General and Logical Theory of Automata" predicted,[3] computers have taken on their own reproduction—which, as CAD, does not simply offer imitation or partial reproduction of the "parents," but augments complexity itself from generation to generation.

Such complexity, however—its presence and planning—is hardly approachable. The authorities of the Imperium, especially in the Pentagon, have ordered that files or data banks concerning artificial intelligence be treated as classified information. All that remains for historians (leaving aside the blindness they exhibit with regard to data banks) is the classified material of yesterday, which ministries of defense typically make available only after thirty years have elapsed. As a result, Turing's innovations decided a world war in 1945, but only in 1975 were they released from secret governmental archives to reach what is still called the "public sphere": books, lectures, and languages that have not yet been formalized.

To be sure, the computers that decided the Second World War possessed no artificial intelligence in a strictly technical sense. They had no frame of reference to communicate with the Symbolic of natural languages and no programs of pattern recognition to process the Imaginary of forms and images. However, they already met the formal conditions of intelligence that Lacan gleaned from Karl von Frisch's experiments with bees. As everyone knows, bees give other bees—through a dance, for which instinct provides the code—information about where a particular blossom is located; they do so in terms of the angle of sunlight, on the one hand, and distance, on the other. This code "is distinguished from language precisely by the fixed correlation between its signs and the reality they signify." In contrast, "the form in which language expresses itself in and of itself defines subjectivity. Language says: 'You will go here, and when you see this, you will turn off there.' In other words, it refers to discourse about the other [*discours de l'autre*]."⁴

In still other words (which are not Lacan's), bees function as projectiles, whereas human beings operate as remote-control weapons. A dance gives the former objective data about angles and distance, and a command gives the latter "free obedience" [*freien Gehorsam*]—which ever since the creation of national armies during the French Revolution has been synonymous with the highest of virtues. In 1943, when Turing's colleagues in the Secret Service succeeded in building a calculating device that could, while performing an assigned task or interim calculation, decide for itself about the commands that followed—that is, determine their future—free obedience finally came to function automatically. Ever since IF-THEN commands ceased to be a privilege of the human being, all philosophical debates about the death of the subject have been settled, simply because weapons have become subjects themselves. Today cruise missiles steer their own flight paths with optical sensors and television memory [*Fernsehbildgedächtnis*]. An onboard computer—artificial intelligence—tells the drive unit [*Antriebsaggregat*]: "Fly this way, then that"—say, to the Ukraine—"and when you see this and that"—say, Kiev—"make a turn and detonate."

Norbert Wiener, the inventor of cybernetics, admitted that "the deciding factor" for such progress was called "World War II."⁵ Both etymologically and in actual fact, cybernetics is a matter of steering, and steering is a matter of military chains of command. With Turing's computers and Wiener's automatic artillery cannons—which predicted the future point in space of an enemy aircraft on the basis of its chance location in the past, and which foresaw where bombers and mortar rounds would merge in an explosion—the history of war ended and Star Wars could begin. Yet even though literature about computers grows daily—in professional journals, social critiques, software advertising, and popular writings—silence prevails about their military history.

Of course [*selbstredend*], in the Beginning was the Word. The Word was with God and tried, for seven days and seven nights, to set up binary distinctions, that is, bits: Day and Night, Heaven and Earth, and Sun and Moon—to say nothing of Good and Evil. These days before the Day, which were repeated sequences of digital codings, literally created nothing—nothing that did not already exist under a title of infamy: *tohu wa bohu*. After all, Heaven and Earth, Land and Sea, did not need to wait for Elohim's inscription. What the administrators of Holy Writ call God's "creation out of nothing" was, instead, the creation of Nothing: pairs of oppositions, code words, signifiers. For this same reason there arose the fundamental difference between differences, on the one hand, and white noise, on the other—and for this same reason a signal-to-noise ratio now separates the order of command from the Chaos of old.

"There is little in our technological or physiological experience," wrote John von Neumann, the mathematician of the Second World War, "to indicate that absolute all-or-none organs exist."[6] Neither *tohu wa bohu*s nor nervous systems operate digitally—as God the Creator does. However, this deficiency of nature has in no way prevented all-or-none organs from being indispensable in strategy, if only to distinguish between commandments and prohibitions. Ever since Genesis, the language of Supreme Command has only operated with "yes" and "no." So as not to admit the smallest possible misunderstanding, which might disrupt the chain of power with the inevitable noise of channels, the language of Supreme Command overcodes even its own codes. Thus, for example, all telegrams from the German general staff since the days of Schlieffen made a fundamental distinction between "Western" [*westlich*] and "Eastward" [*ostwärts*]—instead of "Eastern" [*östlich*]—because, after all, in a war on two fronts, the difference between West and East is just as fundamental to generals as the distinction between Heaven and Earth for gods of creation. When Colonel General Jodl, the last commanding officer in a glorious short story, wanted to sacrifice this binary overcoding to civilian language [*einem zivilen Alltagsdeutsch*], the consequences exceeded mere misunderstanding: the officer corps of the Wehrmacht, in the Holy Spirit of its code, registered "general and universal indignation [*eine allgemeine helle Empörung*]"[7] about the noise that resulted—*tohu wa bohu* that had not been sufficiently digitalized.

Even the conquest of Troy by Agamemnon could not occur without sound. However, and as Villiers de l'Isle-Adam already remarked, for want of means to record it, the Homeric rhapsodies did not transmit the slightest bit of sound; rather, they passed along only the word "sound." What is more, the storied optical telegraph—which, according to Aeschylus, ran from Asia Minor, over the Sporades and various mountain peaks, all the way to Agamemnon's Mycene—operated with a simple fire signal; its absence signified the failure of the expedition, and its presence success. This binary economy of signs seemed made to order for war as a zero-sum game between two adversaries, yet it went missing again along with the twenty-six letters that, after 1794, remotely guided the fourteen revolutionary armies of France by means of the same telegraphic principle. Only with electrical cable telegraphy did it return—at the latest in

1848, when Friedrich Clemens Gerke switched the texts of all dispatches into Morse dots and dashes, two minimal signifiers that constitute a system of exemplary economy.[8]

Since Gerke, military communications have ceased to operate as the same postal system (or literature) that functioned up to the time of Napoleon's dispatches. To be sure—via the mediation of the first general staff to exist in military history—the Emperor would dictate up to twenty letters in the hours preceding combat to autonomously operating corps and divisions of the army (which for their part had been schooled in free obedience). The volume of correspondence represented only a numerical difference between the supreme command of the military and the executive commands [*Durchführungsbestimmungen*] of contemporary literature—that is, epistolary novels such as Goethe's *Werther* or Rousseau's *Nouvelle Héloïse*.

But no modern-day Aeschylus has ever been able to compete with the semiotechnics that broke a two-thousand-year connection to literature by means of codes, channels, and networks. When told by a writer that his invention would now enable novels to be written automatically, Charles Babbage—who had proposed the first calculating machine employing IF-THEN specifications to the British Navy—coolly replied that the book fair at Leipzig and the pig auction at Padua were one and the same zoo [*Menagerie*] in his eyes.[9] The monopoly of writing foundered on discrete steps of calculation and recursive functions. A little later—three years after Gerke's telegraphic (that is, equally discrete) minimal signifiers—George Boole devised the means to symbolize binary logic; for all that, the invention recorded Boole's so-called laws of thought less than it yielded digital switches made of relays, tubes, and ultimately, transistors. Whereas Leibniz had simply proposed a new number system (base two instead of ten) with his binary arithmetic, Boolean algebra sacrificed all counting and place values to binary decision making. Its symbols did not have arithmetical values, but rather logical or strategic ones, as if made to order for games theory and computer simulations—which, as everyone knows, have taken the place of sandboxes, maps, briefings, and so on in general staffs.

In 1898, this kind of two-party zero-sum game occurred between the old-fashioned colonial power of Spain and the United States. A few islands—Cuba and the Philippines—changed their white masters simply because one nation, which had made telegraph wires fit for mass production, effortlessly prevailed in a conflict that the *Proceedings of the US Naval Institute* fittingly christened "the war of coals and cables."[10] Admiral Cervera's Spanish colonial armada, cut off from all Morse information originating in Madrid, did not learn the address of the Cuban coal supply for its steamships, and so it went under. That there is no energy without information—no Boltzmann without Shannon—had been proven. And for the first time, the law of war faced the problem of defining an immaterial blockade.[11]

Fear of such blockades—of cable monopolies of this kind—gave rise, at the turn of the new century, to competition for wireless telegraphy and telephony, and soon for radio as well. This mobilized fleets and armies, but at a price. Guglielmo Marconi, the Fascist senator who invented the technology, described it in a phonographic recording broadcast hours after his death via Radio Roma:

> forty-two years ago, when I achieved the first successful wireless transmission in Pontecchio, I already anticipated the possibility of transmitting electric waves over large distances, but in spite of that I could not hope for the great satisfaction I am enjoying today. For in those days a major shortcoming was ascribed to my invention: the possible interception of transmissions. This defect preoccupied me so much that, for many years, my principal research was focused on its elimination.
>
> Thirty years later, however, precisely this defect was exploited and turned into radio—into that medium of reception that now reaches more than 40 million listeners every day.[12]

The name of the anonymous instances that spurred Marconi to seek a pure contradiction—radio without the possibility of interception—is not difficult to guess. It was (in Eisenhower's parting words) the "military-industrial complex" that finally admitted a compromise between state radio security and civilian mass reception.

On the one hand, new decrees emptied out [*plombierten*] or castrated the two-way devices [*Wechselsprechgeräte*] of the Signal Corps [*Nachrichtentruppen*]—which in the trenches of 1917 had transmitted music for the first time—so that they could no longer broadcast but only receive government entertainment. German civilian radio emerged after the First World War, in order to put a stop to the revolutionary games of left-wing ex–radio operators in the army; that is, to preempt civil wars.[13] On the other hand—and on approximately ninety percent of the wave bands available—the military broadcasters/receivers continued to function, uncastrated. The Reichswehr permitted competition from civilians only after *Poststaatssekretär* Dr. Hans Bredow had brought the happy news that a novel machine had been invented. This device encrypted radio exchanges automatically and banished the dangers of eavesdropping [*Abhörgefahren*] that had been Marconi's nightmare. As the machinic replacement for cryptographic work by hand on graph paper, it fully merited the name it was given: "Enigma."

A partial solution of this enigma or riddle lies in the trenches of the First World War, which had solved all problems of technical storage. Soldiers' sepulchral existence for four years simply prolonged the encryption of sequences of images and sounds. Film and gramophone, Edison's two great inventions, had broken the immemorial monopoly of writing and (according to Paul Virilio's thesis) made technical war possible.[14] At the same time, the newly minted typewriter—by making anonymous, or indeed, by feminizing the production of texts—had completed the modern trinity of the Real, the Imaginary, and the Symbolic (in other words, the three storage possibilities of photography, cinematography, and dictated standard command).

What the storage media of the heroic age [*Gründerzeit*] still lacked, however, was connection to technically adequate media of transmission—radio and television (and their military twins: sonar and radar). According to an analysis by Lieutenant General Rudolf Schniewindt, the mobile war that Field Marshal Schlieffen planned had failed simply because at the Marne, proper capacity for telecommunications was wanting; Captain Heinz Guderian, as the chief of an armored radio station, could have attested as much. But even in the static warfare that necessarily ensued, there was no viable way to control soldiers in the trenches remotely—especially not in the catastrophic command system of the British, which operated without any back coupling between troops and staffs.

From 1917 on, General Erich Ludendorff, in order to overcome the immobility of the trenches, experimented with pedagogical remote control and free obedience. However, success remained modest even in the most famous of cases: whenever Lieutenant Ernst Jünger caught sight of the enemy between clouds of gunpowder—an enemy who was *per definitionem* invisible—he confused the foe with his own cinematographically stored doppelgänger.[15] Technical feedback became more and more urgent on land, in the air, and at sea. The first tanks were intended to pave a way through enemy trenches for the infantry—if only it had been possible to steer them from afar. But because, as a rule, the outer antennae of these armored vehicles broke off in the barbed wire of no-man's-land, the carrier pigeons of old—that is, of the Franco-Prussian War—had to take the place of radios.[16]

In 1939, to counter such atavism and aporia, the gospel of the *Blitzkrieg* announced a new technology that promised "speed of attack through speed of communication" [*Angriffsgeschwindigkeit durch Kommunikationsgeschwindig-*

keit]. VHF transmitters—long before they became the vehicles of consumerism in the postwar era (conveying everything up to and including rock music)—steered the three mobile weapons systems of the Second World War: tanks, submarines, and, Stukas (as well as, from 1944 on, twenty percent of the guided missiles from Peenemünde). In 1934, the Weaponry Office [*Heereswaffenamt*], in agreement with Marconi but counter to expert opinion, had managed to prove that VHF exchanges between staffs and combat tanks operated without interference from optical hindrances such as forests and hills.[17] Therefore, in 1939 the Wehrmacht—the sole army of the time to do so—was able to take the field with ten remote-controlled and autonomous tank divisions. Guderian had learned his lesson at the Marne. The *Blitzkrieg* replaced static warfare in the trenches and immobile storage media with transmissions media, making "cauldron battles" and tank maneuvers possible.

Both innovations—Guderian's tanks and Fellgiebel's messaging networks[18]—necessarily relied on the Enigma encryption system. Connecting an army of five million and forging feedback networks via radio amounted to maximizing the risk of interception that had distressed Marconi. And so, after 1939 war itself coincided with its information network; it provided—as soon would be the case for every computer—the organigram of all addresses, data, and commands. However, electrical networks need to run over electrical networks. Therefore, even though—and because—it was based on transmission devices, the *Blitzkrieg* armed itself with the least spectacular of all storage media. All commands, data, and addresses ran over typewriters, both the Enigma and cipher machines [*Geheimschreiber*] which were veiled in even greater secrecy.

Unlike what occurs in the tedium of our everyday, academic lives, these apparatuses could surprise their users in tanks, submarines, or Stukas. Instead of the unambiguous correlation of keyboard activity and system output (where the only consolation comes from typos), the Enigma offered all the joys of discrete—that is, combinatory—mathematics. The twenty-six letters of the alphabet ran over electric power lines into a distribution grid composed of three (and later, four or five) drums [*Walzen*] that switched between ever-changing coding possibilities. With every tick [*Anschlag*], the drums (like clock hands indicating seconds, minutes, and hours) moved 1/26 of a rotation; they returned to starting position only after 26^7 or eight billion letters of text had been generated. As a result, interception yielded a pure mishmash of letters, which only an Enigma machine operating in reverse at the other end could decode.

For five years—from the first day of the war to the last—Wehrmacht Supreme Command placed absolute trust in cryptography that, finally, had been automated. Even if enemy cryptanalysts had intercepted individual radio messages and deciphered them, the uncoded text—after many millions of years of mathematicians' lives—would have arrived far too late for the fighting troops. As everyone knows, only real time analysis counts in technical wars.

For all that, German military command was not in full possession of its senses. Any machine steered by an algorithm can be beaten—indeed, surpassed—provided that the enemy device has a superset of algorithms at its disposal. That is precisely what decided the Second World War. Computers were, and are, the strategically decisive counteroffensive in a media war.

In short, World War II occurred simply as combat between two typewriters. On one side stood the Enigma and cipher machines, which did not encrypt just single messages but a telecommunications system in its entirety. On the other side stood apparatuses called "Bombe," Eastern Goddess," and "Colossus," which merited their prophetic or gigantic names because of their capacity to decode this same system (after relatively simple radio interceptions). The most important factor for the end of the war was the fact that British intelligence set up the first operational computers in history (and thereby brought about the end of history). Even though and because...

Even though and because, in the 1930s, the eccentric and homosexual son of a colonial official had attended public school—or rather, private school—at Sherborne. The teachers at the venerable institution could hardly forgive Alan Turing for his chaotic habits and messy writing. His brilliant tests in math received bad grades simply because the penmanship was "the worst ... ever seen."[19] To this very day, schools are just that true to their old mission of conditioning neat, coherent, and personal handwriting in order to produce "individuals" in the literal sense of the word. Turing, however—a master at undermining education and the caste of intellectual priests—came up with a means of evasion: at the age of eleven, his letters "described an exceedingly crude idea for a typewriter."[20]

Nothing came of Turing's childish construction plans. But when, on Grantchester Meadows near Cambridge (the meadows of all English lyric poetry, from Romanticism up to Pink Floyd), the idea for a Universal Discrete Machine dawned on him, his schoolboy dreams were fulfilled and transformed. A typewriter stripped down to bare essentials became the prototype for all computers that can be conceived.

Turing machines are exceedingly more primitive than the typewriter designed at Sherborne. All they consist of is a roll of paper that simultaneously contains commands, data, and addresses—input and output, a program, and results. Turing machines do not need the many redundant letters, figures, and signs of a typewriter keyboard. They make do, loosely following Boole, with one sign and its absence: one and zero. By sampling this binary information on the basis of an IF-THEN specification that constitutes the entirety of their artificial intelligence, they operate automatically: the roll of paper moves either not at all or just a bit to the left or the right; that is, it moves just as discretely as typewriters do with their space bars and backspace keys. The difference is that the way they read—because Turing also, and especially, made do without human secretaries—determines what is subsequently written. It depends on the sign (or as the case may be, its absence), whether Turing machines leave the mark standing or delete it—or conversely, whether they leave the empty space standing or replace it with the sign. After this simple operation, the program loop jumps back to reading, and so on, ad infinitum.

That is the whole of it. No computer that has ever been built—or will ever be built—can do more. Turing proved as much mathematically in 1936. The artificial intelligences of today run more quickly, and with more parallel operations, but the principle is no different.

With the Universal Discrete Machine, the media system achieved completion. Together, storage and transmission media yield a switching principle that can simulate all other information machines—simply because it stores, transmits, and calculates in every single program loop. An inhuman bureaucracy assumes all functions that are sufficient and necessary for the formal definition of intelligence. In today's standard processors, a bus administrates the transmission of addresses, a silicon memory the storage of data, and an arithmetic logic unit—as the combination of Leibniz and Boole—the binary calculation of commands.

But to what end? . . . Initially, in 1936, Turing's Universal Discrete Machine was simply a thought experiment for solving, in negative fashion, the *Entscheidungsproblem* posed by the great Göttingen mathematician David Hilbert. Hilbert's "program" of 1928 had called for mathematics to be complete, consistent, and decidable. It was a matter of demonstrating, then, that theorems could be either proven or disproven, that they not be derived by contradictory means, and finally, that they be resolved in a determinate and finite set of steps. As is well known, Gödel disproved the first point of the program; on the basis of the incompleteness theorem at which he arrived, he affirmed, yet again, the superiority of human intelligence. Turing's machinic thought experiment disproved the second point, but it led him to the opposite conclusion. The fact that there are theorems that machines cannot decide in a finite number of steps defines computability in general, according to Turing. "Computing"—which until 1936 had referred to a human capacity—assumed the new and technical sense that, ever since, has made world history. Artificial intelligences, instead of still being measured by what they cannot do, mastered everything they could do. This was not a matter of science, but of strategy. The very fact that finite-state machines had an advantage over the physical or neurophysiological universe—namely, the fact that they were predictable—qualified them for war. Turing replaced nature with the enemy, an analog system with a binary one, physical laws with encoding technology, observable phenomena with intercepted messages, and natural constants with cryptographic keys. His justification: "The subject matter of cryptography is very easily dealt with by discrete machinery, physics not so easily."[21]

And so it happened. Turing had hardly disproven Hilbert when he wrote to his mother about "a possible application" for the new mathematics at which he was working, which seemed worlds away.

> It answers the question "What is the most general kind of code or cipher possible," and at the same time (rather naturally) enables me to construct a lot of particular and interesting codes. One of them is pretty well impossible to decode without the key, and very quick to encode. I expect I could sell them to H.M. Government for quite a substantial sum, but am rather doubtful about the morality of such things. What do you think?[22]

Instead of his mother, the government itself provided the answer. Germany's "Enigma machine was the central problem that confronted British Intelligence in 1938. But they believed it was unsolvable."[23] Until, that is, the Government Code and Cipher School took Alan Turing into its service three days after the war had erupted—all moral doubts notwithstanding.

Bletchley Park, the bombproof country seat of British cryptanalysis in wartime, occupied a uniquely good starting position: young mathematicians from the Polish secret service had already constructed, on the basis of intercepted Enigma signals, a deciphering machine that had been christened "Bombe." However, in December 1938, when Fellgiebel's communications network for the Wehrmacht increased the number of drums to five, even Bombe could no longer compete. The 150,738,274,937,250 ways of electronically interconnecting pairs of letters surpassed the machine's ability to calculate—at least in the real time on which *Blitzkrieg* and countermeasures depended. The overtaxed Poles gave their documents to the British and to Turing.

Alan Turing and Gordon Welchman made the primitive Bombe into a machine that the chief at Bletchley Park did not baptize "Eastern Goddess" by chance. It was a fully automated oracle for interpreting fully automated secret radio messages. From May 1941 on, the enemy (loosely following Goebbels) listened in on Enigma commands with only twenty-four hours' delay. The fact that Enigma was a machine also made mechanical cryptanalysis possible. As a pseudo-random generator, the cipher-typewriter produced nonsense only with respect to systems whose period fell short of its own. Turing's Goddess, however, discovered regularity in the jumble of letters.

In the first place, the Enigma had the practical advantage—or the theoretical disadvantage—that its ciphers, in terms of group theory, were self-inverse. So that they could be encrypted and decrypted on the same machine, letter pairs had to be exchangeable. Therefore, if the Supreme Command of the Armed Forces—*Oberkommando der Wehrmacht*, or "OKW"—coded its O as K, K yielded an O. This "particular feature" meant, secondly, "that no letter could ever be enciphered into itself."[24] The OKW, then, was incapable of writing its own name. Turing subjected this scant but traitorous information to sequential analysis, which could assess—and therefore steer—all probable solutions. With an automated faculty of judgment, the Eastern Goddess cycled through permutation upon permutation, until letter salad became plain language again. War of the typewriters.

And because "15 up to a maximum of 29 percent"[25] of German telecommunications ran through more than 30,000 Enigmas, the war of espionage reached a new level: interception "capture[d] not just messages, but the whole enemy communication *system*."[26] The midlevel of command—from army and division staffs all the way down to particular weapons of the *Blitzkrieg* on land, in the air, or at sea—divulged their addresses (which, all spy novels notwithstanding, are more telling than data or messages). Sixty different Enigma codes and three thousand secret radio transmissions per day, plus all the details of transmitters and receivers, made the war look like one big typewriter the size of Europe. Under high-tech conditions, strategy coincides with its organigram. This was reason enough for the Government Code and Cipher School to mirror systematically, as a miniature of the Wehrmacht, the enemy himself.[27] Turing's famous game of imitation became a historical event.

It is only a step from the flowchart to the computer. Finally, the addresses, data, and commands that, in the Wehrmacht or its British simulacrum, were still circulating between human beings and typewriters could become hardware. In 1943, the Post Office Research Station in Dollis Hill performed the final step for Bletchley Park. Fifteen hundred repurposed tubes, instead of amplifying analog signals in as linear a fashion as possible (as occurs in radio), simulated, through overmodulation, the binary play of Boolean algebra. (Transistors were first invented in 1949.) The Universal Discrete Machine—with data entry, the possibility for programming, and greatly improved internal storage[28]—was implemented for the first time. Turing's successors could find only one fitting name: Colossus. Logically, the secret strategies of the Führer's headquarters in the "Wolf's Lair" [*Wolfsschanze*] could only be cracked by the monstrosity that was, and is, the Computer.

Colossus went into action to decode a further forty percent of German telecommunications—everything that, for security reasons, did not pass through Enigma and radio, but rather went via the cipher machines built by Siemens. As a teleprinter operating on Baudot/Murray code, this souped-up typewriter also took care of burdensome manual operations and human error; its strictly digital signals consisted of "yes" and "no" in series of perforations that, with the binary addition of plaintext and pseudo-random generators, enabled much more efficient encryption than the Enigma. Moreover, radio interception was only possible if, exceptionally, the signals passed through a radio link instead of telegraph wires.[29] Instances of high command are just that exact when shopping for typewriters.

It goes without saying that Colossus beat binary addition through binary addition. For all that, however, the first computer in the history of science (or in the history of war) would only have been a thousand-pound version of the typewriter manufactured by Remington—albeit enhanced with calculating features[30]—if it had not obeyed conditional jump commands [*Sprungbefehlen*].[31]

In 1938, in Konrad Zuse's private residence in Berlin, conditional jumps— first foreseen for the Analytical Engine that Charles Babbage left uncompleted in 1835—reached the world of machines that ever since has been one with the Symbolic itself. Without success, the autodidact Zuse offered his binary calculator as an encryption machine surpassing the (supposedly) infallible Enigma.[32] The opportunity that Wehrmacht Signals Communications [*Wehrmachtnachrichtenverbindungen*] missed was seized only in 1941, by the German Experimental Institute for Aeronautics [*Deutsche Versuchsanstalt für Luftfahrt*], in order to "calculate, test, and check remote-controlled missiles [*Flugkörper*]."³³

On all fronts—from top-secret cryptanalysis up to spectacular offensives that occurred by means of futuristic weaponry—the Second World War switched over from human beings or soldiers to machine subjects. Indeed, Zuse's binary calculators, instead of meeting the fate of the V-2 under the rocks of the Harz mountains,³⁴ almost programmed the flight paths of rockets from the inception. The "array of assignments" with which the Weaponry Office at Peenemünde had charged German universities rather clairvoyantly included—in addition to devices for measuring Doppler effects, calculators of flight mechanics, and sensational acceleration integrators—what Wernher von Braun called "the first attempt to perform electronic digital calculation."³⁵ The weapon as a subject (or *subjectile*, as Artaud soon put it) needed a brain to match.

And so, Colossus begat child after child—each one even more colossal than its secret father. According to the Ministry of Supply, Turing's postwar computer ACE calculated "shell, bombs, rockets, and guided missiles." The American ENIAC "simulated[d] the trajectories of shells through varying conditions of air resistance and wind velocity, which involved the summation of thousands of little pieces of trajectory." EDVAC, built according to the findings of von Neumann, solved "three-dimensional 'aerodynamic and shock-wave problems . . . shell, bomb, and rocket work . . . in the field of propellants and high explosives.'" BINAC worked for the U.S. Air Force, and ATLAS for cryptanalysis. Finally, MANIAC—if only this wonderfully promising name had been implemented in time—would have optimized the shock waves of the first hydrogen bomb.³⁶ Annihilation in nanoseconds called for suitably automated mathematics.

On the manifest level, then, the movie looks as if everything involving the "marriage of two monsters"³⁷—which von Neumann arranged between German guided missiles and American atomic payloads in 1954 by dispensing with conventional amatol and equally conventional bomber pilots—had completed the decisive step from the *Blitzkrieg* to the strategies of the present day. It was not entirely so, however, because both guided missiles and nuclear weapons had passed, with altogether strange ease, through curtains of bamboo and iron, in part through espionage and in part through the transfer of technology. The same did not hold for the machinic subject itself—the unprepossessing but fully automated calculating typewriter [*Schreibrechenmaschine*]. Stalin, with the thunderbolt of a theory that is omnipotent because it is true, condemned the bourgeois perversion of cybernetics. It seems that once revealed, the secrets of mass destruction in the duels of intelligence agencies—like rocket exhaust and the lightning flash of detonation—had blinded historical materialism.

Annihilation still counts as decisive in war. Now, after forty years, it is gradually emerging from secret archives that, among all candidates for the title of grand destroyer, Bletchley Park was the most qualified. The Second World War witnessed the triumph of materialism that had materialized mathematics itself. "Intelligence had won the war,"[38] Turing's biographer has written of Enigma and Colossus—and with British verbal precision that makes no distinction between understanding, intelligence, and information machines [*Verstand, Geheimdienst und Informationsmaschine*]. This, however, is exactly what had remained a secret of state. During the war, an entire organization emerged for the purpose of transmitting the results of fully automated cryptanalysis to front staffs exclusively in disguised form. Otherwise, the greatest secret of war (through looted papers, defectors, or treacherously accurate countermeasures) might have leaked through to the Wehrmacht and Enigma would have fallen silent. Accordingly, the final historical task for intelligence officers was to invent an array of glorious espionage tales to conceal the fact that interception and automated calculating typewriters had made secret services and agents superfluous. (Which spy novels do to this very day.) The mysterious "Werther" who is supposed to have transmitted so many plans of attack from the Wolf's Lair, via Swiss double agents, to Moscow—a party who cannot be located historically, however—could be one of the simulacra that Bletchley Park systematically concealed from the Red Army.[39] In such a case, at any rate, Stalin's theory would have had a material basis: information blackout [*Informationssperre*].

On 28 August 1945, three weeks after Hiroshima and four weeks after Potsdam, President Truman issued a secret order about secret radio interception—an information blackout on information machines. Cryptanalysis, which had decided the war, became the epitome of top secret—for past and present, in terms of technology and method, concerning success and failure, and for both Bletchley Park and Washington, D.C.[40] Hereafter, the same war, which now went "cold," could start again: in the wake of Truman's order, Colossus and its American replicas learned Russian instead of German. Perfectly hidden, "the legacy of a total war, and of the capture of a total communication system, could now be turned to the construction of a total machine."[41] The machines that had been developed during world war for world war "did not disappear with the armistice. For decades thereafter (and even today to some extent), the same machines remained in use. Their ghosts appear today in computer systems such as the UNIX operating system and in numerous commercial 'black boxes.'"[42]

In the period immediately following the war, there were two whole computing centers, one for British intelligence and the other for its American counterpart. If what the most recent research in England supposes is true—that machinic cryptanalysis of Soviet radio transmissions first uncovered the tracks of the nuclear spies Rosenberg—there is great probability that Turing, even though he was supposedly designing computers only for civilian and academic purposes, continued his intelligence work. And if the track of the Rosenbergs led to the famous trio of homosexuals in the British Secret Service, then Turing would necessarily have stumbled on the names of former friends from Cambridge—Alastair Watson, for example—in radio transmissions that had been intercepted. Ultimately, of course, the discovery of Turing's homosexuality made him an absolute security risk himself.

But British media power was coming to an end anyway. One of Truman's first decrees was that all wartime cryptanalysis should be kept secret; one of his last decrees founded the National Security Agency. The NSA—the most secret of the three branches of American intelligence—took over, with its 80,000 employees, Turing's innovation through the European-American technology transfer that promptly occurred. The inventor of the computer had to bite into Snow White's cyanide apple so that the fruits of his labor could assume global domination.

According to one of its rare public-relations campaigns, "NSA . . . certainly hastened the start of the computer age."[43] However, if only because the economy of the American private sector during the postwar period (all rumors notwithstanding) viewed computer development as a "profitless venture,"[44] it is surely more accurate to affirm that "NSA became the world's leader in the development of computing equipment—pushing far beyond the publicly available technologies of the day."[45] Since then, its spy satellites have intercepted telephony, telegraphy, and microwaves—mail from all corners of the globe, that is; its computers have deciphered messages that are potentially coded, scrambled, and so on, stored the transmissions automatically, and trawled through them (just as automatically) for suspicious keywords. And so, 0.1 percent of all telecommunications on this planet are absorbed by the NSA's artificial intelligence. What then happens with them, no one knows. As a rule, orders for secrecy are lifted only after thirty years. Perhaps they will no longer be necessary at all three decades from now. The Word that was in the Beginning is vanishing into computer data banks anyway. When all that is said by the inhabitants of the earth has disintegrated into bits, Alan Turing's Universal Discrete Machine will be perfected.

14   Unconditional Surrender

The warlords of the three unions—the U.S.A., U.S.S.R., and U.K.—had planned a further conference for January 1943 to coordinate military objectives and postwar plans. Ultimately, Stalin did not agree to Soviet participation, and for two good reasons. For one, he could not personally leave Moscow, because (at least in his lofty military fantasy) he had to guide the deathblow that General Chuikov's 62nd Army was in the process of dealing to Colonel General Paulus and his 6th Army (cornered at the city named after Stalin himself). Second, the Soviet commander in chief was as yet unwilling to discuss operational details—he would only have been able to voice, once more, the call to open a second European front. A telegram sufficed to make this one-syllable point.

The Casablanca Conference, then, was not exactly strategically decisive. Rather, it was dictated by Roosevelt's wish for "fresh air."[1] The American president—whose party had almost lost control of Congress in the most recent election—felt drawn to Humphrey Bogart's cinematic city, which had just been half liberated.[2] In the courtyard of a villa encircled by military police and barbed wire, Roosevelt discussed Stalin's demands with the British prime minister and accepted that victory must occur in Europe before it happened in Japan. Accordingly, a division of labor was decided upon: British air raids would occur by night on area targets and American air raids by day on point targets[3]; there would also be two landing operations: first in Sicily (which belonged to Cosa Nostra[4]) and then, a year later, on the coast of Normandy.

These operational decisions already raised the political question about the conditions of surrender to be placed on the Wehrmacht. At a press conference at noon on January 24, Roosevelt declared that the future United Nations would only accept "Unconditional Surrender"—that is, "the total elimination of German military power." Churchill, who for his part "would not have used" these words, "backed the President up" because journalists were present and "it had now been said."[5]

Roosevelt had no intention of snubbing his British allies—much less did he mean to provide the German propaganda machine with an argument for verbal counterattack. (Indeed, "Total War" was not a response to "Unconditional Surrender," but had already been envisioned by Goebbels on 17 January 1943.[6]) Roosevelt—at any rate, this is what he said later on—was simply quoting from school history books that every American (but alas, not every European) knew. According to these textbooks, General Ulysses Simpson Grant, who was destined to become equally famous in word and in deed, had fought on the side of the Union. In 1862, Grant had besieged the Confederate general Bricker at Fort Donelson. When the surrounded forces found themselves obliged to acknowledge defeat, Grant discovered a little pun: the "Unconditional Surrender" he demanded had the same initials as his own, slightly megalomaniacal name (Ulysses Simpson) and the country he represented (the United States).[7]

Literally, then, "Unconditional Surrender"—this formulation without a basis in international law[8]—signifies capitulation to America as such. What the words say in a more technical sense is explained by America's great postwar novel. In *Gravity's Rainbow*, a bomb raid of the 8th U.S. Air Fleet against a German chemical factory inspires the black protagonist—a former colonel in the Waffen-SS from Major General Dr. Walter Robert Dornberger's military research facility at Peenemünde—to decipher matters as follows. While paranoid, his view is only too plausible in historical terms:

> if what the IG built on this site were not at *all* the final shape of it, but only an arrangement of fetishes, come-ons to call down special tools in the form of 8th AF bombers *yes* the "Allied" planes all would have been, ultimately, IG-built, by way of Director Krupp, through his English interlocks—the bombing was the exact industrial process of conversion, each release of energy placed exactly in space and time, each shock-wave plotted in advance to bring *precisely tonight's wreck* into being. [...] If it is in working order, what is it meant to do? The engineers who built it as a refinery never knew there were any further steps to be taken. Their design was "finalized," and they could forget it.
>
> It means this War was never political at all, the politics was all theater, all just to keep the people distracted [...] secretly, it was being dictated instead by the needs of technology [...] by a conspiracy between human beings and techniques, by something that needed the energy-burst of war, crying, "Money be damned, the very life

of [insert name of Nation] is at stake," but meaning, most likely, *dawn is nearly here, I need my night's blood, my funding, funding, ahh more, more.* [...] The real crises were crises of allocation and priority, not among firms—it was only staged to look that way—but among the different Technologies, Plastics, Electronics, Aircraft, and their needs which are understood only by the ruling elite.[9]

Germany's ruling elite had started the war on the assumption—which was largely correct, thanks to the "utmost acceleration of our armament"[10] that Hitler had pursued—that the country enjoyed a technological head start of two years. The principle of acceleration (which would also experience technical implementation in the V-2) had made it possible to make this leap forward without sizable "deep armament" [*Tiefenrüstung*]—that is, it prevented Total War from merging with social revolution (as had occurred during World War I).[11]

Given the belated (albeit massive) beginnings of deep armament in Great Britain and the United States, German victory would have to occur before 1942. That is why the ten tank divisions of the Wehrmacht—which were inferior in number but deployed with VHF transmitters[12] that were unique at the time and made the *Blitzkrieg* possible in the first place—had to barrel over Poland and France. That is also why Hitler, as a matter of due (if mistaken) course, having already won half the war in September 1940, gave orders to stop all long-term military-technical research projects. Officially, at least, liquid rockets, 50-cm radio equipment, and long-range bombers ceased to be priorities. Instead, all efforts focused on developing multipurpose artillery, "Würzburg giants" [*Würzburgriesen*], and tactical bombers. To be sure, the commander in chief of the Wehrmacht himself—whose "astonishingly far-reaching view of technology and tactics" had already made him "the creator of modern military armament"[13]—continued to propose, almost weekly, innovations for these tactical (or all-too-tactical) weapons systems. For all that, however, it is a matter of record that Hitler had no clue about higher mathematics.[14]

Only when Operation Barbarossa stumbled and before the gates of Moscow the *Blitzkrieg* came to an end, period, did the regime change its economy of weapons and butter. According to Lieutenant General Jodl, head of Wehrmacht Supreme Command, Hitler had recognized the defeat of his strategy "sooner than anyone in the world."[15] In view of the situation, the Führer appointed Albert Speer, his former architect, to direct the newly founded Reich Ministry for Armament and Munitions [*Reichsministerium für Bewaffnung und Munition*], which in September 1943 changed its name (just as aptly as ambiguously) into the "Ministry for Armaments and War Production" [*Ministerium für Rüstung und Kriegsproduktion*].[16] And indeed, Speer nearly managed to "produce war" even though the enemy's domination of airspace increased—the result of the centimeter-wave radar used by the Western Allies (a technology that Hitler had prohibited further research on).[17]

Producing war simply required that warfare no longer be left to warriors. After the bitter experiences of 1914—when Schlieffen's plan for mobilization had mastered only massive armies but not the massive production of gunpowder—the Reichswehr and Wehrmacht had tried to achieve technological and economic control [*Kompetenz*]. By founding bodies such as the Army Weapons Office [*Heereswaffenamt*] and Wartime Economy Office [*Wehrwirtschaftsamt*], they sought to draw the consequences of global warfare—that is, recognize the fact that soldiers now depended on weapons systems, and weapons systems on the availability of raw materials.[18]

With the exception of modernizing office generals, however, warriors are conservative by nature—if only because they teach their men the art of death instead of learning about switching-technology from machines. And so, Speer's wartime "economic miracle" required that the same technical-economic offices of the Wehrmacht lose power.[19] In 1940, the Army Weapons Office had been obliged to surrender control over arms production to Speer's predecessor in order to preserve—if only on paper—control over weapons development. Having lost face, its head committed suicide.[20]

It was only when the military lost power that war could become the playground of engineers, who provided numerous unrequested innovations.[21] Defying explicit orders from Hitler, Wilhelm Emil Messerschmitt developed the Me 262, the first jet-powered fighter aircraft ready for mass production.[22] Alexander Lippisch personally tested his ramjet-powered design, whose commercial use may yet begin in the 1990s.[23] Hellmuth Walther designed submarines that finally deserved the name, and on 3 October 1942, Wernher von Braun launched the first V-2 into the sky over Peenemünde. For these engineering newcomers—who from 1942 on were also permitted to belong to supervisory boards[24]—Speer's appointment meant exactly what Pynchon has described. The new order eliminated the one-sided military finalization of arms production, shifted the priorities from Hitler's materialism of natural resources toward high technology, and invited capital-rich companies to compete freely with each other.[25] As a result, the Germans triggered—out of sheer necessity—a second wave of innovation. According to an internal report of the U.S. Navy, they "were already building the weapons of tomorrow, today."[26] In spring 1945, some of these arms were even deployed.

For all that, the weapons of tomorrow—that is, jet fighters without gasoline supplies, rockets without launching bases, and night-vision devices without armor—could not change the fortunes of war. Instead, they changed the infrastructure of both Germany and the European fortress that—and not just according to the *New York Times*—is supposed to have anticipated the economic order of 1992. "Ironically," the historian of the Reich's forced laborers has written,

> Germany's labor experience was a factor in the preparation of European postwar integration. Hitler and his brutal Gauleiter Sauckel have the distinction, along with Jean Monnet and General George Marshall, of being the founders of the Common Market.[27]

Through the systematic exploitation of industrial capacities and human raw material in occupied lands, the military sector increased, between 1941 and 1944, from sixteen percent of the German economy to forty percent. New technical elites and—because Allied superiority in the air forced movement into the country or even beneath the earth—a reindustrialized hinterland laid the foundation for Professor Erhard's future Economic Miracle.[28]

Defeat, which had been in sight since Stalingrad, could not halt the technological breakthrough. On the contrary, the Ministry of Economic Affairs [*Reichswirtschaftsministerium*],[29] which Speer made partially inoperative from 1943 on, shifted over to long-term postwar planning. It encouraged, among others, an independent scholar [*Privatdozent*] in Erlangen and a journalist in Frankfurt to plan projects of reconstruction. The scholar was named Ludwig Erhard, and the journalist Erich Welter[30]; indeed, by founding the *Frankfurter Allgemeine Zeitung* after the war, the latter did in fact contribute to the economic reconstruction of the Federal Republic. No Western welfare state after 1945 would have been possible without a warfare state beforehand.[31] To this extent, a "zero hour" never existed.

Given his competitors' foresight, Speer, the erstwhile architect, could not just stand back and watch. Hardly had cities or factories—actually, cities more than factories—fallen in rubble when teams of his young architects (as if to prove Colonel Enzian's paranoia) raced from Berlin down the autobahns of the Reich to examine the high-modern architecture of ruins. On site, they noted the clearing operations [*Aufräumarbeiten*] performed by Allied bombing coalitions and the catastrophic consequences that narrow, medieval street planning had once again had for populations in flight. Consequently, any postwar reconstruction would have to work on the premise that city highways and "green zones" should explode the city centers of old.[32] The green zones would provide the side benefit of serving recreational purposes—and the main purpose of offering flight zones during bombing in the Third World War. And so it was. Speer's city planners concretized their principles—above all in Düsseldorf, Hamburg, and Hannover—into postwar architecture. As their historian put it, "total war" must "not only be understood as the end of the Third Reich, but also as the prehistory of German reconstruction."[33] Pynchon's Colonel Enzian would add that carpet bombings probably belong to the same prehistory.

The only problem was that the Reich—supposedly so totalitarian—actually consisted of a highly entropic balance between competing subsystems of power and bureaucracies, and the only countervailing instance was the so-called Führer-principle. For this reason, all plans for reconstruction—lying in filing cabinets in anticipation of the post-Hitler era—met with an absolute enemy on the German side.

Hitler, in his personally optimized concrete bunkers, lost all confidence in one power subsystem after another. The first victim, given Allied air superiority, was the Luftwaffe, and the second—after Stauffenberg's assassination attempt—was the Wehrmacht; its technologies of the future (e.g., the special-purpose Army Rocket Corps [*Raketen-Armeekorps z. b. V.*]) came under the command of the Waffen-SS from July 1944 on. Finally, in March 1945, when even the *Reichsführer SS* could no longer defend the Oder-Neisse Line—Berlin's last natural defensive barrier—the number of possible candidates for Final Victory had been cut. Hitler duly declared that the whole, treacherous population of Germany would rightly be defeated by "the stronger people to the East,"[34] and in imitation—or escalation—of Stalin's initial defensive tactics (which had, after all, made Western Russia into scorched earth), he issued the so-called Nero Decree:

> On 19 March 1945, the Führer issued the following order:
> Re: destruction measures in the territory of the Reich.
> The struggle for the existence of our people demands [*zwingt*], also within the territory of the Reich, to exploit all means that will weaken the fighting power of our enemy and hinder his further advance. [. . .] Therefore, it is an error to believe that undestroyed transportation-, communications-, industrial, and supply facilities, or those that have only been damaged [*gelähmt*] for the short term, could be made operational for our own purposes upon recapture of lost territory. Upon retreat, the

enemy will leave behind only scorched earth for us and abandon all concern for the population.

I therefore command:

1. To destroy all military transportation-, communications, industrial, and supply facilities within the Reich that the enemy might in any way use, immediately or in foreseeable time, to continue fighting. [...]

3. This command is to be conveyed to all troop leaders immediately. Directions to the contrary are invalid.[35]

If there were any doubts about the purpose of scorched earth, they vanished at the latest with the "enforcement directives" [*Durchführungsbestimmungen*] issued by General Albert Praun, who succeeded Fellgiebel as head of the Wehrmacht Signal Corps after the latter was executed in July 1944. It was impossible to speak of a Reich—that is, a media system—after the "edict of total destruction," which concerned not just the army but also "the wires and installations of the post office, the railroad system, the waterways, the police, and electric-power transmission lines," as well as "all stocks of spare parts, all cable and wire, even the switching diagrams, cable diagrams, and descriptions of equipment."[36] As in Borges's short story, the land and its maps were set to implode into a single ruin.[37]

Clearly, then, the scorched earth order was not directed against external enemies. It concerned, first, postwar reconstruction plans in the economic and defense ministries, and second, the Wehrmacht's strategy of ceasing combat in the West and redirecting as many units as possible from Soviet areas to areas under control of the Western Allies. When, on April 10, a "British document on the administration of the Reich after occupation, including corresponding maps" was "captured,"[38] *Anschluss* to the future Western Europe could certainly be counted on. And so Speer, who had unsuccessfully lobbied against the scorched earth policy, contacted Guderian, the chief of staff, and friendly industrialists, with the effect that the order declaring directives counter to the Nero Decree null and void was hardly observed. Reconstruction efforts (and the populace) were spared having to start again without any infrastructure.

At the same time, the stage had been set—with the rescue of circuit diagrams and technical manuals—for a grandiose transfer of technology, which would soon constitute the international postwar order. When Hitler decided, on April 22, to remain in Berlin and die, because *Reichsführer-SS* Himmler had betrayed him politically by participating in negotiations for surrender,[39] the system was free to fall apart into its subsystems, and these subsystems could now merge with their corresponding surroundings. At the latest after General Bradley and Marshal Konev met at Torgau, the erstwhile Reich became a fractal of American, Soviet, British, and—last but not least—French zones, where, for one final moment there survived islands of the Wehrmacht, the Navy, and even the Waffen-SS (as in the high-tech center at Nordhausen).[40] And because fractals, as mathematical patterns, tend toward self-similarity, the formation of zones was repeated on multiple geographical scales:

Führungsgruppe B of Wehrmacht Supreme Command had reached the far south on the last autobahn of the Reich that still was open, where the Supreme Command of the Luftwaffe under Göring—who when taking leave of Hitler had traded all his fantasy *Reichsmarschall* outfits for the simple "olive-drab of the American uniform"[41]—awaited General Patton's tanks. *Wehrmachtführungsstab* A and Naval High Command had departed for the extreme north, that is, for an area sure to be under British control, where Himmler dreamed of negotiating a partial surrender with Montgomery. Only Hitler and other prospective suicides like General Krebs—whom perfect knowledge of Moscow and the Russian tongue had qualified to be the last army chief of staff (after Guderian)—remained in the Berlin *Führerbunker*, that is, in the kettle of the Red Army. With that, all the subsystems and the vanishing center of the Reich had recognized, and exhausted, their options (except for the French one).

So that the former center might preserve the Soviet option, fractalization was repeated on a lower—that is, operational—level. Notwithstanding all the Wehrmacht's westward marches, General Wenck's newly formed 12th Army received orders to disengage the American enemy and assist Berlin from the southwest. At the same time, Busse's 9th Army (after breaking out from encirclement) was to attack from the southeast while the so-called Army-Group Steiner moved in from the north. SS General Steiner—who had already proved unable to prevent the "secret" nighttime departure of the V-2 Rocket Corps from the Oder Front "to the south"[42]—found just enough time, in a telephone exchange with Krebs, to call the attack orders "inoperable and senseless" before the last line to the bunker went dead. It was precisely this feedback loop with a "phantom" of battle, "which only existed in the fantasy of the Führer's headquarters,"[43] that triggered the final act: the deaths [*Todesarten*] of Hitler, Eva Braun, and Blondi—of the smallest fractal, that is, or (as Hitler had been saying since autumn 1943), the "only loyal ones."[44]

That was reason enough for the three commanding generals—Wenck, Steiner, and Busse—to continue their armies' withdrawal to the west, which

Hitler's orders had interrupted. It was also reason enough for Goebbels, just one day after Hitler's suicide—on May 1—to send General Krebs with a white flag into Chuikov's command post in Berlin, where the last army chief of staff explained, in fluent Russian, that only two states in the world had honored their workers by setting aside a day for them: Germany and the Soviet Union.[45] Finally, that was reason enough for SS General Kammler, the commander of the Special-Purpose Army Corps, to pile the high-tech blueprints from the restricted area of Mittelbau into a car and (according to "a report that could not be confirmed in detail") bring them to interested researchers in the southeast.[46]

When the center disappeared, the technology transfer could begin, in keeping with the options of the individual subcenters and their new surroundings. The documents of "Unconditional Surrender" were first signed secretly in Reims before the Americans and British, and then in Karlshorst, with all four Allies present, in the name of the Wehrmacht Supreme Command. They included the following specifications:

> 2. No ship, vessel, or aircraft is to be scuttled; nor may ships' hulls, machine installations or devices, machines of any kind, weapons, apparatuses, or any technical means of continuing the war in general be damaged. [...]
>
> 6. This declaration is made in English, Russian, and German. Only the English and Russian version is binding.[47]

And so, the countermeasures undertaken by Speer and Guderian against the Nero Decree—which would have turned Stalin's scorched farmland into fried technology—perfectly matched the Allied prohibition on destroying any kind of military technology. "Unconditional Surrender" meant technology transfer.

In the Soviet Occupation Zone, the selfsame antiaircraft searchlights whose blinding radiance had started Marshal Zhukov's final campaign[48] now enabled the nighttime dismantling of weapons plants. As Pynchon observes, "The roads heading east [were] jammed day and night with Russian lorries, full of materiel. All kinds of loot. But no clear pattern to it yet, beyond strip-it-and-pack-it-home."[49] For all that, a few factories, concentration camps, engineers, and Waffen-SS training instructors remained at work, ultimately providing uranium experts for the Red Army, MiG-15s for the Korean War,[50] and cadres for the future National People's Army of the German Democratic Republic. Legend holds that *Sputnik* won the race against the U.S. *Explorer* because Peenemünde's assistant researchers went to Kazakhstan and the only full professor to White Sands . . .

The technology transfer to Great Britain occurred along lines that were less strategic, and more like colonial trade. The 5-decimeter waves of dismantled Würzburg Giants, brought to Jodrell Bank, led to the invention of radio astronomy. Wind-tunnel measurements and engineers from the Aeronautical Research Institute in Volkerode led to the construction of the Concorde,[51] and tape recorders from navy supplies produced the sound at Abbey Road Studios—and thus of the Beatles.[52]

Walther's submarine designs for Vickers-Armstrongs,[53] on the other hand, simply reflected a fallen empire's turn to defensive strategies. In general, Great Britain seems to have been in the uncommon position of losing more by technology transfer than it won, at least after Truman and Churchill agreed in Potsdam not only to keep intelligence about German technologies from Stalin, but also, and especially, to do the same with their own innovations.[54] As is well known, British prototypes of digital computers—because they were able to decipher, in real time, the Wehrmacht's entire system of command, from the operational level of Enigma transmissions up to the strategic level of Siemens cipher machines [*Geheimschreiber*]—proved decisive for combat in the Atlantic, in Africa, and likely in Europe as well.[55] When Alan Turing, the inventor of the computer and a cryptographer for British intelligence, came to Ebermannstadt on a final assignment in July 1945, there was nothing left to dismantle at the German center of cryptanalysis. On the contrary, he could only pity the technological gap of his enemies (or colleagues), who had failed to have themselves replaced by machines.[56]

The German-French technology transfer went much more smoothly. Likely because neither one of the two languages was binding in surrender documents, mutual hermeneutics continued without disturbance. Just as German press officers had, during the Occupation, authorized printing paper for Sartre's *L'être et le néant* because it exemplified heroic nihilism, French officers in turn invited a Freiburg philosopher—Martin Heidegger—to "think" technology as such. It seems that empirical technologies promised greater "success," which in fact was simply the "inheritance of [four years] of cooperation in war": the Mirage and the Airbus were designed by engineers commandeered in wartime,[57] when Fortress Europe was united economically.

But because "Unconditional Surrender" means, in concrete terms, capitulation to America, transatlantic technology transfer surpassed all others. Internal statistics of Air Force Intelligence—according to which 17 percent of all German military scientists worked in the Soviet Union, 12 percent in France, 11 percent in England, and only 6 percent in the United States—were "obviously false"; the figures were provided to other government agencies only to accelerate the transfer operations "Overcast" and "Paperclip."[58] In 1945, the Intelligence Committee at the Joint Chiefs of Staff summarized its personnel needs as follows:

> Unless the migration of important German scientists and technicians into the Soviet zone is immediately stopped, we believe that the Soviet Union within a relatively short time may equal the United States' developments in the fields of atomic research and guided missiles and may be ahead of U.S. development in other fields of great military importance, including infrared, television, and jet propulsion.[59]

But preparations had already been made in Germany. Already at the end of August 1944, it was known in the Reich Central Security Office [*Reichssicherheitshauptamt*]—from "reliable" reports provided by agents abroad—that plans existed, "in the event of a German collapse," "to transfer at least 20,000 German engineers to the United States."[60] This and nothing else, then, had prompted Major General Kammler to order his rocket technicians to the deepest south, that is, into American territory.[61] Otherwise, however—and up to the day of its so-called prohibition (which likely represented the beginning of planning for the postwar Federal Army [*Bundeswehr*])—Wehrmacht Supreme Command and general staffs continued the strategy of transferring, in addition to one and a half million soldiers, as many technicians as possible from east to west,[62] in order to meet American personnel demands.

Since then, the results of Operations Overlord and Paperclip have made history: Wernher von Braun's rocket scientists and Professor Strughold's space-travel physicians—whose test subjects only concentration camps could have provided—completed a nuclear deterrent whose foundations, as everyone knows, were laid by emigrants from Hitler's Europe.[63] John von Neumann, the mathematician of all atom bombs and of the computer architecture named after him, saw to it, at the Pentagon, that the "marriage of two monsters"[64]— a payload from Los Alamos and a carrier rocket from Peenemünde—became the strategic standard.[65] *Pax Americana* rests on what Eisenhower called the "military-industrial complex." Thanks to higher mathematics, it has moved beyond personnel-heavy world wars like the First and material-heavy ones like the Second.

But for the same reason, civilian and entertainment electronics (with the sole, and notable, exception of transistors) remained stuck at 1945 standards. At the same time, the Second World War—which replaced tubes, coils, and condensers wired in improvised fashion with circuit boards—has formed the platform of the world we perceive [*Merkwelt*]. VHF radio for tanks, which the Wehrmacht introduced in 1934 and Bell Labs provided for the U.S. Army in 1940,[66] became the secondary medium for entire populations, which was then complemented by a storage medium when magnetophones once employed for defense were repurposed. In the postwar, the primary medium has been the very same television whose development the BBC and Reichspost put on ice when conflict erupted—simply because exactly the same imaging electronics in radar equipment received the highest military priority. When Walter Bruch—to whose PAL system half the world owes its color television—was not training his TV camera tube on a V-2 as it rose from Test Stand VII at Peenemünde,[67] he spent his time testing devices that are today employed in every cruise missile. Bruch equipped flying bombs with television cameras and self-guidance mechanisms. Launching "pleasure boats" that had seen better days to sail across the Müggelsee—"without passengers of course"—he then sought to optimize the feedback loop between TV and servomotors until the bombs found the pleasure cruisers all on their own.[68] At the behest of the National Defense Research Council, Norbert Wiener and Claude Shannon, the two mathematicians of the coming theory of information, pursued research following the same principle. For the Battle of Britain, they developed automatic antiaircraft control systems[69]—without which, Wiener said, his later cybernetics would never even have been conceivable.[70]

And so, the self-guided weaponry of the Second World War did away with the two fundamental concepts of modernity—causality and subjectivity—and inaugurated the present as the age of technical systems. That said, only Shannon and Turing—that is, neither Wiener nor the engineers of the Wehrmacht (except for Zuse)—performed these calculations digitally or made the decisive step from radio waves and differential equations to the pulse technology of radar or the algebra of computers.[71] *Pax Americana* is based on a solid technological foundation.

But whether they are digital or analog, technical systems are always self-guided. "The build-up of negative impulses, each reinforcing the other," wrote Defense Minister Speer in the final paragraph of his memoirs, "can inexorably shake to pieces the complicated apparatus of the modern world."[72] In his final words at court in Nuremberg, he told the victors, for the record:

> Hitler's dictatorship was the first dictatorship of an industrial state in this age of modern technology. [...] Telephone, teletype, and radio made it possible to transmit the commands of the highest levels directly to the lowest organs. [...] To the

outsider this state apparatus may look like the seemingly wild tangle of cables in a telephone exchange; but like such an exchange it could be directed by a single will. Dictatorships of the past needed assistants of high quality in the lower ranks of the leadership also—men who could think and act independently. The authoritarian system in the age of technology can do without such men. The means of communication alone enable it to mechanize the work of the lower leadership.[73]

"Men who could think and act independently." This, ever since Kant or Gneisenau—and before the development of self-guiding weapons—has defined subjects. Consequently, the *singulare tantum* called "man" also disappeared along with the fifty-five million human beings who died in the world war. As Pynchon maliciously puts it, "the mass nature of wartime death ... serves as a spectacle, as diversion from the real movements of the War," and conceals what is really happening: battles of contingencies, priorities, and technologies.[74] This is also why Germany's fractionalization through occupation zones, technology transfer, reconstruction, and five-year plans forms part of the logic of technical systems. In 1943, after a film of the V-2 had finally convinced Hitler—"the greatest cineaste of all time" (according to Syberberg)—that self-guided space weapons were viable, the Führer is supposed to have declared that states "are now, and for all the future, too small."[75]

And so nothing and no one—not even the *Führerprinzip*—was able to stop the technology transfer. After all, technology transfer means that communications technologies fulfill their definition and become transmissible communications themselves. If empires are media, and media are postal systems,[76] then their destiny [*Schicksal*] must involve "dispatches" [*Verschickung*]. When Zhukov's artillery shot down the last tethered balloon connecting the last radio link between the *Führerbunker*, under the Reich Chancellery, and Army Group Steiner,[77] nothing ended. It was all just getting started.

Besides America, Japan—the technology empire of tomorrow—was waiting for Unconditional Surrender and military technicians. Although blueprints for the Me 262 and the Heinkel 117 had already reached the Far East (when one of two submarines succeeded in breaking the American blockade), an official "Japanese plan to introduce German experts" is first recorded for August 1944.[78] On 30 April 1945, Supreme Command drew up the "outlines of measures to be taken in the event of Germany's capitulation"; "the interests of German citizens in East Asia" were to be "generously preserved," even while forces resolutely pursued the "noble aims of the Great East Asian War." In due consequence, Japan declared "all agreements with the German Reich" null and void on May 9.[79]

Unconditional Surrender

The unthinkable occurred only after Hiroshima and Nagasaki. For the first time in their history, the Japanese heard—as if Sony's media empire had already begun—the voice of a *Tenno* on record and over the radio. In his classical, and therefore nearly incomprehensible, Japanese, Emperor Hirohito declared the end of a world war. He made no reference to "Unconditional Surrender."

15 Protected Mode

AirLand Battle in 1991 demonstrated yet again that of all the postmodern strategies of illusion [*Schein*] none proves so effective as simulating that software really exists—until, that is, the contrary was demonstrated in combat, when computers made it perfectly clear that they were hardware for destroying enemy hardware. Meanwhile, advertising brochures and press sessions disseminated the fairy tale that software was being developed which would become increasingly gentle, user-friendly, spiritual, and intelligent: one day in the not-so-distant future it would effectively amount to German Idealism—that is, software would become human.

And so, software—a billion-dollar enterprise based on one of the cheapest elements on Earth—used everything at its disposal to prevent said "humans" from having access to hardware. With Word 5.0 on a generic AT 386, running "under" (as it is so aptly put) the operating system Microsoft DOS 3.3, one can write entire essays on these same three entities without even suspecting that one has been duped by strategically produced illusions. After all—and the "under" already says as much—one is writing as a subject or underling of the Microsoft Corporation.

Such a worm's-eye view did not always prevail. In the good old days—when microprocessor pins were still big enough to operate on with a simple soldering iron—even literary critics could do whatever they wanted with Intel's 8086 Processor. Because of the lack of difference between RAM and ROM, through the misuse of both stack registers as universal registers, given the absence of all interrupt vectors and the possibility of repurposing the wait input, and so on, even standard chips—which at that time still took 133 clock cycles just to multiply integers—still had to be brought up to the processing speed of primitive signal processors. Because the von Neumann design does not differentiate between commands and data, the silicon chip could compete in stupidity with its maker or user. This same user, in order to make a program run, had to forget everything still spooking about in his head from school days—including elegant proofs and closed solutions. Indeed, one had to forget all ten fingers and translate all the decimal numbers factoring into the program into monotonous columns of binary digits. The task as such was forgotten when one pored over data sheets in order to translate the commands—IN, OUT, and so on (which were already formulated in English, of course)—into the operation code. Alan Mathison Turing, once the Universal Discrete Machine he designed in 1936 was finally available, thanks to a world war, is the only party who has ever been said to have preferred such activity to mnemonic aids and higher programming languages.[1]

Anyway, once such exorcism of the Spirit [*Geist*] and Language had been completed, the machine's stupidity equaled its user's: it worked. To be sure, so-called machine language ran a million times faster than the pencil with which the user had pieced together the zeros and ones from Intel's data sheets. To be sure, the "flip-flops" that cover silicon chips with patterns repeated ad infinitum took up a million times less space here than on paper. But with that, the differences between the computer and the "paper machine"—as Turing had rechristened humanity[2]—had already been exhaustively tallied.

Those good old days are gone forever. Since then, with keywords such "user interface," "user-friendliness," and even "data protection," Industry has condemned human beings to remain human beings. The evolutionary potential of "man" to mutate into a paper machine has been blocked with great cunning. In the first place, Microsoft's data sheets have switched to presenting assembler abbreviations as the outer limit of what users might understand or want of machines [*maximale Zumutbarkeit oder Maschinennäherung*]—and that means, no operating code is made public at all anymore.[3] Second, the relevant professional journals "promise us"—and this is a quote—"that even under the best circumstances, one would quickly go crazy from programming in machine language."[4] Third and last, these same publications already consider it inexcusable "to write a procedure for calculating *sine* in assembly language, of all things."[5]

At the risk of having already gone crazy long ago, the only thing one can conclude from all this is that software has gained in user-friendliness to the same extent that it has approached the cryptological ideal of one-way functions.[6] The higher and more effortless programming languages become, the more unbridgeable the gap grows between them and the hardware that still does all the work. In all likelihood, this trend cannot be satisfactorily explained either in terms of technological progress or through the formalizations of type theory; instead, and like all matters of cryptology, it has strategic functions. On the one hand, it remains possible, in principle, to write user software (i.e., cryptograms) with knowledge of codes or algorithms. But on the other hand—a matter that is hidden out of "user-friendly" considerations—it is well nigh impossible to determine, on the basis of the end product, what its conditions of production are (or for that matter to change these conditions). Users fall victim to a mathematical ruse that is said to have driven Ralph Hartley, the erstwhile head of Bell Labs, to despair in old age: the fact that operands can no longer be observed in so many operations.[7] The sum conceals the addends, the product the factors, and so on.

For software, of course, this mathematical ruse is made to order. In an age that has long since said good-bye to the phantoms of creators or authors, yet at the same time holds fast, by copyright, to these same historical ghosts for financial reasons, the ruse has become a gold mine. At any rate, the subjects of the Microsoft Corporation did not fall from the sky. They had to be produced in the first place, like all of their media-historical predecessors: the readers of books, film audiences, and TV viewers. The only problem now is how to conceal from these subjects the fact that they are subjugated, so the global victory march may proceed.

As far as the politics of knowledge is concerned, the answer follows a proven recipe for success in modern democracies. In the technical arena, the hardware of microprocessors themselves is changed. Perhaps only the engineers at Siemens can tell it like it is. In the *80186-Handbuch*, Klaus-Dieter Thiel writes:

> Today, modern 16-bit microprocessors increasingly take on tasks that belong to the typical application range of classic mini-computers.
>
> And so, in multi-user systems, it is necessary for the programs and data of individual users to be separated, just as the operating system must be protected against users' programs. In order to give every individual user the possibility to run his software independent of numerous other users, and in order to give him the impression that the computer is there for him alone, it is essential to divide the CPU among individual programs through multi-tasking, which, however, can only remain hidden from the user if the CPU is extremely powerful.[8]

In keeping with the Siemens approach, which also has currency at IBM-Germany, Intel did not raise the operating frequencies of the 80286 and the 80386 to levels between 12 and 33 megahertz in order to meet the standards of professional users—or even the Pentagon's specifications for electronic warfare[9]—but rather in order to entangle civilian users in an impenetrable simulation.

Protected Mode

Like the tortoise in the fairy tale, multitasking is supposed to feign, for users, that only a single tortoise (i.e., process) is running, and above all that this "race" [*Lauf*] involves only one hare (or user). That is the same tune that novels and poems, ever since the Age of Goethe, have played to readers—especially female ones: they promise that they exist for them alone. It is also the same song by which modern politics subjugates vast populations in a capacity wholly opposite to the way they are actually treated: as individuals.

In contrast to traditional simulations (which all met their absolute limit in the power [or powerlessness] of everyday language), electronic simulation—according to which every microprocessor is there for a single user alone—also has hardware at its disposal. From the 80286 on, Intel's processors have featured a Protected Mode that (in the words of the Siemens engineer above) guards the operating system from users, which is what makes it possible to "illude" them in the first place. What began as a simple switching possibility between *supervisor stack* and *user stack* in Motorola's 68000[10] (a rival system of which no mention is made, naturally) achieved systemwide implementation in the separation between Real Mode and Protected Mode. Different command sets, different address possibilities, different register sets, and even different command execution times henceforth split the wheat from the chaff, the system design from the users. And so, in the selfsame silicon in which the prophets of a microprocessed democracy-to-come have placed all their hopes, the elementary dichotomy of modern media technologies returns. Once upon a time, civilian radio was permitted to exist in Germany when the postal system could credibly promise the Reichswehr that the consumer radios of 1923, already gutted and stripped of any potential for transmission, would never be able to disrupt military-industrial radio exchanges—because an automated encryption device (which Turing's proto-computer put out of commission in the Second World War) had just been invented.[11]

The innovation of Intel's Protected Mode consists simply in having transferred such logic from the military-industrial sphere into the realm of information systems [*Informatik*] themselves. The distinction between the two operational statuses is not just quantitative, as holds, for example, for the varying ranges of operating temperatures in commercial, industrial, and military silicon chips (in this telling order of rank). Rather, the CPU itself works with priorities, prohibitions, privileges, and handicaps, of which it constantly keeps a record—in Protected Mode, of course. That such controls, which themselves take time after all, do not exactly promote the general goal of increasing data output, goes without saying. In Protected Mode, the same *interrupt* requires up to eight times as many cycles as Real Mode would require. Evidently, high technology can only be passed along to end users and "nontrustworthy" programs (as Intel calls them) if and when signal processing—the military-industrial dimension of computers[12]—has been braked by bureaucratic data processing. There are no longer written tablets of prohibitions assuring an imbalance of power; rather, the binary system as such encodes what counts as a command and what counts as data, that is, what is permitted to the system and what, conversely, is prohibited to user programs. Von Neumann's classic computer architecture—which made no distinction at all between commands and data and, indeed, had no need to do so at a time when all the computers in existence were government secrets—has vanished beneath four sequentially numbered levels of privilege. With due irony, then, the most incorruptible of Germany's computing magazines has observed: "Despite all the abundant talk about privileges, higher privileged code segments, privilege violations, and so on, you are not reading the political manifesto of a former functionary of the [East German] Socialist Unity Party, but an explanation of the security-concept of the 80386!"[13]

Political manifestos, as the name indicates, played out where everyday language governed. And so, the privileges to which they lay claim are now—and have been for some time—null and void [*gegenstandslos*]. Intel's so-called flagship—a CoCom List transferred into the innermost binary number system—has probably contributed more to the liquidation of politically based privileges than the constant stream of television across Eastern European borders. Carl Schmitt once wrote a short text, *Dialogue on Power and Approaching the Ruler* [*Gespräch über die Macht und den Zugang zum Machthaber*], that culminates in the thesis that power amounts to its conditions of access: the antechamber, the office, or more recently, the front office consisting of a typewriter, a telephone, and a (female) secretary.[14] With and by means of such instances of power, dialogue could in fact still occur. Technologically implemented levels of privilege, however, derive their power from the efficacy of silence [*aus stummer Wirksamkeit*]. In order to finally have access to memory reserves beyond DOS—as if by some kind of posthistorical metaphysics—the 80386 user installs one of the "user-friendly" utilities on offer and loads the debugger with a do-it-yourself program [*Eigenbauprogramm*] that still ran yesterday without any problems; promptly, he discovers that the new installation not only administrates storage space, as promised, but also, and without any warning, has locked all privileged commands.[15] As Mick Jagger put it long ago, the user gets, instead of what he

Protected Mode

wants, only what he needs (and as determined by industry standards, at that).

The consequence for the analysis of power systems—the vast task bequeathed to us by Foucault—is twofold. For one, one should no longer seek to understand power, as conventionally happens, as a function of so-called society; instead, and conversely, one should seek to reconstruct sociology from chip design [*Chiparchitekturen*] up. Initially, at least, the task is to analyze the privilege levels of a microprocessor as the truth of the same bureaucracies that commissioned its design and brought it to be deployed en masse. It is no accident that Motorola separated Supervisor Level and User Level, and Intel split Protected Mode from Real Mode, at the exact time when the United Sates of America went about constructing an airtight two-class system. (The Embedded Controller is used in every hotel room lock of the better sort in New York.) Not for nothing are the input and output commands of the 80386 protected by the highest privilege level: in an empire where the populace sees the rest of the world only through the blur [*Mattscheibe*] of the television news, even thinking about foreign policy is a governmental privilege. That is probably also the reason why the latest antics of systems theory simply deny, at the highest level of abstraction, the fact that information systems possess [*verfügen über*] input and output. All in all, this would provide good reason for computer scientists from other countries—somewhere between Japan and Europe, that is—to oppose other, possible bureaucracies to the American one hidden [*versenkt*] in silicon. Whether they would be better or not is debatable; at any rate, they would still be bureaucracies. All the same, competition between different systems and power structures would, simply by occurring, give the subjects of MS-DOS a breath of fresh air [*aufatmen lassen*].

So long as IBM compatibility continues its victory march, strategy is in order more than sociology. In moving from reception offices and everyday languages into micrometer range, the procedures and targets [*Verfahren und Angriffsflächen*] of power have also changed. The brusque "no" of denying information/access [*Auskunftsverweigerung*] is not an option for binary code, simply because the entire type hierarchy of self-similar program levels—from the highest programming language down to elementary machine code[16]—is, in the material dimension, flat. To borrow the words of Lacan, there can be no "Other of the Other" in silicon itself,[17] and therefore no protection of protection. Even the hidden segment descriptors that keep a record of access rights to all programs in a system must be accessible in order to operate. Even the fact that the CPU sets these registers to zero[18] when privileges are violated—all possible and explicit commands notwithstanding—leaves legible traces. At the level of the machine, then, protection mechanisms wind up in the awkward situation of having no hiding place that is absolutely secure. To the same extent that microprocessors remain usable for users—that is, to the same extent that they are supposed to be able to communicate with them—Intel's Protected Mode experiences a classical dilemma of power.

According to the *Programmer's Reference Manual*, even tasks performed by the operating system do not enjoy the privilege of freely opening tasks at a lower privilege level. Simply because the exchange over the stack runs symmetrically or through "direct democracy" [*basisdemokratisch*]—that is, because the program that has been opened "gets busy" [*poppt*]—the task with a lesser privilege level might be tempted not to give back control when it is completed, but rather to infiltrate the higher operation level by way of a simulated return occurring through the program [*programmtechnisch*]. Accordingly, Intel's engineers have deemed it safer to take the fundamental Boolean concept of the "gate" and replace it with bureaucratic access control.

What such prohibitions strikingly demonstrate, however, is simply the impossibility of perfect access controls. In the good old days of microprocessors, when the difference between the system and applications was literally burned into the silicon and resided there—the system in ROM, applications in RAM—nothing could disturb it. But ever since the difference has been rendered programmable, it has also stood open to all manner of circumventions.

Approximately 170 times—that is, at every single 80386 command—Intel's *Programmer's Reference Manual* repeats the warning [*Drohung*] that Interrupt 13 is triggered in Real Mode as soon as any one of the command operands comes to lie outside the effective 20-bit address field. In other words—but still those of the company itself—the 80386 runs in Real Mode only as a faster AT.[19] In the event of noncompliance, a draconian rule applies: "all violations of privilege that do not trigger a more specific exception" occasion a verbal monstrosity called "General Protection Exception."[20] But neither the 170 warnings repeated in the manual nor their innumerable rewritings in other computer books on the market—which seems to offer mostly partial reworkings of manuals done robotically and published under pseudonyms—make this threat any truer. That is, a single subordinate clause in the manual gives away the fact that the address boundaries in Real Mode are nothing more and nothing less than presets programmed into the system start-up. Needless to say, these same words disappear in all the translations, summaries, popularizations, and user handbooks—simply so that the subjects of Microsoft may be kept in the dark about its logical inversion: the fact, namely, that presets can be easily changed.[21] Instead of the default values that the CPU automatically loads into the hidden parts of its segment registers in switching back to Real Mode, programs can also set entirely different values. Every 386-AT enters each of the four possible operating modes with one hundred lines of code: Protected Modes with 32- or 16-bit segment width, and Real Modes with the same segment width. Real Mode with 32-bit segments would produce the most compact and, accordingly, the fastest code by far, yet no mention is made at all in data sheets and manuals of this even being a possibility[22]—to say nothing of it occurring in operating systems of the 80386 as they actually exist.

Therefore, one hundred lines of assembler—and nothing else—solve the problem of postmodern metaphysics. At the risk of going crazy, they lead, under MS-DOS, beyond MS-DOS. In a dramatic paradox, it is precisely the most backward of all operating systems that enables a way out. Intel's built-in barriers—which would engage immediately under more complex operating systems, identify these same hundred program lines as illegal commands, and deny them accordingly—stand powerless before stupidity.

And so, a machine can do both less and more than its data sheets admit. The 80386 has at least two "undocumented commands" that the data sheet purposefully conceals.[23] And in 32-bit Real Mode, it has at least one operating mode that it passes over to no end at all. This chaos does not prevail at the most up-to-date levels of computer science, where it is said debates concern computability and predictability in Finite State Machines in general. Rather, it occurs at the simple, empirical level tailored to engineers. Because, to quote Christian Morgenstern freely, "what may not be cannot be," mere presets are sold to users as if they were absolutes. It was not much different in the early 1920s when the *Reichspost* saw to it that only detection devices—and no tube equipment—were sold to consumers; otherwise, listeners would have been able to broadcast, too, and interfere with military-industrial radio exchanges.

It seems, in other words, that computer science [*Informatik*] is facing internal information blackouts [*Informationssperren*]. In the space of the codes that it must employ in actual fact—even if theory could (and should) devise entirely different models—working against the wishes and without the knowledge of code developers is just as possible as it is rare. Clearly, long after the end of the print monopoly and authorship, the phantom of "man" [*des Menschen*] has seen to it that opinions and even copyrights [*Schutzbehauptungen*] continue to be copied, instead of codes being cracked. A work project [*Arbeitsprogramm*] would need to occupy precisely this site—one for programmers, first of all, but also, and in principle, for machines too. Just as it is possible, and meanwhile, viable, to set programs generated by chance against each other according to purely Darwinian rules, the empirical switching behavior [*Schaltverhalten*] of machines should, on the one hand, be decoded and, on the other, compared with the data sheets.

At least to a literary scholar, it seems that this military-strategic division [*kriegslistige Sparte*] of computer science, as it were, holds a grand future in store. On a strictly technical field of operations, one would proceed with methods similar to those that Foucault proposed for discourses and texts. Instead of inquiring about the meaning of a chain of signs, as occurs in interpretation, or about the rules behind it, as occurs in grammar, discourse analysis simply and straightforwardly concerns chains of signs insofar as they exist and do not *not* exist. It is an idle question whether meanings are not simply a pedagogical-philosophical fiction, or whether grammatical rules are comprehensive or fully comprehensible. That the two words—"grammar" and "rule"—have occurred in a discursive context is and remains a matter of fact.

Over thirty years ago, Johannes Lohmann, the great linguist and Indo-Germanist, already suggested that one seek the historical grounds enabling programming languages in the fact that English, and only English, has verbs like *read* and *write*—that is, verbs that, unlike the Latin *amo amas amat*, and so on, have discarded conjugation forms. Context-free verbal units [*Wortblöcke*], which according to Lohmann go back to the historically unique confusion of Norman and Saxon in old England, can do nothing to prevent being translated into context-free mnemonics and ultimately into computer operating codes. As everyone knows, the endless litany of "read," "write," "move," and "load" is called *assembly language*.

Such discourse analysis—with elements involving not just words but also codes—would, of course, level the sacrosanct distinction between everyday languages and formal ones. In view of the lovely orthogonality that, for example, Motorola's processor series has displayed from the 68000 on, the undertaking would amount to heresy. All the same, the history of Protected Mode—as a half-compatible, half-incompatible continuation of the standards of the good old days—might teach us that codes yield the same opacity that everyday languages do. It is well known that the 8086 featured more than a few commands that were synonymous with other commands, which surpassed them only in the speed at which they were executed. It made a significant temporal difference whether a universal register or the accumulator wrote its contents into the memory. But ever since the new generation produced by Intel "optimized away" precisely this advantage of speed while retaining synonymous commands for reasons of compatibility, code has achieved the level of redundancy that everyday language has always displayed—which Frege illustrated with the pretty example of "evening star" and "morning star."

Such redundancy can only increase over time if machine codes are to remain compatible from generation to generation. In contrast to everyday languages—and especially to German, which places no limit on the length of words or the length of word combinations—all elements in an instruction set possess a finite length and, accordingly, a countable quantity. As a result, there would no longer be any space for further commands on the 80386 (for example), if it did not permit excess code length [*Überlängen*]. And with that—no matter how economical or orthogonal the initial design may have been—codes begin to proliferate and approach the opacity of everyday languages that, for millennia, has subjected people to these same languages. The lovely phrase "source code" names the literal truth.

Of course, discourse analysis can neither tame nor debug such proliferation. Quite possibly, it would be more efficient simply to count on it. Turing's old idea of having machines themselves come up with [*entwürfeln*] their codes may in fact have already secretly become reality. Precisely because "the complex function of highly integrated circuits (aside from memory ICs) can no longer be checked by testing all possible combinations of signal inputs—as occurs in the case of a simple, logical connection"[24]—tests independent of the producer are called for. Objections, which U.S. Patent Law has practically made the norm [*nachgerade zum System erhoben*], should not prevent any number of measurements, modifications [*Patches*], and techniques of circumvention (of which official documents make no mention) from being disseminated publicly. Whether for peaceful purposes or not, it would be information about information science.

Hugo von Hofmannsthal once ascribed the ability to read "what has never been written" to the "wonderful being" called "man" [*Mensch*]. In the chaotic welter of codes that has begun now that everyday languages have abandoned power for the benefit of a Universal Discrete Machine, cryptanalysis of the same kind should be performed universally—and by machines, at that.

16   There Is No Software

"The Eastern world is exploding," Barry McGuire sang. The first time, he did so in the wild 1960s—to talk all his friends (via vinyl or eight-track tape) out of the belief that we are *not* standing on "The Eve of Destruction." The second time, after a brilliant electronic remake that made his vinyl track into the digital chart-topper of the Armed Forces Network in Dharan, his words (via ultra short waves) sought to talk the warriors of Operation Desert Storm out of the belief that they—or we—still face destruction . . .[1]

McGuire (or rather the digital signal processor that erased his phonographically immortalized negation without a trace) was still right the second time, but only because explosions did not count at all. It is unimportant whether oil derricks or Scud missiles (those direct descendants [*Reichsunmittelbare Enkel*] of the V-2) fly into the air. The East can go ahead and explode. All that matters is what happens in the Western world at the present moment: first and foremost, the implosion of high technologies—and as a result, the implosion of a set of signifiers [*Signifikantenszene*] that otherwise would still be called "World Spirit" [*Weltgeist*]. Without computing technology, there would be no deconstruction, Derrida declared in Siegen. Writings and texts no longer exist in perceptible times and spaces, but rather in the transistor cells of computers. And since, in the last three decades, the heroic deeds of Silicon Valley have managed to reduce the dimensions of transistor cells to the submicron level (that is, to less than a micrometer), our present-day scene of writing can only be described by way of fractal geometry: as self-similarity of letters over some six decades that extend from corporate billboards the size of a house down to a bitmap the size of a transistor. At the alphabetic beginnings of history itself, a mere two and a half decades separated a camel and the Hebrew characters designating the animal. Now that all signs have been miniaturized to a molecular scale, the act of writing has vanished.

There Is No Software

As everyone knows, even if no one wants to say it, nobody writes anymore. Writing—that peculiar kind of software—long toiled at the incurable confusion between use and reference [*Gebrauch und Erwähnung*]. Up to, and including, the time when Friedrich Hölderlin composed his hymns, the mere mention of lightning was evidence enough that it might be used for poetry.[2] Today, in contrast, after the transformation of this same lightning into electricity, human writing occurs through inscriptions that are not just burned into silicon by means of electron beam lithography, but rather—and in contrast to all writing implements of history—are themselves able to read and write.

The final act of writing in history, then, may have occurred in the late 1970s, when a team of Intel engineers, under the direction of Dr. Marcian E. Hoff, spread a few dozen square meters[3] of drawing paper on the floor of an empty garage in Santa Clara and drew up the hardware architecture for the first integrated microprocessor. In a second step—mechanical this time—the manual layout of two thousand transistors and their connections was reduced to the size of a thumbnail on a real chip. Third, electro-optical equipment wrote the design onto silicon. Fourth, after the end product—the 4004, which has provided the prototype for all microprocessors ever since—was employed in the new adding machine of Intel's Japanese client, our postmodern scene of writing could begin.

Meanwhile, given the complexity of hardware in present-day microprocessors, manual design techniques no longer have a chance at all. In order to develop the next generation of computers, engineers do not use drawing but rather Computer-Aided Design: the geometrical capacities of the most recent generation of calculators are just enough to map out the topology of their successors. In this way, "the feet" of those who "will carry you out also" once more "stand at the door" (as the biblical phrase has it).

All the same, Hoff's primitive blueprints provided an almost perfect example of a Turing machine. Since Turing's 1937 dissertation, every act of calculation—whether performed by human beings or by machines—can be formalized as a countable quantity of commands that work through an infinitely long band of paper and its discrete signs. Turing's concept of this kind of paper machine,[4] whose operations encompass only writing and reading, movement forward and backward, has proven the mathematical equivalent of all calculable functions and seen to it that machinic, literal meaning [*der maschinelle Wortsinn*] has pushed aside the innocent professional designation of "computer."[5] Universal Turing machines need only be fed with the description (the program) of any other machine to imitate this machine in its effects. And because ever since Turing it is possible to abstract from the differences in hardware between two devices, the so-called Church-Turing conjecture amounts, in its strictest—that is, physical—form, to declaring nature itself a Universal Turing Machine.

As such, this affirmation has doubled the implosion of hardware with the implosion of software. Ever since it has been possible to build computers—since 1943 with vacuum tubes, and since 1949 with transistors—the problem has existed of somehow describing and reading these universal writing-reading machines, which are in fact illegible. As is well known, the solution is called "software," that is, the development of higher programming languages. The ancient monopoly—whereby everyday languages functioned as their own metalanguages, therefore admitting no Other for the Other—has collapsed and given way to a new hierarchy of programming languages. This postmodern Tower of Babel[6] now extends, in the meanwhile, from simple command codes, whose linguistic extension is still a configuration of hardware, over assemblers, which are the extension of these same command codes, up to so-called standard languages [*Hochsprachen*], whose extensions—by way of innumerable detours through interpreters, compilers, and linkers—are also called "assemblers." Writing today, as it occurs in software development, is an infinite series of the self-similarities discovered by fractal geometry; except that, in contrast to the mathematical model, it remains mathematically impossible, in physical/physiological terms, to have access to [*erreichen*] all these layers [*Schichten*]. Modern media technology—ever since the invention of film and gramophones—is fundamentally arranged to undermine sensory perception. We can simply no longer know what our writing is doing, and least of all when we are programming.

The situation may be illustrated in everyday terms—for example, with the word-processing program from which the words here derive. May the *genius loci* of Palo Alto, which produced both the first and the most elegant operating systems, forgive a subject of the Microsoft Corporation for limiting the examples to the dumbest of all operating systems.

In order to process texts—that is, to become oneself a paper machine on an IBM AT under Microsoft DOS—the purchase of a commercial software package is the first item of business. Secondly, some of the files in this package must have the extension .EXE or .COM; otherwise, word processing cannot start under DOS. Executable files, namely, entertain a singular strange relationship with their proper names. On the one hand, they bear magniloquent titles such as "WordPerfect"; on the other hand, they are more or less cryptic acronyms (because vowels are missing) like "WP." For all that, the full name serves only the advertising strategies of software manufacturers; the latter still employ everyday languages as a matter of course because the Disk Operating System, aka DOS, cannot read file names with more than eight letters. That is why unpronounceable abbreviations freed of vowels—acronyms that revoke the elementary innovation of ancient Greece—are both necessary for postmodern writing and perfectly up to the task. Indeed, for the first time since the alphabet was invented, these abbreviations seem to have endowed it with magical powers. "WP," that is, does exactly what it says. Unlike both the signifier "WordPerfect" and empty old-European names like "Spirit" or, indeed, "Word," executable computer files comprehend all the routines and data that are necessary for them to achieve realization. The act of writing—typing "W," "P," and "Enter"—does not make the Word perfect, but it does make WordPerfect run. Software affords many small triumphs of this sort.

The more or less inflationary literature that accompanies software—so as not to fall too short of the command line—doubles these magic powers. Typically, software manuals, because they must bridge the abyss between everyday languages, electronics, and literature, present the program package as a linguistic agent with the power to command, absolutely, system resources, address spaces, and hardware parameters of the computer in question: WP, activated by command-line argument X, switches the screen from mode A to mode B, starts in setting C, finally returns to D, and so on.[7]

However, all the actions that, according to what is written on paper, Agent WP performs are entirely virtual, because each one of them has to run "under" DOS (as it is so aptly put). But in factual terms, only the operating system—or more precisely, its shell—is at work: COMMAND.COM searches the keyboard buffer for an 8-byte file name, translates the relative addresses of the file it (perhaps) finds into absolute ones, loads this modified version from external bulk memory into silicon RAM, and finally assigns the execution of the program (which occurs for a limited time) to the first lines of code belonging to a slave named "WordPerfect."

At the same time, the same command-line argument can also be turned against DOS, because, in the final analysis, the operating system works as a simple expansion of a basic input/output system called BIOS. No single application, nor even the underlying microprocessor system, could ever start if a few elementary functions—which have been burned into silicon for security reasons and therefore form part of the indelible hardware—did not, so to speak, possess Baron von Münchhausen's ability to pull themselves up out of the marsh by their own hair [*Zopf*].[8] Each material transformation of entropy into information, from a million dozing transistor cells into electric voltage differences, necessarily presumes a material event called "Reset."

In principle, this descent from software to hardware—from higher to lower levels of observation—could run over as many orders of magnitude as one wishes. Even elementary code operations, notwithstanding their metaphorical promises (e.g., "call" or "return"), amount to strictly local manipulations of signs and therefore (more's the pity, Lacan) to signifiers of varying electric potentials. All formalization—as defined by David Hilbert—effectively abolishes theory, simply because "the theory" at issue "is no longer a system of meaningful propositions, but one of sentences as sequences of words, which are in turn sequences of letters. We say by reference to the form alone which combinations of words are sentences, which sentences are axioms, and which sentences follow as immediate consequences of others."[9]

When meanings shrink down to sentences, sentences to words, and words to letters, then no software exists either. Or rather, it would not exist if computer systems did not need—at least until now—to coexist with an environment of everyday languages. This environment, however, has consisted, ever since a famous, twofold Greek invention,[10] of written characters and coins; that is, of "letters" and "litter." In the meanwhile, compelling economic reasons have fundamentally done away with the modesty of an Alan Turing—who, during the Stone Age of the Technical Era, preferred reading machine output in binary numbers to decimal computations.[11] The so-called philosophy of the so-called computing community places all its stock in hiding hardware behind software, and electronic signifiers behind human/machine interfaces. In a duly philanthropic spirit, handbooks for high-level programming languages warn of the mental breakdown that would result from writing trigonometric functions in assembler code.[12] In all compassion, BIOS (Basic Input/Output System) guidebooks (and their professional authors) take it upon themselves to "hide the particulars controlling the underlying hardware from your program."[13] Taken to a logical conclusion—and without much difference from the gradations in medieval hierarchies of angels—operating-system functions such as COMMAND.COM would hide the BIOS altogether; application programs like WordPerfect would conceal the operating system entirely; and so on.

Now fundamental changes in computer design (i.e., in the way the Pentagon conceives science) have led this whole system of secrecy to be perfected. First—and on an intentionally superficial level—graphic interfaces were developed for use that, because they conceal operations [*Schreibakte*] necessary for programming, deprive users of the machine as a whole. The IBM-authorized compendium of computer graphics does not even pretend that its user interfaces make system programming faster or more efficient than simple command lines would be.[14] Secondly—in immediate conjunction with Ada,[15] the Pentagon's programming language, but also on the microscopic level of the hardware itself—a new operating mode called "Protected Mode" was developed. According to Intel's *Microprocessor Programming Manual*, it has the sole purpose of keeping "untrusted programs" and "untrusted users" from all access to system resources such as input/output channels and the core of the operating system.[16] In a technical sense, however, all users are untrustworthy; in Protected Mode (as it prevails under UNIX, for example), they are not permitted to control their machines at all anymore.

The uninterrupted victory march of software represents a strange reversal of Turing's proof that there can be no problems calculable in a mathematical sense that a simple machine could not solve. At the precise location of this machine, the physical Church-Turing conjecture, by equating physical hardware with the algorithms for calculating it, created a blank that software could successfully occupy—and from whose obscurity it benefits.

After all, high-level programming languages—the higher their Tower of Babel grows and the more everyday it becomes—operate just like the so-called one-way functions of the newest mathematical cryptography.[17] In their standard form, such functions may be calculated with a justifiable time investment; for example, when the length of operations [*Maschinenzeit*] only increases for polynomial expressions of functional complexity. On the other hand, however, the time cost for the inverse operation—that is, the matter of calculating input parameters on the basis of a function's results—would increase in exponential, and therefore untenable, relation to the function's complexity. In other words, one-way functions protect algorithms from their own results.

This cryptographic feature is, as it were, made to order for software. It offers a comfortable way to avoid what Turing's proof shows: that the concept of intellectual property has become impossible—and especially where algorithms are concerned. Yet the very fact that software has no existence independent of machines has only increased insistence on the commercial (or American) quality of the medium. All licenses, dongles, or patents that have been registered for

WP—or WordPerfect—prove the functionality of one-way functions. American courts, in contempt of all traditions of mathematical honor, have even confirmed copyright claims to algorithms.

And so, it is not surprising that recently and on the highest level—at IBM, that is—the hunt has opened for mathematical formulas that might determine the difference in complexity (the Kolmogorov equation) between an algorithm and its output. In the good old days of Shannon's theory of information, maximum information coincided with maximum noise to some degree.[18] In contrast, the new IBM measure of logical depth is defined as follows:

> The value of a message ... appears to reside not in its information (its absolutely unpredictable parts) nor in its obvious redundancy (verbatim repetitions, unequal digit frequencies), but rather in what might be called its buried redundancy—parts predictable only with difficulty, things the receiver could in principle have figured out without being told, but only at considerable cost in money, time, or computation. In other words, the value of a message is the amount of mathematical or other work plausibly done by its originator, which its receiver is saved from having to repeat.[19]

IBM's measure of logical depth, in its mathematical rigor, might also replace the antiquated, necessarily imprecise everyday terms of "originality," "authorship," and "copyright"—and thereby make them legally enforceable as well. Unfortunately, however, the algorithm for calculating the originality of algorithms in general is incalculable even by Turing's methods [*Nur leider ist gerade der Algorithmus zur Originalitätsberechnung von Algorithmen überhaupt selber turing-unberechenbar*].[20]

In this tragic situation, penal law—at least in Germany—has given up the concept of intellectual property for software (which is just as immaterial as "the Law" itself) and, instead, defined software as a "thing" [*Sache*]. The ruling of the Federal Constitutional Court [*Bundesgerichtshof*], according to which no computer program would ever run without corresponding electrical charges in silicon circuitry,[21] proves yet again that the virtual undecidability between software and hardware is hardly based—as systems theorists would so gladly believe—on a change of the observer's perspective.[22] On the contrary, there are good reasons to find for the indispensability, and therefore the precedence, of hardware.

A machine with unlimited resources of time and space, with an endless paper supply and unlimited calculating speed, has only ever existed once: in Turing's paper "On Computable Numbers with an Application to the Entscheidungsproblem." In contrast, all physically constructible machines are limited by strict parameters within their very code. The inability of Microsoft DOS to recognize file names longer than eight characters (e.g., "WordPerfect") does not just illustrate, in its own trivial and obsolete way, a problem that has entailed more and more incompatibility between the different generations of 8-bit, 16-bit, and 32-bit microprocessors. It also points to the impossibility, as a matter of definition, of digitalization—that is, of calculating the body of real numbers, which used to be called "nature."[23]

That means, however, in the words of the Los Alamos National Laboratory, that

> We use digital computers whose architecture is given to us in the form of a physical piece of machinery, with all its artificial constraints. We must reduce a continuous algorithmic description to one codable on a device whose fundamental operations are countable, and we do this by various forms of chopping up into pieces, usually called discretization. Using finite differences, elements, or some similar scheme, an algorithm with an operation count belonging to N is constructed and is translated into some high level language. The compiler then further reduces this model to a binary form determined largely by machinic constraints.
>
> The outcome is a discrete and synthetic microworld image of the original problem, whose structure is arbitrarily fixed by a differencing scheme and computational architecture chosen at random. The only remnant of the continuum is the use of radix arithmetic, which has the property of weighing bits unequally, and for nonlinear systems is the source of spurious singularities.
>
> This is what we actually do when we compute up a model of the physical world with physical devices. This is not the idealized and serene process that we imagine when usually arguing about the fundamental structures of computation, and very far from Turing machines.[24]

And so, it is no longer a matter of further pursuing the Church-Turing hypothesis and "inject[ing] an algorithmic character into the behavior of the physical world for which there is not evidence."[25] If the world does not arise from God playing dice, the algorithmic behavior of rain clouds or waves in the sea does not exclude, but rather includes, the fact that their molecules operate as computers of their own activity. On the contrary, it would all be a matter of calculating the "price of programmability" itself. This decisive capacity of computers clearly has nothing to do with software; it depends only on the degree to which a given item of hardware can house something like a writing system.

In 1937, when Claude Shannon—"in the most momentous [*folgenreichste*] master's thesis that was ever written"[26]—provided proof that simple telegraph relays could implement Boolean algebra as a whole, such a recording system [*Aufschreibesystem*] was established. And when the integrated circuit, derived from William Shockley's transistor in the early seventies, combined, on one and the same chip, the controllable resistance of silicon with its own oxide as a near perfect insulator, the programmability of matter could, as Turing had prophesied, "take control."[27] And so, software—if it even existed at all—would simply be a billion-dollar business revolving around one of the cheapest elements on earth. Connected on a chip, silicon and silicon oxide yield almost perfect hardware. On the one hand, millions of switching elements work under the same physical conditions—which is decisive, above all, for the critical parameter of chip temperature and prevents exponentially increasing deviations of transistor resistance. On the other, these millions of switching elements remain electrically isolated from each other. Only this paradoxical relation between two physical parameters—thermal continuity and electrical discretization—makes it possible for integrated digital circuitry not just to be finite machines [*Automaten*], like so many other devices on earth, but to approximate the Universal Discrete Machine that has long since swallowed up the name "Turing."

This structural difference can be illustrated quite easily. For example,

> a combination lock is a finite automaton, but it is not ordinarily decomposable into a base set of elementary type components that can be reconfigured to simulate an arbitrary physical system. As a consequence it is not structurally programmable, and in this case it is effectively programmable only in the limited sense that its state can be set for achieving a limited class of behaviors. In contrast, a digital computer used to simulate a combination lock is structurally programmable since the behavior is achieved by synthesizing it from a canonical set of primitive switching components.[28]

Switching components, however—be they telegraph relays, electron tubes, or finally, silicon transistors—pay a price for their decomposability [*Zerlegbarkeit*] or discretization. Apart from the trivial (i.e., discrete) case of word processing—which all but fades away in view of all the other scientific, military, and industrial areas where computers are employed—digital calculators, as the sole "all-or-none organs" in the strict sense of the word,[29] continue to face a continuous environment of clouds, waves, and wars. This avalanche of enormous, and real, numbers, as Ian Hacking would say, can only be mastered by adding more and more switching elements—until the 2,000 transistors of the Intel 4004 have turned into the 1.2 million of the current Intel flagship, the 80486. However, it can be mathematically demonstrated that the growth rate of possible connections between these elements—that is, the computing power as such—has a square-root function as its upper limit. The system, in other words, "cannot keep up with polynomial growth rates in problem size,"[30] to say nothing of exponential rates. The same isolation between digital or discrete elements that secures its ability to function—at least under conditions that are not tropical or arctic—also limits the dimension of possible connections to the local environment of a given chip. Under conditions of global interaction [*Wechselwirkungen*], however, which digital chips experience only in thermal terms, connectivity "according to current force laws"[31] and following combinatory logic could rise to an upper limit set by the squared value of all elements involved.

Precisely this optimal connectivity—on the other, physical hand—distinguishes nonprogrammable systems. On the basis of their global interaction, such systems, whether they are waves or beings, can display polynomial growth rates in complexity; therefore, however, they could only be calculated by machines that would not have to pay the price of programmability themselves. Clearly, this hypothetical—but sorely needed—type of machine would be pure hardware: a physical apparatus working in an environment of any number of physical devices and subject only to the same limitations of resources to which they are also subject. Software, in the conventional sense of an abstraction that may be realized, would no longer exist. The procedures for such a machine, even though they would remain open for algorithmic scripting, would essentially have to operate on a material substrate whose connectivity would permit continuous reconfiguration of its cells. And although "the substrate can also be described in algorithmic terms, by means of simulation," its "characterization is of such immense importance for ... effectiveness and so closely connected with the choice of hardware"[32] that programming it would have little in common with that of approximated Turing machines.

Such badly needed—and none-too-distant—machines, which are already being discussed in current computer science and have already been approximated by the chip industry,[33] might tempt the eyes of some observers to discern the familiar visage of "man" [*das vertraute Antlitz des Menschen*], whether evolutionarily disguised or not, in them. That may be. At the same time, however, our equally familiar silicon hardware is already following many of the demands placed on highly connected, unprogrammable systems. Between its million transistor cells, a million times a million interactions have always already been occurring: electron diffusion and quantum-mechanical tunnel effects occur over the entire chip—except that the manufacturing technology of today treats such interactions as system limitations, physical side effects, sources of interference, and so on. All this noise, which cannot be avoided, is at least to be minimized: that is the price that the computer industry must pay for structurally programmable machines. The inverse strategy—maximizing noise—would not just lead back from IBM to Shannon; it would also offer the sole path to that body of real numbers, which once was called "Chaos."

"Can't you understand what I'm tryin' to say," goes the original version of "Eve of Destruction."

17 *Il fiore delle truppe scelte*

In memoriam H.M.

> In phases of rupture and change, German critique of the military has always continued to combat the mistakes and inadequacies of the past, while ruling elites and the military were already creating a new order of command.
> —Michael Geyer

Napoleon always ordered his elite troops to attack last. Only when a battle was as good as won (Austerlitz) or as good as lost (Waterloo) did the Emperor put his old guard at risk. In the century now in course, elites play the opposite role: they are always first to receive the order to attack. And so they represent "another way in which truth grounds itself": "the essential sacrifice."[1]

Italy's *Arditi*—"the flower of elite troops"[2] according to their battle hymn, but simply the "company of death" for the rest of the army—were called into being by Colonel Cristofaro Baseggio on 1 October 1915, just two months after war was declared on the German Reich.[3] The following year, in April, after the volunteer company had lost ninety percent of its officers and enlisted men in the assault on Sant'Osvaldo, it was dissolved.[4] It was no different for the first German assault division [*Sturmabteilung*], which Major Calsow had put together in March 1915 from former Pioneer Companies: in July, it was already wiped out.[5]

Yet the "forlorn hope" [*der verlorene Haufe*]—which took the stage of the First World War as revenants of the earliest modern infantries—made history, and not just military history. Arditi were three-quarters of the troops with which D'Annunzio occupied Fiume for sixteen months. Likewise, "Sturmabteilung" was not a name invented by Corporal Hitler or Captain Röhm for the SA or *Saalschutztruppe*; rather, it referred to an elite force awarded a state budget in all divisions of the field army by order (15 April 1916) of Lieutenant General von Falkenhayn.[6] Today—after an arduous path leading from world-war armies to storm battalions, *Freikorps* and SA, and on to the Waffen-SS—peacekeeping measures and rapid deployment forces have finally become synonymous. *Il fiore delle truppe scelte* now serves in Bosnia and so on.

1.

When the armies, crowned with flowery garlands, boarded railway cars in August 1914 to hold to exacting mobilization plans, there was as yet no talk of Sturmabteilungen or Arditi. Ever since—at the latest—Prussian war minister von Roon and Field Marshall von Moltke had expanded Lazare Carnot's revolutionary *levée en masse* into universal, obligatory conscription, it seemed the war machine had become one with that of the nation-state. Even imperial armies observed the principle of giving the same weaponry to all infantry companies (if not to all troops). Democracy began with the breechloader. Storm troops, on the other hand, surely represented a unique occurrence in the autopoietic process that produced a new machine in and out of the First World War. Storm troops were invented neither by a ministry nor by a general staff. Rather, they emerged when positional warfare (which, after six weeks, Schlieffen's great plan of attack had become, or collapsed into) sought a way out.

The matter was plenty difficult. Trench systems—which extended from the English Channel down to the Jura after October 1914, and from the Isonzo up to Tyrol after June 1915—had created a no-man's-land between the fronts that cost every attacker his life.[7] In the "competition between technology and tactics," as Hans Linnenkohl called the First World War, it was weapons technology that had won: nests of machine guns and field artillery liquidated all the firing lines that had the audacity, or orders, to venture into the "zone of annihilation."[8] In 1914, the machine gun—this "irrefutable object"[9]—had become untrue to its original calling to hold back black-, red-, or yellow-skinned hordes.[10] Now it set its sights on the infantries of its own inventors (which neither Kitchener nor Schlieffen, the victors at Omdurman and Waterfontein, had ever anticipated). Even the Poisson distribution of preparations for barrage fire that lasted for weeks could not guarantee that, somewhere among shell craters, a single enemy machine-gun unit had not managed to survive; consequently, the fourteen British divisions advancing on the Somme in July 1916 lost one out of every two men.[11]

The shock troop emerged from mass death in the trenches. Ernst Jünger's *Der Arbeiter*—the literary universalization of the First World War—rightly stresses that

> those carrying a new fighting force only become visible in the late parts of the war; their different nature [*Andersartigkeit*] is evident in proportion to the disintegration of armies formed according to the principles of the nineteenth century. Above all,

one meets them where the characteristic of the age [*Eigenart des Zeitalters*] finds particularly clear expression in the use of resources [*Anwendung der Mittel*]: among the land- and air squadrons, among the shock troops, where the collapsing infantry, worn down by machines, gains [*gewinnt*] a new soul.[12]

Calsow's shot-up storm detachment found its new soul under a new commander on a new front. Captain Willy Martin Ernst Rohr (1877–1930) could look back on a career from the picture book of elite sociology. The erstwhile teacher at the Infantry School for Marksmanship and company commander in the Guards Rifle Battalion Lichterfelde loaded the lost hope on railway cars; they did not stop before reaching the wine villages of Oberrotweil, Bischoffingen, Niederrotweil, Oberbergen, and Schelingen. Only the Kaiserstuhl (literally, "Emperor's Chair") was good enough to serve as training grounds for revolutionary infantry tactics. While twenty kilometers to the southeast a revolutionary philosopher was just starting his military service in the censor's office at the main postal office in Freiburg,[13] Colonel Bauer's prize captain was granted permission to change the ravines and loess mountains of the Kaiserstuhl into a landscape of trenches.

The Flanders simulated on the Upper Rhine created a war machine that no longer had any use for the mass armies of nation-states. The steel helmet replaced the *Pickelhaube*. A light carbine rifle, which Rohr's shock troops simply slung over the shoulder,[14] took the place of the Gewehr 98, which infantrymen had carried as if on parade.[15] Above all, Colonel Bauer—who was only responsible for storm troops in Falkenhayn's high command (and not, as later under Ludendorff, for the war economy, period)—gave his elite troops weapons that no infantry had ever wielded before: the flamethrower perfected by Colonel Reddemann,[16] lightweight machine guns looted from the Czar's army,[17] 3.7 cm assault cannons,[18] and—last but not least—short-range mortars [*Minenwerfer*] developed from Rhine metal. Captain Rohr's Kaiserstuhl had them all.

And with that arrangement of weapons technology, the "infantry" vanished. The erstwhile Pioneer Battalion with its infantry training and the weaponry of the field artillery just needed to be organized as something more than a multimedia conglomerate of specialists with different arms.[19] Therefore, Rohr's Assault Battalion [*Sturmabteilung*]—a labor force [*Arbeiterschaft*] not in Bebel's sense, but certainly in Jünger's—consisted of Headquarters [*Stab*], two Pioneer Companies, a Vehicle Company, and an Artillery Division. "Auxiliary weapons included: Machine-gun Squad [*Maschinengewehrzug*] 250 (6 weapons), one

Il fiore delle truppe scelte

Minenwerfer troop (4 short-range mortars), and one flamethrower troop (4 small devices)."[20]

Napoleon's campaigns had been able to cover Europe so completely only because they carried out the military reforms of 1792 thoroughly.[21] The armies were all divided into corps that, because they each had their own infantry, artillery, and cavalry, were able to operate independently. The storm divisions of 1915 elevated this differentiation from the operational plane to the realm of tactics. Each individual shock troop fought as an autonomous war machine whose parts or weapons systems were coordinated by wristwatches[22] and elaborately planned scenarios. Portable *Minenwerfer* took enemy foxholes and positions secured by barbed wire[23] under the high-angle fire that circumstances required; machine guns held the enemy's firepower in check; and flamethrowers rendered remaining patches of resistance inoperative. All this occurred just so that a couple of marksmen with carbines and hand grenades (that is, grenadiers in the literal sense of the Second World War[24]) could traverse the annihilation zone of the no-man's-land and survive.

And so, the "narrow and deeply articulated" [*schmale tiefgegliederte*][25] shock troop liquidated the whole order of the infantry. The dense firing line of equally armed companies—which had been prescribed since 1906 and, at the latest, under the mythical machine-gun fire of Langemarck, had turned into the "self-destruction of the attacker"[26]—died a theory-death, too. In October 1915, Rohr's division needed to move only fifty kilometers over the Rhine, adapt their Kaiserstuhl maneuvers to the situation in the Vosges, and begin the storm. Already after the seventeenth mortar round—the day before Christmas—the French occupying forces on the Hartmannswillerkopf capitulated.

It is no wonder that Rohr's storm division achieved the status of storm battalion, and that the storm battalion was the favorite troop of his commander in chief, Crown Prince Wilhelm of Prussia. The front itself had devised a new elite and tactics that could be fed back into itself as a doctrine. Already because "the undertakings of the storm division, with minimal losses, almost always met with success, the troops themselves expressed the desire to be trained in this mode of combat."[27]

On 15 March 1916, Falkenhayn let it be known through official channels:

In war, the "Storm Battalion" serves in operations against difficult attack sites [*Angriffsstellen*], and in times of peace, as a training force. To make the experiences of the Storm Battalion in matters of arrangement [*Gliederung*] and deployment generally useful, two experienced officers (captains or older lieutenants and four non-commissioned officers [*Unteroffiziere*]) are to be commandeered from all armies on the Western front, for a period of fourteen days, to the storm battalion. [...] After the return of the officers and non-commissioned officers from the commando, storm detachments [*Sturmabteilungen*] are to be formed within individual units. These detachments are to be expanded gradually, so that every division, in the course of time, is in the position to put together a core troop of select and specially trained officers and men for difficult attack missions.[28]

With that, and for the first time in German military history, a chief of general staff had changed infantry tactics in the middle of war. Directly or indirectly, the instructional storm battalion trained any number of storm battalions; Captain Rohr trained any number of lieutenants, whose names then entered literary or military history: Ernst Jünger of Infantry Regiment 73, Erwin Rommel of the Württemberg Mountain Battalion, Felix Steiner of Machinegun-Sharpshooter Unit 46, and so on, and so on. At any rate, the ruling class of World War $N+1$ had been recruited.

Therefore, already in October 1916, Crown Prince Wilhelm could predict what occurred to the War Ministry only in August 1918:

A thoroughly trained infantry, supported by Pioneers and equipped with machine-guns and grenade-launchers, must ultimately be able to do without the allocation of particular troops from the Storm Battalion.[29]

But for all that, a thoroughly trained infantry remained a pipe dream for as long as Supreme Command, along with Colonel Bauer—the point man for storm troops—stood under Falkenhayn. Only when Hindenburg and Ludendorff took over in August 1916 was the tactical realignment in the middle of the war followed by its necessary prerequisite: weapons-technical realignment. "The Third Supreme Army Command [*3. OHL*] completed the transition to mechanized warfare and, thereby, to the industrialization of warfare in Germany with a radicalness that may rightly be called unique."[30] The reason was simple: the radical gesture that qualified as unique—and enthuses the U.S. Army in its resolute adherence to C4 (Communications, Command, Control, Computers) to this very day[31]—was formed by Ludendorff's endless telephone calls, which lent an ear to every request of the field army. The quartermaster general himself performed technical feedback, the same kind that all wireless army radios had performed since Dr. Alexander Meissner had applied positive feedback to vacuum tubes.

The so-called Hindenburg Program—a name whose seeming loyalty to the Kaiser in fact masked the revolutionaries Ludendorff and Bauer—ordered a military economy in the literal sense: the output of munitions doubled, that of machine guns tripled, and the output of *Minenwerfer* is supposed to have increased a hundredfold.[32] With that, the weapons systems that the crown prince and his 5th Army had foreseen for an infantry trained as a single, great storm troop entered into mass production[33] and headed to the front. On 4 June 1917, the War Ministry ordained a new organizational structure, "which fused the infantry with machine-gun weaponry without remainder by forging together the smallest infantry unit, the company, and the machine-gun troop."[34] And with that, the business secret of Rohr's storm battalion—"the thoroughgoing substitution of machines for human beings"—now formed the "core of the new principles of deployment for the German field army."

> Weaponry itself—the "war machine," as one said in the First World War—became the means of, and point of departure for, military deployment and the organization of military units. The movement of the lowest unit, the group, was determined by the firing properties and the protection of the machine gun. Its deployment, in turn, depended on the interplay of associated weapons—that is, on artillery and the infantry assembled around the machine gun. Nowhere was this change in deployment clearer than in the new training regulations. Drills and exercises fell almost completely into disuse. Weapons-training moved—in direct reversal of Wilhelmine practices—to the absolute fore.[35]

War that relied on personnel gave way to war based on materiel.

For Captain Rohr, who had triumphed all down the line, happy times began. Ludendorff traveled, just a month after assuming command, to Army Group *Kronprinz*, where the assault battalion—"the favorite troop of the Crown Prince"[36]—received him in battle uniform,[37] and the quartermaster general "for the first time" saw "a closed formation in storm-uniform with the altogether practical steel helmet [*mit dem so überaus nützlichen Stahlhelm*]."[38] A little later, all storm battalions (with Rohr's prototype as the significant exception) passed from the Pioneers to the Infantry.[39] Whatever tactical innovations had been developed on the front, the Operations Division of Supreme Army Command, with Captain Hermann Geyer leading the charge by pen, transferred to general combat regulations. Above all, the innovations now held a status that prohibited sending their inventors to the slaughter. Under Falkenhayn, Rohr's storm battalion, still in March 1916, had been deployed—with little success and heavy losses—to attack the "blood vacuum" [*Blutsaugpumpe*] at Verdun.[40] Under Ludendorff, however, this "fundamental thought," that troops are arranged for the purpose of dying, went "missing more and more." The change noticeably

Il fiore delle truppe scelte

chagrined Colonel Bauer: storm battalions "were deployed as elite shock troops and then immediately withdrawn. But since the real work [*das dicke Ende*] for the attacker came only after the position had been taken—counterattacks with the heaviest artillery fire—this mode of deployment gave rise to much bad blood among the other troops."[41]

The Third OHL bureaucratically consecrated and granted all the demands that Rohr had made since his deployment at Verdun. While Italian Arditi had orders to dig themselves in on the terrain they had stormed until (if matters worked out) the common foot soldiers could follow their lead,[42] German storm troops were "removed immediately after completing their missions, so that they would stay freshly preserved and ready for use. Therefore," Rohr's *Instructions for the Use of a Storm Battalion* concluded, "they [are] intended only for attack, and not meant to hold positions."[43] Just like their strategic successors in the Second World War—the divisions and armies of the Waffen-SS—Ludendorff's tactical fire brigade had the most up-to-date weapons, the longest periods between deployment [*Etappenzeiten*], and the choicest rations.

> We who were in assault battalions had a general advantage in this respect, for we always received a bonus that consisted of cheese, sausage, or tinned meat. Whenever parts of the battalion were deployed, the whole battalion received a combat bonus, and when the whole battalion was deployed, it received, in addition, an "extraordinary combat bonus." [...] Besides the rations provided, food and sundry items [*Lebens- und Genußmittel*] could be purchased. When the battalion sat idle, as occurred in Beuville, the supply of foodstuffs and sundries from Belgian border towns met the demand. Moreover, during this time, the companies were able to improve their provisions significantly by making use of the gardens they had been assigned.[44]

It is no wonder, then, that from 1916 on the staffs no longer needed to rifle through their companies for able bodies to man shock troops. Plenty of soldiers volunteered for a troop that was the envy of all the others, although they were required to be single and no older than twenty-five. The *gestalt* of the soldier specialist [*Facharbeiter*]—just as the project of multiplying multiplicators had foreseen—entered a series that was already working toward World War $N+1$. In 1941, on the left bank of the Pruth, Italian war correspondents and German tragedians no longer doubted that tank crews, whether from Essen or Charkow, all spoke the same jargon [*Schraubenziehersprache*].[45]

And so, it is also no wonder that the storm battalions of 1916—in addition to the pleasures of leave—enjoyed the privilege that had the brightest future: they were motorized. A troop that, as an operational fire brigade, was thrown into unpredictable sites of conflict [*unvorhersehbare Brennpunkte*] and withdrawn immediately after performing assignments exploded the logistical framework of the First World War by definition. Since Moltke's innovations of 1866 and 1870–71, only railroads had lent war movement. Therefore, the "mobilization" that occurred in 1914 deserved its title. This is why Ludendorff's strategic castling in winter 1917–18 from the Eastern to the Western Front succeeded. But when the troops had been unloaded, and the transition from mobile strategy to equally mobile tactics should have occurred, everything froze into stationary warfare once again. In the impassable combat zone of trenches, grenade launchers, and barbed wire, attacks were abandoned simply because the carriers and horses available could convey neither reinforcements nor artillery to the fighting lines fast enough. The railway network—which still provided the basis for the First World War logistically—only extended to railheads in the rear echelon. Tracks in the combat zone, where they were needed most, had long since been damaged or destroyed by the enemy.

Therefore, the switch from rail war to motorized war began—in a movement wholly parallel to the founding of storm battalions—in the trenches of the First World War. It involved more than just the tanks that became famous when, in the Battle of the Somme, they emerged for the first time. Altogether unprepossessing trucks debuted, too. On the Allied side, their number ultimately reached into the hundred thousands. On the German side, after great efforts undertaken by Ludendorff's Third OHL, they ultimately counted 40,000.[46] Events foreshadowed the *Blitzkrieg* of 1939—an effort to switch an entire logistics from the rail system to motorization that was just as systematic as it was halfhearted.[47]

All the motorization afforded by the First World War benefited assault troops first and foremost. A (training) *Park-Kompanie* had hardly been established when Captain Rohr already laid claim to all its trucks to "replenish equipment, munitions, and provisions."[48] At any rate, the railways cars that had still transported Rohr's battalions to the Kaiserstuhl could be retired. It was no different for the first Assault Armored-Vehicle Division [*Sturm-Panzer-Kraftwagen-Abteilung*] that was equipped with looted British tanks and German replicas. Crown Prince Wilhelm, as usual, saw to it that the brand-new weapon was "shipped to Assault Battalion Nr. 5 (Rohr)."[49] Finally, in summer 1918, when the deployment of the American Expeditionary Forces on the front sealed Allied superiority and chased storm battalions and fire brigades from flare-up to flare-up, their arrival and departure in "rapid carrier vehicles [*raschen Lastautos*]" had already become a matter of course.[50]

Il fiore delle truppe scelte

Wagner's Valkyries—the first assault troop in military or opera history[51]—owed their superhuman speed to the optical trickery of a *laterna magica* that, at the original production in Bayreuth, projected cloud-steeds onto the horizon of the stage. The Valkyries achieved technical positivity, however, when the internal combustion engine was invented: a form of locomotion developed on the model of storms put an end to marching as the one-thousand-year epitome of the infantry. Assault battalions belonged—and belong—on tanks, trucks, or jeeps.

2.

Arditi—as *Comando Supremo* defined them in a secret memo that specified, in its heading, that it was never to reach the foremost lines of battle—are, as a matter of principle, "thrown." This pitch [*Wurf*] takes them exactly where the words that command them must not arrive. "Offensive missions"—which, more innocuously, can also be read as "expeditions"—"cast them into varied positions on the front, but preferably into the flanks or even the back of enemies" who had penetrated Italian territory.[52]

And so, in their thrownness, Arditi, just like German assault battalions, necessarily turn into the crews of combat vehicles [*LKW-Besatzungen*]. Let the common infantry toil away, practicing parade and forced marches on maneuvers and enduring the opposite—sheer stasis—in the trenches. Arditi have always already transcended archaic modes of movement.

On 10 November 1918—the day of the great Allied victory festivities—Mussolini only needed, in order to celebrate the Arditi as the "wondrous, warring youth of Italy," to board one of their "trucks." (The transport vehicle that would carry his corpse to Milan in 1945 had not been built yet.) As Mussolini's motorized rostrum rolled through Milan from the monument to the glorious Five Days of 1848 to the Garibaldi statue, the later architect of the first highways in Europe declared the united will of Fascism and *Arditismo*: "All wretched parties blocking the way to a Greater Italy" were "to be destroyed" with the signature weapons of "bombs and daggers."[53] And as if to translate the metaphor of the "way" into the plain language of motorized vehicles, there spoke, from the same truck platform as Mussolini, the honorable Edoardo Agnelli, lord of the Fiat factories.

During the war, the fronts had already been clear. Arditi—with privileges exempting them from all service in the trenches and extending from special rations to model barracks—stood on one side. The infantry and the wretched

Il fiore delle truppe scelte

Figure 7. Armored vehicles in Fiume.

stood on the other. In contrast to the German Empire—which the likes of Ludendorff, Bauer, and Rathenau had systematically undermined through their revolution from above, after all—the equally young Kingdom of Italy preserved its old-fashioned political structure. Consequently, it was violent criminals above all who volunteered for Arditi companies.[54] During the retreat from the Isonzo, after the discipline of the Italian Army had collapsed anyway, they summoned their professional aptitudes once again. At the end of 1917, the laments of despoiled peasants in the Veneto left *Comando Supremo* no choice but to take a stand against its own creation and to bar prisoners an Ardito career—at least as a rule.[55]

And so, "the wretched," against whom Mussolini's salute had mobilized the victorious Arditi, were ipso facto the powers of the state: from the prime minister down to the military police. Their powerlessness to pave the way to a Greater Italy followed from the simple fact that Arditi were, by definition, motorized. The (in)famous incident—when four Arditi, en route to the front, fired their carbines at *Carabinieri*, and the authority of the military polices literally collapsed[56]—necessarily presupposed that these valiant individuals were not marching but rather "racing to the front line in a truck."[57]

Thus, the *Marcia su Ronchi* simply represented the logical continuation of

Il fiore delle truppe scelte

well-rehearsed [*eingespielt*] logistics. While poets like Marinetti found reason to voice full-throated complaints about the "arduousness" of their way to Fiume ("between forests and Mediterranean coasts"), D'Annunzio's Arditi once again made the journey in trucks, armored cars, and tanks. Guido Keller—a flying ace and the so-called *segretario d'azione* of D'Annunzio—after the midnight arrival in Ronchi, heard that the trucks urgently needed for the quick nocturnal transport of the troops had not arrived; "he disappeared with a handful of others [i.e., other Arditi] into the night, before returning, a few hours later, with twenty-six vehicles that he had stolen from a fleet of vehicles located a few kilometers away."[58] And because the railway line from Trieste to Fiume lay idle, a strategic operation occurred—perhaps for the first time in military history—on the model of blitzkriegs to come and tank divisions.

However, the operation goal did not cooperate. The harbor city of Fiume was not a trench system as in Flanders, nor was it a network of tunnels as in the Dolomites. As soon as the "army of liberation" had done its name proud and made Fiume a liberated city, nothing was left to storm. Arditi and legionaries were reduced to festive idleness and to the rule of engagement [*Gefechtsvorschrift*] that prescribed to them (in contrast to German assault battalions) that they dig in in secured territory until reinforcements arrived from the infantry. This regulation did not anticipate that the infantry might not arrive as relief but rather as an adversary. On 24 December 1920, during the *Natale di Sangue*, the Arditi therefore encountered their inveterate enemies: their own country's naval artillery and *Alpini*. Guido Keller—because he did not accept the bureaucratic order of even his own *Comandante*[59]—fought against the advancing Alpini with a bamboo stick, the sole weapon permissible in civil war.[60] The Arditi of the commander's bodyguard encircled D'Annunzio's palace with improvised trenches, barbed wire, and barricades, until all assault-troop tactics had turned into hopeless positional warfare. Bloody Christmas ended with 203 legionaries dead.[61]

In other words, Fiume froze *Arditismo*. A whole army congealed into the figure of a world war that was over. D'Annunzio's great promise—that "the victorious army that had been undermined by traitors and agents of corruption" would, in Fiume's ten legions, "reconstitute and heal itself, rise up, and burst into flame"[62]—was fulfilled only too literally.

"Plan for a New Order for the Army of Liberation," written by Captain Giuseppe Pfiffer and signed by D'Annunzio, represents the monument to such immortalization. Just like assault-troop Lieutenant Ernst Jünger, whose 1922

Il fiore delle truppe scelte

instructions for training prepared the Reichswehr Infantry for everything but the blitzkriegs to come, D'Annunzio's orders for the army consolidated the tactical and weapons-technical arrangements of 1918. They took for granted a situation that had frozen into positional warfare; whether on the Isonzo or in Fiume, the only task was transition into movement. The assumption, then, was that the entire army should be transformed into a single assault troop. Even though the number of Arditi in the Italian Army during the First World War never exceeded 50,000, the Plan made the 7,000 fighters D'Annunzio had at his disposal in 1920[63] all into Arditi.

Because storm troopers—in contrast to the infantry of old—are specialists in weapons systems, the Plan foresaw an order that, going beyond the Futurist love for metal, grouped each unit around its respective equipment. And because the combat mission represented a shock-troop undertaking on a massive scale, the sniper companies [*Schützenkompanien*] of all legions were to be armed like the Arditi: with machine guns and hand grenades, automatic pistols and flamethrowers (to say nothing of the omnipresent dagger, the Ardito trademark). At the same time—as D'Annunzio observed with equal poetic and etymological acuity[64]—"legion" already means "elite"; and elite forces, especially generalized ones such as were established by Fiume's military order, must as a matter of course bring forth other elites. His talk of an "auxiliary company" is pure understatement, for this company reproduced, in fractal repetition, the autonomously operating legion and all of its weapons systems; the only difference was that it did so "as a 'forlorn hope' that throws itself into ruin to turn the tide of battle."[65]

Arditi, as has been noted, are "thrown." And so long as no *Natale di Sangue* threatened ruin, this thrownness excluded infantry marches in some measure. The very existence or readiness-to-hand [*Zuhandenheit*] of a fleet of vehicles saw to it that, of all the kinds of sport that the Plan imposed on its legionaries,[66] there was only one that still called for company marches.[67] All other forms of athletic activity followed training principles that from 1917 on were tested for Arditi in Sdricca di Manzano and Borgnano; they called for locomotion only as an existential "borderline situation": sprinting and cross-country,[68] climbing and jumping, rowing and swimming.[69] In just the same way, "weekend warriors" (as Heiner Müller called them) pursue their athletic activities to the southeast of Rijeka to this very day.

Above all, however, "the legionaries," as Captain Host-Venturi boasted of his forces, "gave such impulses to a particular kind of athleticism that," in Fiume,

Il fiore delle truppe scelte

"it ascended to a municipal institution: soccer."[70] Two hundred years after the extinction of Florentine *calcio*,[71] team sports were once again the order of the day. Soccer—according to the insights of military science of 1939—"is less a matter of individual achievement than the concerted efforts of the entire team and the subordination of individual interests to those of the team."[72] The ball was not just a round object, then, but a weapon around which storm troopers or Arditi must assemble. Not for nothing—as the counterespionage services of the French 6th Army recognized—did Rohr's assault battalion "indulge" every other evening in soccer matches "that officers joined."[73]

And so, all so-called team spirit—including a form of address between the ranks[74] that was just as familiar as it was necessary [*das ebenso notwendige wie "vertrauliche Du"*]—may well have originated at D'Annunzio's Fiume or at Rohr's Kaiserstuhl. At any rate, men who were not considered soldiers the less for it, whether in the March on Rome or during the so-called skirmishes of the *Freikorps* east of the Elbe, hardly wore the character armor of the drills that is said to have been the rule at *Gymnasia* and cadet schools at the turn of the century. The discipline that the trenches instilled in a new elite—the systematic training, involving weapons technique and sports in equal measure, which promoted "the oft-invoked 'comradeship of the front,' the unity of combatants that seemed to transcend class"—"made too great an impression for one to explain it simply in socio-psychological terms, in terms of 'male fantasies.'"[75]

"The applied physical exercises of the assault battalions," their National Socialist historian decreed,

> formed the foundation for the paramilitary sports movement [*Wehrsportbewegung*] after the World War. Namely, in chronological perspective, exercise occurred on the same basis in the paramilitary sports formations of the *Stahlhelm* and the *Bund der Frontsoldaten*, in the *Reichskuratorium für die Jugendertüchtigung* and in Party organizations. The athletic badge of the SA is based on the same principles.[76]

In still more general terms—decreed Felix Steiner, the commanding general of the 3rd SS Tank Corps—the shock troops of 1917 "replaced masses with elites" and the "idea of spontaneity, rapid attack, and automatically working together."[77] For this same reason, assault divisions, in the precise sense of Hitler or Röhm (whose ideas were destined to produce such great effects), were excluded from all claims of "succession." Only "the assault soldiers of the Waffen-SS" might be considered the "rebirth" of the idea, "seventeen years later."[78] "Comradeship" between officers and troops, "track and field" in obligatory basic training, "teamwork in the assault detachment [*Stoßtrupp*], whereby

Il fiore delle truppe scelte

machine-gunners, marksmen, mortar operators [*Gewehrgranatschützen*], and hand grenade launchers worked together as a well-oiled machine"[79]—all these things were regulations of the Waffen-SS; their instructors had no difficulty discerning their historical derivation from shock-troop tactics because they themselves had received the same training in 1917.[80]

"Bright sunshine"[81] fell upon the Munster Training Area when Steiner's regiment, *Waffen-SS-Standarte Deutschland*, tested emergency scenarios on 19 May 1939.

> After twenty minutes, heavy fighting began. Hitler had been requested to install himself in a concrete bunker, since all the weapons would soon begin firing around him. When he categorically refused, he was conducted to a site in front of the bunker, which offered him and his attendants at least makeshift cover.
>
> Now, the artillery divisions began to barrage the target—a deep system of trenches that lay three hundred meters from the standpoint of the observer. The heavy infantry weapons joined in. Heavy machine guns firing indirectly reinforced the preparatory fire, and light machine guns assumed appropriate positions and held down the target-enemy [*Scheibengegner*] in the trenches; meanwhile, under the protection of this dense "fire bell" [*Feuerglocke*] and through the gaps of fire from the light machine guns, the first wave of some sixty shock troops advanced to the wire obstacles, which had already been torn apart; with explosive charges and Bangalore mines [*mit Sprenglatten und gestreckten Ladungen*], they made alleyways through which they swept into the foremost trenches; they smoked the enemy out with hand grenades and were relieved by a second wave of shock troops that then—with automatic weapons, hand grenades, and flamethrowers—penetrated the depth of the position while the curtain of fire from the artillery and infantry's weapons hailed down right in front of them.

And so, less than four months before the beginning of the Second World War, the Waffen-SS was still following the storm-troop tactics of the First.[82] (Guderian's tank and radio technologies were not represented at all.) But none of that appealed to the erstwhile field messenger [*Meldegänger*] who had achieved the position of "observer." "Hitler, who had been surrounded by all the fire, spoke not a word. He recognized full well by what had been demonstrated to him that his picture of a conventional guard-troop" had been "destroyed."[83]

3.

On 14 August 1916, Rohr's Assault Battalion tested, at its training grounds in Beuville, tactics that in March 1918 would become the strategy of the entire

Il fiore delle truppe scelte

Ludendorff offensive. The Kaiser's personal press agent, Walter Bloem—before seeing, after the war, the light of the cinema's soul[84]—stressed only the following:

> It involved an attack by the assault battalion on an enemy marked with targets, in the course of which heavy firing occurred—also on the part of the artillery that had been made available for the occasion. The combination of both weapons was to be demonstrated—the forward displacement of the "creeping barrage" [*Feuerwalze*], the task of the assault infantry to follow it closely, even at the risk of losses through their own explosives. This task was performed by the select troops that had been trained for weeks for such special purposes and to have such nerve that, in fact, a few injuries occurred.
>
> Then, Captain von Rohr[85] presented a number of his officers and troops, who had distinguished themselves in their last deployment, and the Kaiser distributed in glowing mood iron crosses of both classes.[86]

1916 and 1939, then, witnessed the same scene—one and the same maneuver. But what had put the warlord of 1916 into wonderful spirits soured the mood of his counterpart in 1939. Wilhelm II gladly saw the initial approach to the "creeping barrage" upon which the German field army placed its last—unsuccessful—hope one and a half years later. Hitler saw, instead of a bodyguard that was just as old-fashioned as it was dreamed up, the same thing.

The notion of creeping barrage goes back to General Robert Nivelle, the French commander in chief of 1917.[87] The creeping barrage as a matter of combined tactics, however—and like so many innovations of the twentieth century—went back to the Russo-Japanese war. Perhaps Italy's Arditi had not imported Japanese martial arts[88] (and bamboo rods, like Keller) in vain. What the creeping barrages of the artillery impose on their own infantry is kamikaze. One third of all Japanese losses at Port Arthur came from "friendly fire."[89]

And yet (to speak in one voice with Captain Geyer), it was only a creeping barrage that enabled *attack in positional war*. In an initial step, Colonel Georg Bruchmüller, the officer responsible for artillery in Supreme Army Command,[90] synchronized the barrage with the weather report, in order to maximize the effect of gas grenades. Second, the barrage began—just as had occurred in Steiner's maneuver—immediately before the assault-troop attack, so as to do away with the advance warning that a day's worth of continuous fire [*Trommelfeuer*] provided as a matter of course.[91] Third, the forward push of the creeping barrage, unlike the continuous fire of the British at the Somme, did not occur on the timetable of the general staff but rather as a variable dependent on field reports (that is, feedback loops) from the storming infantry, transmitted via aerial observers or radio.[92] Fourth, and last, everything was a matter of effectively suppressing enemy machine guns that had survived the artillery fire in trenches or in shell craters during the infantry's attack (which repeatedly failed for this reason).

Accordingly, the key of the entire Ludendorff Offensive was a principle to which the plan of attack returned time and again:

> The principle that *the infantry, when attacking, must run into its own artillery- and mortar fire*—which has been trained into assault battalions with so much success—,must become the shared possession [*Gemeingut*] of the entire infantry. It demands uncompromising courage [*rücksichtslosen Schneid*] and superior morale, because isolated losses through one's own artillery fire must be counted on. However, through this running-forward, close combat with the enemy infantry and its machine guns is made easier. On the whole, then, losses lessen considerably. All means must be employed to make the infantry understand this. It must be possible. *The energy of the infantry attack and its success essentially depend on it.*[93]
>
> In assault [*Beim Sturm*], it is a matter of fully exploiting the effect of the artillery's preparation and the support of its fire. *The storming infantry must, at the same time as the last artillery-shots and mortar rounds, reach the enemy position and, in the further course of events, follow its own barrage immediately*, so that the enemy has no time either to emerge from the shelters that still remain or otherwise ready himself for combat.[94]

And so, the Ludendorff Offensive of March 1918—this *attack in positional war* that was just as impossible as it was audacious—was nothing more and nothing less than the transferal of storm-trooper tactics onto an entire field army. Top-down logic, as it had operated from the autonomously operating Napoleonic corps up to the autarchic assault battalion, ran back, from the bottom up. But with that, the "nerve" that, in 1916, only Rohr's "select troops . . . trained for weeks for such special purposes" had mustered, turned into the universally binding rule of combat [*Gefechtsvorschrift*][95]; now, for the first time, the Operations Division of Supreme Command gave these instructions to troops "down to the slightest detail."[96] Assault Battalion Nr. 5 had "become the teaching master of modern tactics for the entire German army."[97] After half a year of exercises and preparing armaments, 56 of a total of 192 divisions stood ready, as shock troops, to "run into their own artillery- and mortar fire" early on the morning of 21 March 1918. Only after passing beyond this frontier of death did the infantry in fact become a shock troop—that is, a killing machine.

Il fiore delle truppe scelte

After great initial success and breakthroughs reaching fifty kilometers deep, the Ludendorff Offensive came to a halt—that is, to strategic failure. Shock-troop tactics were simply not to be expanded to a country in its entirety. But failure itself had the effect that, starting right away, *Combat as Inner Experience* kept on going. Jünger's narrator dwells on the run-up to the Ludendorff Offensive until the very last page; he does not say a word [*kein Sterbenswort*] about the outcome.[98] It is also the exact position of a philosophy conceived by a party who knew all about such matters because he himself had participated in the Ludendorff Offensive (including the pedagogy behind it, which extended even to the lowliest corporal).

In summer 1918, "Heidegger, as member of Front Weather Observation Unit [*Frontwetterwarte*] 414, was deployed in the field of operation of the 1st Army on the Western Front. The unit stood under the Weather Observation of the 3rd Army; more precisely, it was stationed in the Ardennes near Sedan. The unit's main task in the Champagne-Marne Offensive (which began 15 July 1918) was to cover the left wing of the 1st Army as it advanced on Reims. The meteorological services had been established to support the deployment of poison gas by predicting the weather [*wetterprognostisch*]."[99] Later, in summer 1923, Heidegger—leading "a *shock* troop of sixteen men"[100] against philosophical opponents—answered the call to the University of Marburg. Finally, in spring 1927, the first half of *Being and Time* was published.[101]

As everyone knows, the "*Dasein*" (literally, "being-there") at issue in this Being has given up the name of "human being."[102] It has always already been "thrown" [*geworfen*] into a world that, all the same, it must "project" [*entwerfen*] in the first place. Accordingly, the integrity of Being that being-there is supposed to be remains unrealized—death "is" as the end of Dasein in the Being-of-being-to-its-end [*als Ende des Daseins im Sein dieses Seienden zu seinem Ende*].[103]

If, all the same, *Being and Time* affirms that "the existential edifice of an authentic Being-towards-death must let itself be projected," then philosophy stands before questions that are, quite literally, unheard of:

> Is it not a fanciful undertaking, to project the existential possibility of so questionable an existential potentiality-for-Being? What is needed, if such a projection is to go beyond a merely fictitious arbitrary construction? [...] Does Dasein ever factically throw itself into such a Being-towards-death?[104]

The answer is "yes." To put things as concisely as possible—and to formulate them in terms that are just as unpoetic as they are necessary—all it takes is a Ludendorff Offensive. Death, all philosophy notwithstanding, occurs in the Poisson distribution of historical ways of dying [*Todesarten*]. The "basic state" [*Lage*]¹⁰⁵ confronting Supreme Command in a strategic sense and philosophy in an existential one admits no doubt: before every possible attack there lie a no-man's-land and an artillery barrage that leave the infantry no chance of survival. Therefore, activity on the front [*Frontverlauf*] (which Heidegger also calls "death") "gives Dasein nothing to be 'actualized,' nothing which Dasein, as actual, could itself *be*. It is the possibility of the impossibility of every way of comporting oneself towards anything, of every way of existing."¹⁰⁶

All the same, the project of Dasein—such as Geyer's attack in positional warfare conceived it on paper—must overcome thrownness (or in other words, being-in-the-trenches [*Grabenstellung*]) by anticipating (literally, "running into") the barrage of fire from one's own artillery:

> The more unveiledly this possibility gets understood, the more purely does the understanding penetrate into it *as the possibility of the impossibility of any existence at all*. [...] In the anticipation of this possibility [*Im Vorlaufen in diese Möglichkeit*] it becomes "greater and greater"; that is to say, the possibility reveals itself to be such that it knows no measure at all, no more or less, but signifies the measureless impossibility of existence. [...] Anticipation [*Das Vorlaufen*] discloses to existence that its uttermost possibility lies in giving itself up, and thus it shatters all one's tenaciousness to whatever existence one has reached. In anticipation [*vorlaufend*], Dasein guards itself against falling back behind itself, or behind the potentiality-for-Being which it has understood. It guards itself against "becoming too old for its victories" (Nietzsche).¹⁰⁷

Philosophy, instead of falling back behind Ludendorff's failed offensive, takes it up again. Each victory that does not occur remains a task to be performed [*bleibt aufgegeben*].

> The authentic repetition of a possibility of existence that has been—the possibility that Dasein may choose its hero—is grounded existentially in anticipatory resoluteness; for it is in resoluteness that one first chooses the choice which makes one free for the struggle of loyally following in the footsteps of that which can be repeated.¹⁰⁸

The Greeks may have read a debt owed to nature in death, and Christians a punishment of their God. It was the philosophy of polemical obedience [*der kämpfenden Nachfolge*] that first sought not to attribute the end of mortals to alien forces. Enemy artillery and machine guns, ever since a creeping barrage has held them down, are over and done as a cause of death.

The very opposite holds true: because the tactics of one's own infantry and

Il fiore delle truppe scelte

artillery go back to the same strategic design, which has explicitly "taken isolated losses into account," anticipating/running-into the barrage reveals death "*as the end of Dasein*," its "*ownmost possibility—non-relational, certain and as such indefinite, not to outstripped*" [*eigenste, unbezügliche, gewisse und als solche unbestimmte, unüberholbare Möglichkeit des Daseins*].[109] In its torn state [*Zerrissenheit*] of having to be a thrown projection [*geworfener Entwurf*], Heidegger's Dasein executes [*trägt . . . aus*] the competition [*Wettlauf*] between tactics and technology, "in-each-case-mineness" [*Jemeinigkeit*] and the work of the general staff, shock troops, and Supreme Army Command.

Assault Battalion No. 5 did not just teach a lesson to an entire field army; it also gave the war's philosopher his doctrine. And—in contrast to books—it followed the law that had made it arise in the first place. In March 1918, Rohr was promoted to major.[110] In October of the same year (once again, at the instigation of the crown prince) his battalion was transferred to protect the Kaiser personally. Finally—after Armistice, Resignation, and Revolution—Assault Battalion No. 5 transformed into *Freikorps Hindenburg*.

18 Eros and Aphrodite

> Who decided that the Thespians should honor Eros
> most of all the gods, I do not know.
> —Pausanias

My question is, how did it come that people in Europe do not know love (*ta erotika*), but rather love knowledge (*philosophia*)? Accordingly, one should ask, of all the lovers of knowledge, the party who declared that all he knew derived from matters of love: Socrates in Plato's *Symposium* (178d).[1] At the same time, the point of my question is to show an unattained model for this symposium—which in turn represents the unattained model for all symposia that still bring us together here (and elsewhere) as Plato's academic heirs. The matter concerns seriality as such; that is, the point is not to confirm, yet again, the thesis that origins have always already been displacements, replays, or repetitions. On the contrary, my objective involves a history of decline, at the end of which stand sober knowledge and sober discourse about sex.

Symposia today, at least in their official capacity, are no banquets. This is the case for the simple reason that speaking and drinking, as activities involving one and the same mouth, exclude each other. Therefore—and according to Nietzsche's grand analysis of the academic enterprise [*Betrieb*][2]—symposia involve many mouths that neither drink nor speak, but have delegated drinking to ears that, for their part, all suck at a mouth that does not drink, but speaks instead.

The symposium of the year 416 that Plato records is very different. When the poet Agathon invites his friends to the feast, both drinking *and* speaking occur because the real symposium of the previous night has already broken all alcoholic records. From the inception, then, Plato's *Symposium* stands under the sign of repetition, which at the same time represents moderation or sobering up. Agathon and his friends decide that "none of us should drink more than we think is good" instead of succumbing to the laws of compulsive consumption that otherwise yield little to contests between Athenian tragedies. Now free, the guests' mouths can replace the previous night's "competition in drinking" with a competition in speaking—if only because, somewhat ironically, their discourses define discourse itself as "relief" (*anapsuchē*) from alcoholic aftereffects. Not for nothing do the boon companions who lend the work its name count in their number one of those physicians who inaugurated European medicine in general by advocating restraint when consuming potations—if not abstaining entirely (176d). And so the friends, at least for a few hours, until Alcibiades unexpectedly crashes the party, take this medical advice to heart to some degree; consequently, there need be no talk of intoxication. The symposiasts can freely decide on possible themes for oratorical competition. Without a discernible run-up, they arrive at the very topic that symposia, or graduate seminars, pursue to this very day: desire and sexual difference. According to Pausanias, Eros is the order of the day—or rather, night—not because love and intoxication entertain a relationship of substance [*in der Sache*], but rather for purely philological reasons: "Is it not . . . an extraordinary thing that, for all the hymns and anthems that have been addressed to the other deities, not one single poet has ever sung a song [*enkōmion*] in praise of so ancient and so powerful a god as Eros?" (177a).

And so it came that all European discourse on love began with men suffering from hangovers, whose prose, as a media-technical (i.e., recordable) innovation, simply sought to fill in a space that had been left blank in poetry. Even if, in terms of classical philology, this omission is contestable—there are, after all, the pseudo-Homeric hymns—it is outweighed by the fact that ever since, all prose on love (at least up to Nietzsche) carried intoxication [*Rausch*] as a blank space, too. In the *Symposium*, something comes over even Socrates—who, famously, cannot be made drunk by any earthly amount of wine—to explain to his neighbors that prose (aka theory) does not flow (i.e., communicate) between one symposiast and the next as wine pours between one vessel and another (175d). States of intoxication, in other words, are not calculable, not storable, and certainly not communicable—again at least before Nietzsche, invoking Charles Feré, established his law of Dionysian *inductio psychomotrice.*

If only because the participants at Greek banquets reclined and did not dance, states of intoxication—also in the transfer of knowledge from man to man such as occurs in Plato's work as a whole—do not play a further role in the *Symposium*. Everything that intoxication does not perform [*leistet*] is taken over by a form of Eros that the speakers before Aristophanes and Socrates define in terms that are essentially—indeed, fundamentally—pederastic. Therefore, Eros, in contrast to intoxication, is transmissibility itself: he/it wanders from a lover to a beloved, who will himself become the lover of another beloved later on, and so forth, ad infinitum. The power of this chain is attested, via Racine's *Andromaque*, up to Derrida's *Postcard* today; according to Plato's *Phaedrus*, it might forge the most indomitable of all armies (just as it governs the succession of symposiastic situations and discourses).

For all that, the forgotten question—whether Intoxication and Love belong together, or at any rate what the story of their difference is—returns in the *Symposium*, and it does so precisely when matters do not involve the transmission of knowledge between homosexual soldiers or philosophers, but between the sexes. For Socrates' statement that all he knows is erotic was not always already the case. All that Socrates knows and shares at the symposium he has learned from Diotima, the wise woman from Mantinea (201d). Exceptionally, however—and, as Lacan demonstrated in his seminar on transference (of all things)—such knowledge is not dialectical in nature. Whereas he otherwise exploits the coherence of signifiers—in other words, of Greek grammar—here, Socrates passes along, for once, the mythical knowledge of a woman; and no matter what Lacan says, Diotima is hardly just "the woman in Socrates."[3]

As mythos, this knowledge differs from all the other definitions given at the symposium, which have declared Eros—with varying schemes of justification—either to be the oldest or the youngest god. According to Diotima, Eros is a daemonic hybrid [*Zwitterwesen*] somewhere between gods and human beings. The story of his birth tells as much simply and straightforwardly:

> On the day of Aphrodite's birth the gods were making merry, and among them was Poros ["Resource"], the son of Metis [*mētis* = "Ruse," "Craft"]. And when they had supped, Penia ["Need," "Poverty"] came begging at the door because there was good cheer inside. Now, it happened that Poros, having drunk deeply of the heavenly nectar—for this was before the days of wine—wandered out into the garden of Zeus and sank into a heavy sleep, and Penia, thinking that to get a child by Poros would mitigate her penury, lay down beside him and in time was brought to bed of Eros. So Eros became the follower and servant of Aphrodite because he was begotten on the same day that she was born, and further, he was born to love the beautiful since Aphrodite is beautiful herself (203b-c).

At the origin of love, then, lies intoxication—and at the origin of this gathering, which seeks the essence of such love in free [*zwanglos*] drinking and philosophical prose, there lies a very different symposium: one that is divine. The truly Platonic idea of the party that Agathon throws for his thinking buddies is

Eros and Aphrodite

not just the official and wholly drunken symposium that occurred the previous day, but also a birthday celebration of the Olympian gods.

At the same time, however, a distinction exists between Idea and Appearance—the heavenly symposium, on the one hand, and the earthly one, on the other—and in two respects. In the first place, in lieu of Eros—whom Agathon and his friends celebrate as "that god of love who watches over the young and fair" (265c) (as Socrates does in Plato's *Phaedrus*)—the gods extol a female divinity of love. The Olympians celebrate Aphrodite's birth. Secondly, the goddess is not celebrated by means of rhetorical prose and wine, but rather—because neither "wine" nor storage media for words exist among them—through nectar, the divine potation, which they consume by mixing silence and excess in equal measure. And if nectar is simply fermented honey (as estimable research by Victor Hehn suggested long ago), then its alcohol content, at approximately 17 percent, blows away anything Agathon might mix up with wine and water in terms of potency. And so, in an etymological anamnesis of Hehn's "Age of Mead,"[4] every flight of intoxication recorded in Plato's *Symposium*—no matter whether it occurs thanks to the gods' nectar or thanks to the wine of mortals—goes back to the word *methē* (cf. 176e).

Accordingly, when Poros—whom Diotima's myth declares to be a god, unlike the mortal Penia[5]—succumbs to sleep under the influence of nectar (or mead) (203b), he is already anticipating the final scene of Agathon's earthly symposium. After Alcibiades' drunken intrusion has once again invalidated all rules of moderation, all the participants at the banquet fall asleep, except for Socrates (223b-e). Both the divine orgy and its human equivalent have fallen out of a world that, in Greek, is always defined as order (*kosmos*) (223b). "But," as Lacan drily notes, "the joy of parties [*le bonheur des fêtes*] is precisely the fact that things happen which topple the usual order."[6] For this reason, moreover, the heroic figure in both cases is a strategist who—like the protagonist of Poe's "Purloined Letter"—perceives that the others can no longer perceive anything [*wahrnimmt, daß die anderen nichts mehr wahrnehmen können*]: Penia on Olympus, Socrates in the darkened city of Athens. This figure, practicing the ruse of strategically arranging disorder, represents the genealogy of knowledge.

Poros—even though, as the son of (crafty) Knowledge, he has a name that means "way," which he *is*—barely manages to make it to the garden of the gods. And if he gets that far, he lacks sufficient force to make it to a Monte Carlo algorithm. Just like Alcibiades in the *Symposium*, and a certain Anchises of myth, the god loses the ability to stand on his own two legs. Conversely, it is precisely Penia's neediness or resourcelessness (*aporia*) that permits her to make her way over the threshold, where she had been lurking, into Zeus's locked garden. And so, Poros and Aporia exchange names (or attributes[7]); their union proves that although sex can produce Eros it does not require Love itself. The honeymoon needs the proverbial honey (or nectar) for the union to prove fertile, but no desire for eroticism. To this extent, Penia and her sober cunning, on the one hand, and Poros in nectar-induced aporia, on the other, offer the only way out of the aporia that is omnipresent in the *Symposium*—the only way of not always needing to presuppose desire in order to derive (or generate) it. After all, according to the selfsame Diotima, the gods prefer to interact with resting humans (203a), and such company (*homilia*) can mean both sexual intercourse and sleep; according to a grammatical observation that Lacan makes, these are also the names of gods.[8]

However, for all of his understandable praise for Socrates—the fact that the latter replaces the concept of Love with that of Desire (or at any rate defines the matter more exactly) when refuting Agathon dialectically[9]—Lacan is only half right. What makes sex possible between Poros and Penia, between sleep and waking, is another factor that enters into play: Aphrodite's birth. For Eros to be conceived in the first place, Aphrodite must already have been born. This is not because Eros—as Socrates teaches in the *Phaedrus*—is a god and Aphrodite's son (243a), but rather because without Aphrodite's birth no divine banquet would ever have occurred, and consequently there would be no room [*Spielräume*] for erotic parasites like Penia or the philosophical parasites who attend symposia.

The function of birth, in other words, precedes the function of desire. After all, Aphrodite's coming-into-being does not represent an event among others, but rather the emergence of sexual fertility itself. According to Diotima, *Kallonē*—Beauty itself or herself—presides over all Becoming. This is the case in a twofold sense, for Beauty acts in this capacity both as Moira and as Eileithyia, that is, both as death and as the birth of birthing itself (206d). The possibility of Penia bringing a daemon named "Eros" into the world after nine months also depends on this elementary function. Accordingly, in the excess of their orgy, the gods celebrate the birth of an Aphrodite who is Beauty itself, without defect and consequently also without desire. *Ta aphrodisika*—"the things of Aphrodite"—precede and underlie the very fact that Socrates and his friends wish to know about *ta erotika*, "the things of Eros." The philosophical celebration of a desire the poets have forgotten is the oblivion of a birth the gods celebrate.

Accordingly, none of the speakers demonstrates the least interest in the fact that Aphrodite was born on the night of that divine symposium. The explicit distinction that Pausanias made at the beginning of the gathering—between Aphrodite Pandemos and Aphrodite Urania—has already been forgotten. To be sure, the fact that the intoxicated Poros falls asleep in "the garden of Zeus" speaks for the earthly Aphrodite, who, according to Pausanias, is supposed to stem from a union between Zeus and Dione (in contrast to the motherless daughter of Ouranos) (180d-e). Time and again, however, the symposiasts' exchanges and Diotima's narrative allude to the divine battles that led to the emergence of the divine or "foam-born" Aphrodite when Kronos castrated Ouranos. For example, Agathon objects to Phaedrus (and in the same breath to Hesiod's *Theogony* [cf. 178b]) that Eros cannot be the oldest god—and certainly "not older than Kronos and Iapetos"—because as his edifying rhetoric would have it, "if Eros had been among them then, they would neither have fettered nor gelded one another" (195b-c).

Moreover, Diotima's myth of the cunning Penia and the nectar-drunk Poros presumably alludes to the Orphic myth according to which "Zeus," on the "guileful advice" of Night, "pursued Kronos" by "making him intoxicated"—which, pointedly, he did not do "with wine" but "by means of a honey-drink"—so that he might promptly bind the sleeper in chains.[10] In all three situations—Ouranos and Kronos, Kronos and Zeus, and finally Poros and Penia—it is the strategy of weakness (that is, the strategy of women) to overcome strength with cunning[11] or intoxication ("nectarization") in order to "out-Herod Herod." Moreover, in two of the three scenarios—for Ouranos and Gaia, as well as for Poros and Penia—sex plays the role of a lure, which for its part leads to more sex: when Ouranos' "infinitely extending"[12] phallus, to which Gaia's act of seduction has given rise, is cut off and falls into the sea, Aphrodite rises from the foam. The Olympian age, in other words, can begin. Likewise, when Penia sleeps with Poros—"and he perceived not" (to employ a phrase for the wine-drunk counterpart in the Bible [Genesis 19:33])—a male daemon of love arises precisely where Aphrodite stood. The age of classical pederasty can begin.

In this long chain of violent actions, then, it is no accident that Eros is the last-born and nine months younger than Aphrodite. The function of the phallus—whose transfers and castrations are treated by the mythical scheme, after all—becomes his in part, but only partially so. At two points in her discourse,

Eros and Aphrodite

Diotima makes it unmistakably clear what the phallic function involves following the birth of Eros. Whereas for gods like Poros this function is not impeded even by heavy drinking, Diotima says of Eros:

> He is neither mortal nor immortal, for in the space of a day he will be now, when all goes well with him, alive and blooming, and now dying, to be born again by virtue of his father's nature, while what he gains will always ebb away as fast. (203d-e)

Diotima's description of the desire Eros rouses in general is even more drastic:

> when the conceiving power draws near the beautiful it grows genial and blithe, and birth follows swiftly on conception. But when it meets with ugliness it is distressed and frowns; it shrinks into itself, turns away, shrivels up, and does not conceive, but still labors under its painful burden. (206d)

And so Eros, in altogether precise medical terms, incarnates both tumescence and detumescence, erection and atrophy, and potency and impotence—in a word, the penis. Onto the site occupied by the mythical phallus, which made its way from Ouranos to Aphrodite, steps the murky empirical reality of men who, to take Lacan at his word, *are* not the phallus but rather *have* it. Alcibiades—when he tried to sleep with Socrates—could have said a thing or two about the matter [*hätte ein Lied davon singen können*].

The same sad song could be sung by a hero who was associated with Diotima's city of Mantinea like no other: Anchises. In "The Homeric Hymn to Aphrodite," this son of Trojan kings, while pasturing his cattle at the cliffs of Mount Ida, became the secret consort of Aphrodite, who only revealed her divine form to him as he slept. As if Penia and Poros had exchanged places, the union brought forth Aeneas.[13] Despite Aphrodite's prohibition, Anchises boasted of his child's mother; for this reason, he was blinded, or alternately, lamed by a divine lightning bolt. Aeneas had to carry him on his shoulders out of Troy—powerless, or in other words, impotent; finally, the son buried his father (at least in the Arcadian version of the myth) between Mantinea and Orchomenos, at the foot of a mountain henceforth known as Anchisia.[14] And on Mount Anchisia—right by "the tomb of Anchises"—there lay, in Pausanias' day, "ruins of a temple of Aphrodite."[15]

The question why Diotima, who teaches Socrates *ta erotika*, comes from Mantinea has met with any number of replies—when, that is, interpreters have not explained her away altogether (as Wilamowitz did when he called her a Platonic fiction, and Lacan when he dismissed her as a Socratic phantasm). One reading simply points to the etymological, but unexplained, connection between the name of the city and Diotima's prophetic gift, authenticated when she averted the plague. Another reading holds that Diotima represents Plato's political "homage" to "Mantinea, which, between 425 and 423," reformed its municipal constitution under the influence of the Sophists.[16] Either way, however, prophecy and politics have nothing to do with the eroticism that forms the core of what Diotima teaches Socrates. The essential feature of her priestly knowledge is to explain how matters of love (*aphrodisia*) became matters of desire (*erotika*)—how love for the goddess turned into desire for human beings, that is, men. Such knowledge would also be difficult to integrate into the worship of Zeus Lykaios, whom Diotima served as priestess according to a late tradition.[17] Of all the holy sites in Mantinea, the only possible repository for knowledge of the fact that when Aphrodite's birth was being celebrated a daemon named Eros was conceived in the intercourse that occurred between gods and human beings is the shrine erected to Aphrodite and her mortal lover.

And yet, at the site where Socratic dialogue generates knowledge no women are allowed. Eryximachos has banished even the flute girl into the back rooms, where she can "play to herself or to the women inside there" (176e). Only the ignorant among the drinkers—Plato's Protagoras declares—could prefer a woman's voice that makes music to speech from their own mouths.[18]

Hardly has Socrates finished his discourse on the erotic lessons of Diotima when the flute girl returns. Indeed, only thanks to the assistance she offers does the heavily inebriated Alcibiades, bedecked with all the attributes of the God of Wine, even manage to make it to the symposium at all. The inability to walk—the mythical attribute of Anchises—returns for good reason, too. For Alcibiades' political strategy lies buried in the same city where the remains of the Trojan hero rest. If in Plato's *Symposium* Mantinea has connotations other than Aphrodite and Anchises, they involve memories dating back two years, to the decisive battle known by the same name. In 418 B.C., the Spartans were victorious only because (according to Thucydides) they had a faster—that is, more pederastic—chain of command[19] than the allied Mantineans, Argives, and Athenians. The winter following the battle, Alcibiades witnessed, at the agora of Argos, how his strategy for the Peloponnesian War collapsed.[20]

As everyone knows, Alcibiades' erotic strategy does not fare any better than his political strategy. His drunken speech about Socrates continues all the oratory about Eros—except that now a human being occupies the same place otherwise held by gods and daemons. And with that, it involves the last, and worst, displacement that can befall love in its long way from Aphrodite, via Eros, to mortals. Alcibiades, namely, on the basis of the altogether correct assumption that Socrates desires young men like him, once reached the completely mistaken conclusion that the mythical model of Poros and Penia might be repeated. And so he invited Socrates to a symposium, wined and dined him, and finally climbed into bed with him. His erotic strategy, however—and as Alcibiades himself says (first of all, because "drunkards ... tell the truth" [217e], and secondly, because all ears that have not been initiated into pederasty are closed anyway [218b])—yielded utter defeat: he slept all night at the side of Socrates, this "*daemonic* and wondrous man," just as young men otherwise sleep at the side of their fathers or older brothers. This occurred because Alcibiades, notwithstanding wine and other inducements, simply failed to keep Socrates from speaking (219b)—or even more simply, because Socrates actually responded to the rhetorical question, "Are you asleep?" (218sc). With that, discourse takes the place of intoxication, knowledge the place of sleep, and a philosopher named Socrates the place of the god Poros. Finally, someone has mastered sex.

Such mastery, however—and as Alcibiades has realized only too late—is yet another ruse. In a manner that is cunning in its imprecision, Foucault describes the matter as follows:

> In the love relation, and as a consequence of that relation to truth which now structures it, a new figure makes its appearance: that of the master, coming to take the place of the lover; moreover, this personage, through the complete mastery that he exercises over himself, will turn the game upside down, reverse the roles, establish the principle of a renunciation of the *aphrodisia*, and become, for all young men who are eager for truth, an object of love.[21]

And yet, contra Foucault, Alcibiades' plain language [*Klartext*]—which is sober because it has been stripped of illusions—reports that only Socrates, among all the old and ugly lovers out there, managed to achieve the opposite by continuing to play his part: that is, to become the beloved of handsome young men (222b). After the fact, Penia in all her deficiency [*Dürftigkeit*] is revealed as a metaphor for a man and a philosopher who arouses the desire of others despite—or thanks to—his deficiency. And that, as everyone knows, is how academies and universities originated.

Philosophy does not remain as foreign to such ruse as Foucault affirms in the same breath.[22] It is no accident that the god of Delphi announced to Socrates—according to the *Apology*—that he is the wisest of mortals. For Socrates was first to take the declaration [*Leitwort*] of the Oracle at its word: whoever can say that he is not sleeping cannot be recognized by any beloved anymore; instead, he recognizes—knows—himself. On the other hand, in such mantic knowledge, which according to Diotima sprang from the sleep of the gods rather than that of mortals, insight and self-identity [*Erkenntnis und Selbstsein*] remain separate as a matter of principle. Those who came to Delphi with questions were sober and alert, but without any knowledge of what they were or would be. Therefore, they had to rely on a prophetess, the Pythia, who indeed knew their future (albeit at the price of not knowing herself because of intoxication that was systematically induced by bay leaves). Greek "soothsayers," according to a neat formulation in Xenophon's *Symposium*, "have the reputation of prophesying the future of others but of not being able to foresee their own fate."[23]

The very opposite holds for philosophical knowledge: all who do not yet have it fall into sleep and intoxication, whereas he who possesses it drinks all the wine they pour him, even though "it never makes him drunk" (214a). And so, one after the other, "they began to nod, and first Aristophanes fell off to sleep and then Agathon, as day was breaking. Whereupon Socrates tucked them up comfortably and went away" (223d).

Philosophy, in other words, makes states of alcoholic intoxication impossible, because it swallows up all the wine in Athens to no effect. And so there remains only one kind of intoxication that it does not pretend not to know: nectar as the drug of the gods. From the body of Eros—who was after all conceived in the nectar intoxication of Aphrodite's birth—bees suck new honey and, with that, new mead. The last lines of the last poem that has been passed down from Plato's hand return to a garden where, just as in the garden of Zeus, a god lies sleeping:

> But Eros himself, overcome by sleep, lay among the roses;
> A smile danced on his mouth, and over the sweet lips
> Ran the bees, sucking to make honey in their hive.[24]

19 Homer and Writing

According to Mallarmé, there is no prose. The alphabet exists, and then, just as soon, there is verse. Here I am following this affirmation—which inaugurated modern poetry—by assuming that Barry Powell's wonderful thesis also holds for quite some time after Homer. Friedrich August Wolf once robbed us of the singer; Powell has restored him to us.

According to Powell's argument, the Greeks developed their vocalic alphabet, which according to Johannes Lohmann (my most venerable teacher) represented the "first complete and therefore systematic analysis of the sound forms [*Lautformen*] of a language"—around 800 B.C. at the courts of Euboea, when they made a north Syrian system of consonantal notation their own. The Greeks did not do so, however, in shady dealings [*zu schnödem Handel*] with the donors, as scholars have ceaselessly maintained for the last two centuries. Rather, the Greeks created five vowels in order to record and collect the unwritten and blind [*die schriftlos blinden*] songs of Homer. In this sense, and no other, the Greek alphabet has remained our mother tongue: one does not have to understand it in order to read it—a unique event in the world. One need only discern [*erhören*] that singing presumes A, E, I, O, U, women, voices, and song—vowels [*Selbstlaute*], that is. Once the travails and triumphs at Troy no longer stood engraved on the *stelae* of despots who commanded writing—as had been the case in the Near East—they sounded forth, read aloud in mortals' offerings to Muses and Gods. The immortals, as Odysseus learns from Alcinous, the king of the Phaeacians, have made ruin the lot of mankind in order to rejoice those yet to come by song. And so let us hark.

The *Iliad*, for which the Greek alphabet came to be, is a battle cry. Only in the *Odyssey*—that is, ever since vowels have existed—has song received a proper name: the Sirens. Such is the name of those who bind and charm. A single voice flows from two mouths, beautiful and sweet as honey. Whoever hears the pair sing will certainly never *not* return home—that is just a lie of translators *ad usum delphini*; rather, the hearer will experience utter rapture and make his way richer in knowledge. The "god-speaking voice," as Odysseus himself praises it, knows how to tell him everything that the Achaeans, the Trojans, and he himself suffered on Troy's broad plain, all that comes to pass on the much-nourishing earth. And so, on an island abounding in flowers, bees, and drinking water, Sirens are like the nymphs ("brides") that Homer also calls "the Muses." Only when they were born (as Plato wrote of writing, of all things) did we receive the gift of music. The word "music" derives from *muse*, even in Arabic. And so, the words of the Sirens to the hero simply reverse the prayers that the singer of the *Iliad*, when memory failed him, addressed to the Muses: you are goddesses and are present and know everything that we know only by hearsay.

In other words, unconcealment [*Unverborgenheit*] or *alētheia*—this trinity of being-present, being-something, and omniscience—is not a matter of Platonic philosophy, but rather the bequest of the vocalic alphabet; although invented after the *Iliad*, it preceded the *Odyssey*. Less than twenty kilometers from the land of Sirens lies Ischia, the ancient "Island of Monkeys." There the Cup of Nestor attests that in 730 the *Iliad* must have been available to a singer and writer in legible form—and also, "beautifully wreathed in the bed of love," Aphrodite from the *Odyssey*. Accordingly, early vocalic inscriptions did not record death sentences or goods traded, as in the Orient, but rather the voices of music. "The alphabet exists, and promptly there is verse." Or as Richard Bentley put it long ago: the *Iliad* is for men, and the *Odyssey* is for women—more precisely, about them. For it is the voices of women—melodies—that Odysseus's adventure in the Wild West of Greece uncovers: from the depth of concealment [*Verborgenheit*], Hades itself, rise the screams of war widows and orphaned daughters, so powerful that they chase Odysseus to flight. Circe and Calypso dwell [*bergen sich*] in houses in order to sing at the loom and keep Odysseus from Penelope—who for her part weeps at her own island loom for even longer. Only the Sirens sing openly and abyssally [*unverborgen grundlos*] at high noon, when the sea is calm, to announce with bright sound what singing itself means: binding, enchanting with love, knowing.

Being discloses itself therefore—in accordance with the alphabet of Muses—to the hearing first and foremost. In the 62nd Olympiad, Pythagoras came from Ionia to Lower Italy, like Odysseus and his singer. He arrived where there now stood any number of sister cities. Thanks to his doctrine, the area was soon celebrated as "*Magna Graecia*." He was named "Pythagoras" because the god-saying oracle hidden at Delphi had revealed to his mother (to whom the matter was still hidden) that she was carrying him. The name says to say, "Delphi's Python at the agora," even though doing so means bearing witness to horror—to departed souls in the underworld, to daemons trapped in ore, and to gods who are present as a ringing in one's ears.

All this Pythagoras announced both to men and to women—first to the nymphs and ephebes of Croton, then to those of Metapontum.

Pythagoras, who never wrote a single letter, put *acousmata* in his listeners' ears: sayings that are matters of hearing and obeying [*die sowohl zum Hören und Gehören sind*]—as much as they speak of hearing itself. Pythagoras, it is said, invented philosophy in word and deed by asking, "What is it that is?" His riddles [*Fragespiele*] involved posing "*acousmata*" and answering them, too:

> What are (dead) souls? Dust in the air.[1]
> What are the Pleiades? The lyre of the Muses.[2]
> What is ringing in one's ears? The sound of gods [in us].[3]
> What are thunderbolts? So those in Hades will suffer.[4]
> What is the sound of struck metal? The voice of daemons trapped in ore.[5]

And so on and so forth, until everyone else's ears also rang, because *acousmata* for the most part did not locate noise [*Rauschen*] in the field of vision but in that of hearing. If Homer's heroes—according to Julian Jaynes's grand thesis—were guided by divine voices that sparked [*funken*] from the right hemisphere of the brain to the left,[6] Pythagoras spoke as if these distant voices had achieved consciousness [*würden ihrer selber inne*]. For all the ears in Croton that tuned in, amidst tears, the experience led straight to a god [*Das erhebt in all den Ohren, die zu Kroton unter Tränen lauschen, alsogleich zum Gott*].

One *acousma* says why this was so:

> What is it that prophesies/speaks the truth at Delphi? The tetractys. That is all the harmony in which the Sirens [sing].[7]

This *acousma* is the only one in two parts. First, the obscurity of the Pythia turns into the play of combinatory logic [*offenbares Zahlensteinspiel*], which Pythagoras taught as the tetractys—the "square" or "fourfold" [*Geviert*]—to his pupils. Place a one in the first row, a two in the second, a three in the third, and a four in the fourth, until the triangle that has emerged from the tripod records not just counting, but also addition. Whoever has counted from one to four, Pythagoras taught, has already formed the sum without noticing it: the holy number of ten. However, because the question of what it *is* heralds the question of what it *means* (*sēmainei*),⁸ the answer entails something more, too. What speaks the truth at Delphi—and what, moreover, the tetractys is—are the two Sirens, as Odysseus heard them singing together.

In their unison, two mouths are arranged like the harmony of the seven strings of the Muses' lyre; that is, as an octave. Between the first row of stones and the second, the tetractys reveals [*zeigt die Wahrheit auf*] that octaves, as the relation or *logos* extending over all seven strings (*diapasōn*), are equal to half their own length. Accordingly, the philosopher turned *harma* (a yoked team of horses in Homer) and its plural form, *harmonia* (the joints of Odysseus's raft), into the singular form *harmonia*—the joining of two to one—and lent the octave a proper name. Only the Sirens resplendently ascend in harmony from the threatening murmur [*Angrauen*] of noises—in Hades, in one's hearing, in deposits of ore. And Harmony, in the last myth that the Greeks invented, was called the earthly-beautiful daughter of Discord and Love. With the elements of his tetractys, then, Pythagoras recorded [*schreibt . . . an*] the *gamos*—the great marriage (bed) of Kadmos and Harmonia, the marriage of vocalic writing and music. In those days, up in the Kadmeia of Thebes, Gods and Muses visited mortals in dance and song for the last time. *Harma* means, before all else, the intertwining of man and woman.⁹

But as many names as occurred to the Greeks when they celebrated the inventor of their vocalic script—Kadmos and Palamedes, Theut and the Muses themselves—so little did they speak of a second wonder: the fact that the same letters, and only for the Greeks, also stood for numbers. Later, *to gramma*, the inscribed letter, was also called *stoicheion*, "one of a row." This fact makes it clear that another, anonymous adaptor present in the ordinal series of consonants (which had been taken over as a group) permitted figures to be read as cardinal numbers as well. At any rate, in the same year that Pythagoras was born, the first inscription appeared in Lower Italy employing one for alpha, and beta for two. Ultimately, it yielded a system of ones, tens, and hundreds up to a thousand minus one.

Pythagoras—the friend of wisdom—answered the *acousma* "What is wisest of all?" by responding, "Number."¹⁰ And yet he never wrote a thing. "He said so himself," his listeners would declare after the master's death. The house in Metapontum where he died became holy to the goddess Earth, and the path leading to it turned into a grove of the Muses. For all that, the first person to speak of music in numbers—that is, of alphabetic figures [*Alphabetziffern*]—

Homer and Writing

was a pupil who is said to have betrayed Pythagoras through writing during a civil war: Hippasos of Metapontum.

> By the whole [is] like beta to one,
> By five like gamma to beta,
> By four like delta to gamma.[11]

These theorems, though obscure, remain eternally true and can be shown to any child. Hippasos, on his lyre, commanded the play of strings, which unified numbers and music. To its eternal glory, one—as the origin or foundation [*Grund*] of all that exists [*von allem Seienden*]—was written out; two, three, and four, on the other hand, were represented by figures. With that, harmony emerged as what articulates the whole. *Melea*—in Homer, the plural form for "limbs"—gathered into the singular *melos*, a song that played through all the strings in two tetrachords: a fifth above and a fourth below.[12] The dimension of the octave, Philolaos added to Hippasos' observation, equals the fourth times the fifth—*syllaba kai di'oxeian*. $3:2 \times 4:3 = 12:6 = 2:1$. In other words, Hippasos read the master's tetractys as an operator that made not just addition but also music available to reflection. Everyone's ears hear octaves as identities; and the fifth and the fourth, especially after Euler and Fourier, have constituted European high culture. That is why Pythagoreans—who did not just hearken to the master's words in an acousmatic capacity [*wie Akusmatiker*] but also learned the reasons [*Gründe*] underlying them from Hippasos—are called mathematicians. They lend intervals, which are numbers, names that can be spoken and recorded.

"Means," Archytas of Tarentum would later write, "are three in music: first the arithmetic mean, second the geometric, and third the subcontrary—which, since Hippasos, is called harmonic."[13]

The arithmetic mean of the whole that is called an "octave" is the fifth, and the harmonic mean is the fourth. A geometric mean, in contrast, should entertain the same *logos* to the tonic note as to the octave. In modern terms, one to x would equal x to two. Although this can be done with lengths of string, it cannot be recorded numerically. The square root of two remains *arrhēton* and *alogon*—unspeakable, like shameful protrusions [*Schamglieder*]. It is wordless or (in Latin) "irrational." Under conditions where an alphabet limits operations to natural numbers, real numbers do not unconceal themselves [*entbergen sich*] into Being; they found [*stiften*]—as their name indicates—a geometry that is strictly separate from arithmetic. Hippocrates of Chios and Archytas, a general and lord [*Stadtherr*] of Tarentum, founded [*begründen*] geometry by replacing the tetractys with lines and endpoints. Pythagoras's counting stone turned into a duality, which was almost inaudible in the Dorian dialect of Archytas. The line—*hē grammē*—was delimited by points marked, in each case, by a letter: the first point was called "alpha," the second "beta," and so on. And so, almost a century after Hippasos, the Greek alphabet brought geometry into Being and Discourse [*Sein und Sagen*], after all. Joined together, *to gramma* and *hē grammē* made lengths—even if they were a-logical—ready-to-hand for a science of diagrams.

That absolves Hippasos, but it does not redeem him. Transmission or tradition—that is, what Pythagoras once founded and which continues to provide the doctrine of many today [*als Schule west*]—also meant "treason" in Greek. Hippasos declared that not all *logoi* can achieve the status of *logos*, the Word. Moreover, he announced that the beautiful *kosmos* also harbors trouble [*einen Störer*]: *diabolus in musica*, as the Middle Ages would later condemn the square root of two. The traitor experienced the punishment due to what is *arrhēton*: at Metapontum, he swallowed (and was swallowed by) the endless blue sea. In rebellion, the city's populace murdered and expelled his pupils; the Greek alphabet—the only one in history to do so—had led to revolution [*Umsturz*].

Only one man, whose very name promised loyalty, stayed true to the people. Philolaos of Croton, the great teacher of Archytas, survived. The reason is simple: Philolaos was the first to break with the master's instructions and to write books. Hardly had Socrates (who knew no mathematics and never wrote a word) died of drinking hemlock than Plato paid vast sums for these books. In Philolaos's works stands written what is universal about numbers and notes—that is, what interests philosophers. To wit, there are two, and only two, kinds (*eidē*) of numbers: even and odd. From them derive the innumerable forms (*morphai*) between the heavens and the earth—the same as with computers today. The relation between notes [*Tonverhältnis*], on the other hand, unites the two *eidē* because even and odd alternate between numbers, between N and $N+1$ (as in the tetractys). In this way, and only in this way, the art of calculation, which once built advanced civilizations on the Nile and the Euphrates, became the singular science of Europe. Contemplating the *eidos* of uneven numbers, it became possible to discern that something, a being [*ein Seiendes*], remains between them—and that at three, the first uneven number, something lies in the middle. And so, what is odd is also called "limiting"—*perainon*—with a secondary, sexual meaning. In turn, at the center of two, the beginning of even numbers, there is nothing: an *apeiron* gapes, as it did for Hippasos, who first met with the *alogon* and then the sea.

Philolaos taught that this hole offers [*vergibt*] space that three can penetrate, *perainein*. At the origin, all knowledge, to quote Lacan literally, records erotic technique. Conveniently, five as two plus three—like a woman on top of a man—is called *gamos*, "marriage." Therefore, *physis*, all that exists [*das Seiende im Ganzen*], needs, in order to yield beauty (i.e., *kosmos*), the power of that harmony which joins things as different as the two sexes. Were there only even or odd, the "worldliness-of-the-world-as-such"—more than two thousand years before Heidegger—would just have been nonsense [*ein Unfug*]. In this way, however, *ha estō*, "Being" in the language of Philolaos,[14] was identified with the force that occupies the center of Parmenides' sphere of worlds: "the daemon who guides all things," Aphrodite.[15]

We all know that she no longer rules, this *daimōn hē panta kubernā*—cybernetics, back and forth. One might intone a lengthy song of lamentation, then, about what became of ancient wisdom when Socrates, at Agathon's banquet, preferred little Eros to his great mother.

Please permit me an abbreviated account. Plato's *Phaedrus* refers to the inventor of the Greek vocalic alphabet as a daemon from Egypt, as if the many bilingual inhabitants around Naucratis could not distinguish between an alphabet and hieroglyphic script. In the *Phaedo*—the only dialogue that mentions Philolaos by name—the two *eidē* of odd and even numbers become ideas that have absolutely nothing to do with each other. Socrates means to refute Kadmos, that is, numeric writing [*Ziffernschrift*]. And so, he sends the women away from his cell, in order to die alone among beautiful youths. And because he also believes that he has refuted Harmonia, his soul enters the realm of Ideas without knowing anything of music. At this *Liebestod* among disciples, only Plato was not present, because he was pretending to be sick. But in fact he was in Magna Graecia, learning about intervals on lyres; shortly before his death, he invented—in honor of the Sirens—musical intervals in the heavenly spheres. The plaything and piece of evidence [*Spielzeug und Beweisstück*] called the "lyre" transformed into his doctrine of Ideas, which no one can see or hear, and which convinces no one.

Immediately after this death, Aristotle fled from Plato's Academy because it returned to mathematics. The *eidos* abandoned Woman, who turned out to be mere matter, whereas (Aristotle maintained) it is the sperm of men that conceives human beings. That is why Aristotle calls the voice, which people share with animals, the substance [*Stoff*] of *logos*. Now, *logos* does not refer to an interval of sounds, but rather to discourse [*Rede*] as it distinguishes human beings. For human beings do not sing. They read. According to Aristotle, the meaningless sounds of letters—*stoicheia*—form *sullabai* (the "syllables" of modern languages); although devoid of sense, they are speakable. Several syllables together yield meaningful nouns and verbs which when joined, yield a sentence—*logos*, as he calls it. So far, so good.

However, the example that Aristotle has chosen reveals [*verrät*] something else altogether: the first and second letters are gamma and rho, which when read together, make a grunting noise that human beings share with beasts: "GR" like Gryllos, "Grunter"—the name of one of Odysseus's shipmates, according to Plutarch. Instead of turning back into a person, he prefers to remain a pig in Circe's woods, *hulē*. Only when alpha joins as the third letter does the wonder called "vowel" [*Selbstlaut*] occur. Yet "GRA" yields a speakable syllable that suppresses [*unterschlägt*] two things. In the first place, for Presocratics such as Philolaos, *sullabē* did not mean a syllable of language, but a musical fourth.[16] Secondly, "GRA" forms the beginning of the very word that Aristotle suppressed for the sake of *stoicheion*—that is, *to gramma*, "sound" and "letter" in one. Of *eidos*, *logos*, and *sperma*, only a breath [*Sprachhauch*] remained; had Aristotle only read what he himself had written.

And so it has stood with the oblivion of Being [*Seinsvergessenheit*] ever since the time of Socrates, when thinking, writing, music, and numbers ceased to occur together. The only protection is loyal remembrance, the written word [*Dagegen feit nur treues Eingedenken, fester Buchstab*]. We—who else?—even have record of as much.

> Eurytos of Croton, an *auditor* of Philolaos, was told by a shepherd that he had heard a voice from his grave at high noon—the voice of a man who had been dead many years, which seemed to be singing. "By the gods," Eurytos replied, "and what series of notes?" [*kai tina pros theōn eipen harmonian*].[17]

The very first thing we hear in this passage is that sheep and man [*Schaf und Mensch*] are not enemies. Second, we hear that Philolaos does not rest in Croton or Herakleia, but rather where roses, wine, and olives grow: at Italy's highest noon [*in Italiens hellstem Mittag*]. And so, even shepherds will learn what singing, counting, and writing means. For ever since the time when Philolaos died and continued singing, the alphabet of the Greeks has known how to record—in addition to sounds, numbers, and intervals—a certain melody, by the gods: *tina harmonian pros theōn*.

20. The Alphabet of the Greeks: On the Archeology of Writing

Ever since Homer, language has been given to us as the house of Being by the Greeks. But how did this tongue come to itself as the writing [*Schreibweise*] that we come to know when we learn Greek? As is well known, the vocalic alphabet represents the second effort by Greek-speaking rulers, kings, and singers to record their Indo-Germanic language, which had so many more vowels than English or German. As much is readily evident in, for example, the ancient verb form *eaō* ("I let"), which does not contain a single consonant but only epsilon, alpha, and omega. Linear B is a syllabary—a grid with about fifty fields; some nine consonants and a few vowels provide a framework in which one can readily discern "ti," "to," "rho," and so on, even if some of the sounds have not yet been deciphered (that is, have not been assigned determinate theoretical values).

It is no accident that the history of the script's decipherment played out after the Second World War, in the victorious lands of the United States and England. In 1947, Alice Kober in New York had a suspicion: she noticed that many end syllables in words differed only in the final vowel, but not in the preceding consonant. They represented "dry" syllables, then, and accordingly, were to be read as "to" and "ta," "prōton" and "prōta" (for example); in other words, "the first [thing]" and "the first [things]." Kober observed that a fundamental difference between all words—the fact that nouns and adjectives are divided between two sexes, as all human beings happen to have, plus a neuter—is a feature only of Indo-Germanic languages. Thus, she supposed that the Linear B recorded on Crete must have been Indo-Germanic, too. A good example would be *Kirkos* and *Kirkē*, "falcon," male and female—the latter a well-known figure in the *Odyssey*.

This chapter is a transcription of a lecture held in the context of a series entitled *Archäologie als Kulturwissenschaft*.

The Alphabet of the Greeks

In 1952, during his free time, an architect named Michael Ventris also worked on the hypothesis that the ancient inscriptions were written in an Indo-Germanic tongue. Ventris was the first to suppose that ancient Greek had been the language of Crete. He demonstrated the validity of his hypothesis by focusing on proper names. Knossos, Amnisos, and Phaistos were all known sites of archeological digs, but archeologists did not know if these places, when they were built—whether by Greeks or non-Greeks—were in fact called "Knossos," "Amnisos," and "Phaistos."

How did it come that an architect, of all people, had this ingenious idea in his free time? During the war, Ventris had worked as a navigator for the Royal Air Force, decrypting German radio transmissions. That is exactly what Alan Turing was doing at the same time, too—cracking the German Wehrmacht code made by *Enigma*. His efforts helped determine the outcome of the Second World War. Turing entered the proper names of German generals into the scrambled radio transmissions; thereby, they automatically became legible. And so, *mirabile dictu*, computers and Crete are connected. Then, in 1954, Turing had to bite into a poisoned apple; he died right away. And Michael Ventris suffered a strange accident alone in his car—on an empty street, early in the morning in a London suburb. It was not revealed until 1974 that automated decryption processes had contributed to Allied victory over Hitler.

In the *Phaedrus*, Plato has Socrates say there was once a time when the Muses did not exist. And when they were born and entered the world, some men grew so enthused that all they did was sing, forgetting to eat and drink; and so they became cicadas. That is to say, from this passage in Plato we know that writing is a gift and a wonder, and that it *takes place* as a gift from the Muses.

In 750, the people of Euboea set sail with Phoenician merchants and settled in the regions where Odysseus had still felt lost. A little to the north of the Sirens are Capri and Ischia. Ischia was settled with the lovely name of Pithekoussai, "the Island of Monkeys." It seems it still hosted primates then (as one still finds in Gibraltar today). In 1953, archeologists beheld a wonder, to wit, an iamb and two hexameters inscribed on a cup found in the grave of an ephebe. There it stood, in relative clarity and easily legible (apart from slight gaps in the text that were chipped). Written in beautiful, calligraphic script stood: "*Nestoros eimi eupoton potērion*"—"Nestor's cup am I, good to drink. Whoever drinks from here, may the desire of lovely-wreathed Aphrodite seize."

There are some who translate this passage with an objective genitive: "desire for lovely-haired Aphrodite." For the Greeks, Aphrodite was the goddess who bestows desire. She did not need to be lusted after by boozers at a symposium; rather, she awakened desire in the first place. That is why participants at symposia, banquets, bedecked themselves in wreaths—because they were imitating the gods in their finery: *kallistephanos Aphroditē*, "lovely-wreathed Aphrodite." The archeologists recognized right away that the words were quoting Homer. Book XI of the *Iliad* describes an enormous drinking vessel made of gold. It stands on a table, and only the venerable hero Nestor, lord of Pylos, can lift it and drink the wine it contains. Therefore, of course, the little cup on Ischia was not Nestor's chalice but rather a literary reference.

Independently, Ernst Risch and I have taken a look at the last line once again: in Book VIII of the *Odyssey*, Ares and Aphrodite (because the latter's husband, Hephaestus, just happens to be absent) sleep together. Then they get caught. And there one reads, precisely, *eustephanos Aphroditē*—that is, "well-wreathed Aphrodite." Risch and I affirm that what was written on Ischia alludes to the *Odyssey*. And with that, knowledge of a written *Odyssey*, as if by abduction (that is, with the methods of a Sherlock Homes), is proven—and archeologically, at that. Risch goes on to share that Walter Burkert had the honor of taking in hand this chalice of Ischia, Nestor's cup (which naturally is inaccessible to mere mortals like us). On close inspection, he reports, it became clear that the writer of these lines had had a written version of Homer lying before him, which he must have copied.

Another piece of evidence is more or less contemporary. The Dipylon inscription of Kerameikos, now housed in the National Archeological Museum of Athens, presents one complete hexameter and the beginning of a second; after that, there is a little scribbling. Barry Powell, after eyewitness inspection, like Burkert's, has supposed that a second, clumsy hand followed one that was very skilled, one that could write and versify. "Whoever of the dancers now dances most lustily shall" get (we add) "me." The jug itself speaks—it is an *oggetto parlante*, as defined by Jesper Svenbro. The jug stands displayed and exhorts one young person to dance better than the next. The wording is *paizei*: dancers should, literally, "child"—that is, they should play; *paizein* comes from *pais*, "child." A musical agon and a prize stand at issue, and that means that song is offered as the reward for dancing because, until this point, it has also carried the scene [*der Gesang ja auch vorher der Träger in dieser Szene gewesen ist*]. And if the completely meaningless four or five letters in the other, clumsy hand were literacy [*Alphabetisierung*] itself?

One party shows how he can write before he has sung; the other says, "I would like to write, too." Then, the practiced individual tells the inexperienced one: "Just start . . ." That is how one learned to write in Greece—in order to sing and make music. In the *Iliad*, this kind of thing represents an altogether infrequent exception. But the *Odyssey* is different. The *Odyssey* begins with Penelope, waiting for faraway Odysseus. It ends with Odysseus—after a moment of pleasure with his wife—telling her of his adventures. Penelope tells him he has heard the Sirens, too. And then they both fall asleep, at the end of Book XXIII. In between, during this homecoming, which has taken so much time and been so long deferred, the heroes have had to contend with women. After the Cyclopes and Lestrigons have been dealt with, Circe shows up. She sings at the loom, while Penelope cries at hers. Then Calypso appears for a year—another nymph, who also sings and weaves, and therefore entertains a relationship to labor and to work songs.

What is more, there are the countless daughters and widows of the departed heroes of the Trojan War, who wail so loudly that Odysseus flees them at the end of Book XI. He can no longer stand the invocation of the dead [*Totenbeschwörung*], which involves mostly dead women [*die ja eher eine Totinnenbeschwörung ist*]. And to that I say, proclaiming what has never been said before: the spirits are women; they are vowels; they feed the *Iliad*—that is, death—back into love, song, and music [*Das sind die Frauen, das sind die Vokale, das ist die Rückkopplung der Ilias, also des Todes, in die Liebe, den Gesang, die Musik*]. A few times, according to interpreters, exegetes, and philosophers, the hero Odysseus succumbs to this Sirens' song. I do not believe that he yields; rather, I think the Sirens incarnate music, yet again. Book XII, verse 184: "Come hither, now, Odysseus." The Sirens control the airwaves [*ergreifen die Stimme*]; they control the vowels. There are inordinately many vowels in what they sing at the hero:

> Come hither, now, Odysseus, much-renowned, great glory of Achaea; stay your ship that you may hear our voice. Never yet before you has anyone passed by on his black ship who did not hear the honey-sweet voice of our mouths. No, he has taken full pleasure and returns home the wiser. For we know all that, on Troy's broad plain, the Achaeans and Trojans endured at the will of the gods. We know all that comes to pass and many other things, too.

Supposedly, Odysseus sealed the ears of his companions and had himself tied to the mast. It is strange, however, that the following lines simply say, "When we no longer heard the Sirens"; it is not said, "When I no longer heard the Sirens." It also says, "when we left the island," not, "when we had sailed past the island."

On the Archeology of Writing

The wonder of the Sirens—whether it involves eroticism or not—is that they dwell on the island richest in flowers, an island on which Odysseus presumably also sets foot. That means there is fresh water; and that means they are nymphs, for nymphs are freshwater divinities and one does not worship them in a temple but rather where there are no archeological discoveries to be made from the time of the Greeks. And therefore—because of the flowers, the Sirens, and the fresh water—there are also bees, and where there are bees, there is honey, and so on; and songbirds, too, which is why it all sounds so bright and beautiful. (I am attempting an archeology of the text, not the archeology of findings from digs.) And so, although all the verses are beautiful, the most beautiful one declares, "No one passes until he has heard the honey-sweet voice"—one voice, in the singular—"of our mouths" (plural). The two Sirens are two organs, two holes, and yet they sing in unison. This fact never occurs to us. And that is the problem with our theories of Greek music.

Let us remain in this land, Lower Italy, which the Greeks found so beautiful that they all followed Odysseus. Already in 530, the Greeks landed in Messina for the simple reason that Italy abounded in forests—as is still the case today—unlike their native country. In 530, a good reason also prompted the mathematically gifted Pythagoras to relocate from Samos to Croton and Metapontum, that is, to the southern boot, the most beautiful and rose-filled region in Italy. In the early sixteenth century, Gregorius shed tears that a column was still standing on the site where Pythagoras had had his school, which, after all, gave us the university as such—we are all Pythagoreans. There Pythagoras had posed a question, not to one pupil but to many pupils (for such is the essence of the university), so that they would come up with a question themselves, namely, *ti estin*, "What is something?" And then he asked, "What is 'number'?" The answer he expected was "the best." I would say so, too. "What is it that rages in Delphi?" "What is the Oracle of Delphi?" Nobody knows. And then Pythagoras declared: "tetractys"—my magical formula, which is mathematical. He wrote down the numbers one, two, three, and four as one counter [*Steinchen*], then two counters, three counters, and four. Now matters are reversed: we are no longer practicing archeology, but rather writing with numbers—and purposefully so.

We put down one, two, three, four. Pythagoras tells a pupil, "Count!" The student begins slowly: one, two, three, four. "Stop," says Pythagoras. He asks, "What did you just do?" "Count to four." "No, you made the number ten. One plus two plus three plus four: ten." This is the holy number of Pythagoreans.

The Alphabet of the Greeks

And after that, Pythagoras probably thought only of the stones' relation to the first number, the *archē*, the number One. That was the first algorithm—the first operational definition (Bridgman), the first signifier without a signified. With the tetractys one can do everything, yet one cannot say what it is. And that is why, because of this ingenious answer, the disciples devised the *manteion*, the Oracle of Delphi: "Be the tetractys!" Pythagoras himself bore such a title in Delphi: *Pyth-* means "Pytho," the chasm of Delphi; it also means "to rot" and any number of other things. And *-agoras* comes from "agora" and "to speak." Thus, Pythagoras was the one who brought the dark rumor [*Sage*] of the dead (or singers) to light for the Greeks of Lower Italy. That is why he gave his disciples this answer. And then the pupils asked, "And what is the tetractys?" And Pythagoras responded, "*Harmonia!* The harmony in which the Sirens . . ."—end of transmission. It is an "*acousma*," an oral oracle that Pythagoras, who never wrote, pronounced to his pupils.

This can only mean the relation of one Siren's song to that of the other. After all, there are two Sirens: Homer explicitly says so twice, and he employs the dual instead of the plural. The relationship is harmony, as the Sirens sing it, and the harmony that sounds is the Oracle of Delphi—a double explanation. Accordingly, *harmonia* became a key word for the Pythagoreans. The term derives from the word for "chariot" at Troy, *harma*. Then it became "joint," as in what holds rafts together: *harmonia* (not a singular form) refers to the metal hooks that hold together the vessel that Calypso, at divine behest, permits Odysseus to build, so that he may leave her, for love.

Pythagoras's pupils went on to think the matter more radically and expand the tetractys. And so, among them *harmonia* came to signify something entirely new, namely, the fugue—and above all else, the octave. The octave, then, would be the first form [*Gestalt*] of mathematics. And Pythagoras or his pupils recognized that it obeyed mathematical laws. They knew that if one divides a given string on a cithara or phorminx into two equal halves, the octave follows from the tonic note. If one divides the string between two-thirds and one third, it produces the fifth—now the relation of the numbers is no longer 2:1 but 3:2. If one divides the string in the relation 4:3, one obtains the fourth. Then the Pythagoreans could declare the tetractys operationally closed. They recorded as much in the strangest numerical notation, which was overwhelmingly simple and beautiful. Namely, they simply took the sequence of Greek letters for numbers—first ones, and then tens and hundreds—that is, they constructed a mathematical alphabet out of the twenty-seven letters (adding a few old ones, too): alpha to iota for one to nine, kappa to tau for ten to nineteen, and then followed the series of hundred up to nine hundred. And when they wished to say, "The fourth is 4:3," they simply wrote, *delta kai gamma*.

And there it was. And in this way, the music never stopped ending. Now what was most fleeting and most beautiful in the world was not just a script that the singer had sung. Also the cithara, which accompanied it as instrumental music, was situated in a recording system [*kommt ... zu einem anschreibbaren System*]. The Greeks did not come—in the spirit of Enlightenment or during the shift from *mythos* to *logos*—to sing and versify more and more badly; instead, things just became more and more beautiful.

From this theory of music there emerged everything that, ever since, has counted as science—above all, knowledge concerning *physis*, or "nature." It is said, for instance, that Empedocles of Acragas studied among the Pythagoreans. Tetractys is the root of ever-streaming-and-being Becoming, of all *physis*. The mathematical basis of all that is follows from the unity of whole numbers as they appear geometrically and arithmetically.

Empedocles seems not to have believed it. In beautiful Agrigentum, something wholly different occurred to him: he realized that he found himself in one of the loveliest energy systems in the world. Here heavenly fire existed on earth—Mount Aetna, the divine Zeus—and then he beheld the seething sea, the flowering earth, and the sky. In the world picture he devised, Empedocles assigned four divinities to the four elements. His followers subsequently introduced the general name which ultimately—via *stoicheion atomon*, "the indivisible"—led to modern nuclear physics, which is itself an alphabet, just like the alphabet of the Greeks; in the first place, because it calls hydrogen H, oxygen O, and nitrogen N.

This fundamental thought—that all that is has, at its basis, letters in their discrete roundedness [*Abgezähltheit*]—gave the inventor of atomic theory, Leucippus of Miletus, a simple idea, which Aristotle relates: "Tragedy and comedy come from the same letters." That is, heartrending woe and Aristophanic-phallic joy, in the final analysis, are both an alphabet. Very well.

We find ourselves in an acoustic-written realm that I wished to illustrate so fully [*den ich so nahebringen wollte*] so that one might understand that archeology should perhaps be separated from belief. The eyes, it is said, are better witnesses than the ears. I do not believe it. Let us undertake acoustic archeology.

It is commonly held that the song of the Sirens was heard once and for all. I alone dare contradict this belief. The nanophysicist Wolfgang Heckl—*nano*

The Alphabet of the Greeks

derives from the Greek word for "dwarf"—has reflected: a handsome young man, or better, a beautiful young woman, was sitting at a potter's wheel, there in Greece, or maybe in Egypt. She sat and pedaled, and the wheel turned. She put lovely geometrical patterns into the clay. One can draw lines simply by holding steady the needle or comb with which one engraves. And then Heckl observes: the intention of man [*des Menschen*] is one thing, but the physics occurring behind his back is another. Styluses, combs, and hands—when someone is singing or playing an instrument—are exposed to microscopic (or nanoscopic) movements. In short, we are simply making the tiniest traces. And so, why should the voice of the two Sirens, when someone sang of them, not have been recorded, too?

21   In the Wake of the *Odyssey*

We should ask ourselves, above all, why what we ask returns, time and again, to Odysseus. The answer stands in Borges, who wrote that there are only two stories for Europe: in the one, the heroes depart in order to fall in glorious battle for a faraway city; in the other, the hero sets to sea and, after twenty years of war and wandering, returns to his love.

    I think that this *nostos*—the ever-repeated return to the Greeks—gives form not only to lecture series but also to our *Dichten und Denken* in general. As Ernest Renan observed when contemplating the acropolis more than a hundred years ago, progress will only ever occur as the further development of what the Greeks already began. If instead of progress we speak of recursions, this statement can remain our guiding thread. To be sure, I will have to limit my remarks—that is, I will need to forget Hegel, Nietzsche, Heidegger, and Foucault. And we will get back to the *Odyssey* itself only after four other returns to the work: Virgil's *Aeneid* and Dante's *Inferno* (in the *Divina Commedia*) represent two literary wanderings; Jean-Luc Godard's *Le Mépris* and Stanley Kubrick's *2001: A Space Odyssey* offer as many cinematic journeys.

    The very medium that made such fictions and media possible is commonly overlooked: the alphabet in the unique form that the Greeks gave to it—a form, that is, which records vowels and by this means can transcribe any language at all. From the artful language [*Kunstsprache*] of Homer, via Virgil's Latin and Dante's self-invented Tuscan, up to screenplays in French and English, a single writing system is at work.

    Why that is, is an obscure question. The standard answer holds that the Greeks—around 800 B.C., after four centuries without writing—adopted a Semitic alphabet in order to trade with Carthaginians and Phoenicians. For all that, however, it is puzzling why no commercial or political inscriptions have been preserved from archaic times. There are only hexameters, dedicatory in-

In the Wake of the Odyssey

scriptions, obscene graffiti—and Homer. But this same fact puts us on the answer's trail. To recite the hexameters of the *Iliad*, it is necessary to have invented and recorded vowels in writing; otherwise, no singer would know whether the syllables of the verses should be voiced long or short.

And so we assume, with Barry B. Powell—that is, against Joachim Latacz and Walter Burkert—that Homer himself, like his many predecessors, could neither read nor write, and that he dictated his *Iliad* to a user of the alphabet [*dem Alphabet-Adaptor*].[1] Otherwise, the poem's twenty-four books [*Gesänge*] would not have reached us so literally.

The *Iliad* is set around 1200—that is, at the same time when the Greeks and Cretans possessed a syllabic writing system that fell into oblivion as Troy, Knossos, and Mycene burned (and remained preserved thanks to these same fires). This writing system was altogether useless for recording hexameters. Odysseus's wanderings, on the other hand—and in contrast to the wooden horse he invented—take place four centuries later. As Circe tells Odysseus, Jason and the Argonauts have long since discovered the Black Sea; now it is a matter of opening [*erschliessen*] the far west of the Mediterranean in competition with the Phoenicians: the space extending from Libya, over southern Italy, up to the gates of Heracles—today's Gibraltar. Leaving behind the Lotus-Eaters, Odysseus wanders over to fearsome giants, that is, to the megalithic cultures that, long before the Greeks, ruled west Sicily and south Corsica. Then suddenly the tone changes: instead of a world of men (as in the *Iliad*), a foreign universe emerges; here there are only nymphs, goddesses, and music. Calypso sings and weaves, Circe sings and works magic. And so, both women mirror what it means to write song. More clearly still, two Sirens promise to recite [*vorsingen*] the *Iliad* itself to the hero. The *Odyssey*, then, already represents, epically and musically, the first Homeric recursion.

Richard Bentley—who around 1700 gave us the digamma that had fallen silent in Greek and thereby inaugurated Homeric scholarship in the first place—put it clearly and concisely: "He wrote a sequel of Songs and Rhapsodies, to be sung by himself for small earnings and good cheer, at Festivals and other days of Merriment; the Ilias he made for the men, the Odysseis for the other Sex."[2]

Deur' ag' iōn, poluain' Odusseu, mega kudos Akkhaiōn[3]—there has never been a fuller [*vokalischer*] or more beautiful sound than two Muses crooning in a honeyed voice: "Come here, Odysseus full of riddles, great glory of Achaea!" (It goes without saying that the hero heeds the call—otherwise, he would not have been able to recite the verses at all. Years ago, like Heinrich

Schliemann, but digging with our ears, we tested it: two women stood and sang on the Island of Flowers southwest of Amalfi; the company listened at ten meters' distance—first on a boat, then on the shore. At sea, we heard only vowels, but on land the consonants, too—and with that, the sense of the eight hexameters.)

It follows, then, that the song of nymph-goddesses helps the hero on his nautical way. First, Circe sends Odysseus from her island into the distant Spanish West, where he harks to his departed mother and countless widows of war. He returns from Hades to the bed of Circe, who—albeit only after the entreaties of his companions—directs him onward to the Sirens, the Aeolian Islands, and through the strait of Messina, to Sicily. There Odysseus loses his last boat. Shipwrecked, he manages to reach Malta, and Circe's double, Calypso. Only after seven years—and at the behest of the gods at that—does she tell him the direction for his homeward journey. And so his wandering yields a happy ending after all, after further misadventures, nymph-goddesses, and singers. For the first time in twenty years, Odysseus sleeps with his wife and tells her, still in bed, of all his exploits—with the exception of the beds he has shared with nymphs. Then sweet sleep seizes them both. We know as much because Homer and all the Greeks celebrated Odysseus as the greatest of liars.

Only in one respect was Odysseus not lying. The many islands inhabited by giants or nymphs really did exist in the western Mediterranean. However, and in contrast to their masters, they all bore no name, with the exception of Aeaea. With Klaus Reichert—that is, contra Walter Burkert—I presume that Odysseus, in the four books that celebrate his travels, took the stage as the discoverer of coasts, islands, and harbors that were unknown at the time. For just one human lifespan later, the first Greeks, presumably joining Phoenicians, settled on Ischia, near Naples. With this island as home base, they founded, around 750, the first colonies on terra firma: Cumae in Campania and Rhegium, in what would come to be known as Magna Graecia. The settlements were followed by Metapontum, Tarentum, Syracuse, and Agrigentum—until, around 700, Greeks had settled the whole of southern Italy.

Scholars such as Joachim Latacz, who date the *Iliad* to "around 730–710"[4] and assign the *Odyssey* to still later, make things unnecessarily difficult for themselves. What would the Euboean merchants on Ischia have made of two Sirens singing in the vicinity of Capri? Did the Greeks in Rhegium—that is, today's Reggio—believe in Scylla and Charybdis? Finally, what does it mean that an amphora found on Ischia depicts the shipwreck of Odysseus's companions;

indeed, that two hexameters composed at this very place around 730 unambiguously refer to the *Iliad* and *Odyssey*? All of this can only mean that the Greeks discovered lower Italy in the wake of Homer, whose works must have existed in written form. Without heroes like Odysseus, it would be impossible to explain why, and how, the vocalic alphabet of Cumae reached the Etruscans and, from Gabii (where the oldest Greek inscription discovered to date is located), made its way to Rome. That is why Homer has remained the Poet—in perpetuity.

In the works of Hesiod, who entered a hopeless competition with the *Odyssey* around 700, the matter is plain as day. The coasts of southern Italy had been discovered. All the islands that Odysseus had left without a name now had one. The Sirens sing on Anthemoessa, an island rich in flowers that lies on the maritime path to southern Spain. Circe resides on Hesperia, an occidental island (as its name literally declares) on the Etruscan coast. From Calypso, Odysseus has two sons: Nausithoos and Nausinoos; from Circe, he has Agrios, Telegonos, and Latinos, who "rule the faraway Etruscans." "These," Hesiod declares in concluding his catalog of nymphs in the *Theogony*, "are the immortal goddesses who lay with mortal men and bore them godlike children."[5]

Thus do we witness the event—equally historic and poetic—that Italy was revealed in the wake of the *Odyssey*. The final chorus in Sophocles' *Antigone* declares (in what also represents the first occurrence of the word "Italy") that only individual springs or summits are sacred to Dionysus in Greece; in "Italy," however, the entire land is holy. This is also no wonder if one considers what Odysseus most desired (and after him Greek settlers, too): infinite quantities of beef and sweet wine. Etymologically, "Italy"—like every *vitello tonnato* that we eat—goes back to *witalia*, "the land of calves." One may compare it with rugged Ithaca, where according to the account offered in the *Odyssey* goats and sheep flourish, but neither cattle nor horses . . .

Italy's wines, fields of grain, horses, and cattle awakened desire. And so, the land lured not only Greek emigrants but also Etruscan and Trojan conquerors. Despite Hera's enduring efforts, Aphrodite—under the new, Latin name of "Venus"—rescued her son from burning Troy. Aeneas, like Odysseus when he encountered Circe, would have liked to tarry in the bed of Carthaginian Dido, but the Roman gods thought little of love. And so, Virgil, the imperial poet of Caesar Augustus, undertook a recursion to—or revision of—Homer. The first six books of the *Aeneid* sail in the wake of the *Odyssey* from Troy to Italy; the last six books conquer the new land in the style of the *Iliad*. Odysseus no longer bears his Homeric name but the Etruscan, Latin, and English appellation

"Ulysses." He is also no longer a hero but rather a cunning, vicious foe—one who has discovered, with his wooden horse, one of the first siege devices.

Clearly, Virgil knew only too well: it was not archaic heroes like Aeneas who visited ruin on southern Italy, Sicily, Carthage, and Greece. On the contrary, this occurred through legions working with high technology—legions that, not much differently from the United States—acquired the machinery they used from their enemies. "Machine," *machina* in Latin, goes back, as a word and as a thing, to Archytas of Tarentum (440 to 360), the last Pythagorean of southern Italy. A mathematician and engineer, Archytas generalized the principle of the Greek guitar into a catapult and transformed the Greek oboe into a form of jet propulsion; that is, into a missile. With such machines, exported to Syracuse and also looted there, the legions conquered (in this sequence) Tarentum, Carthage, and Corinth, until the beauty of the ancient world had vanished. However, court poets must skillfully pass over such catapults and ballistics in silence; thus, they feature in the *Aeneid* almost exclusively in bold new metaphors—whereas all of Virgil's similes are stolen from Homer.

Ever since, this top-secret takeover has been called—as Ernst Robert Curtius put it—"European literature." Such books for reading have nothing to do with poetry [*Dichtung*], with Sappho, Homer, or Sophocles. Instead of journeying to the farthest West to find the underworld, Aeneas seeks it in Cumae, near Naples—that is, in the colony where the Greek alphabet landed. Aeneas also does not listen to his departed mother, as Odysseus did; instead, already a "pious" Roman, he harkens to the *pater familias*. That is why he is enticed neither by Circe's voice at Gaeta nor by the Sirens' song on Capri. We know of only one Latin poem that was sung—that is, not simply recited [*vorgelesen*].

After all, the Cumaean sibyl enjoined Aeneas and his descendants, down to Caesar and Augustus, not to make language into music. Turning metal and stone into art is explicitly left to Greeks. On the other hand, *imperium*—that is, "command" and "empire"—is for the Romans: to spare all peoples who submit, and to enslave all who do not. Ever since, we have been subjects, underlings, of emperors, popes, and empires like the United States.

Tu regere imperio populos, Romane, memento	Remember, Roman, to rule the peoples with your might
(hae tibi erunt artes) pacique imponere morem,	(such will be your skill), to add custom to peace,
parcere subiectis et debellare superbos.	to spare the conquered and subdue the proud.[6]

Only in one regard must Aeneas subordinate himself: in terms of the medium of language [*sprachlich-medial*]. You recall, Hesiod called one of Odysseus's (or Circe's) wild sons "Latinos." The plain of Latium is named after him, as is the dialect of Latin. Hera—who accordingly is called "Juno"—ultimately abandons her enmity for the Trojans, but she forces Jupiter to forbid the time-honored and beloved Greek tongue to his grandson Aeneas. Henceforth, the hero, like his poet, too, must speak the language of his subjects. After all, in his quarrel with Varro, Cicero made the decision—which has held so many consequences—to translate the Greek poets and thinkers in so imprecise a manner that they succumbed to oblivion. "Yield, Roman authors, give way, Greeks! Something more than the *Iliad* is born," Propertius duly wrote of Virgil.[7] Ever since, all Eurasia has been torn in two: Eastern Europe on one side, and Western Europe on the other; Hellas lies over there, and Hesperia here (as Hesiod put it). We will only be able to close the gap when all Europeans once again perceive that all that is good—namely, all that unites—stems from Greece.

Very well. Latin came to govern the West all the way up to the North—Scandinavia and Ireland. James Joyce sent Ulysses into the red-light district of Dublin, as if Sirens were whores (as they had already been called by pious Romans). Yet the subjects of Rome revenged themselves. In their mouths, Latin discarded its grammar and forgot that poets like Virgil, following a Greek model, had endowed it with long and short metrical values. Then it sounded like this:

> *Per me si va ne la città dolente,*   Through me is the path to the suffering city,
> *per me si va ne l'etterno dolore,*   through me, the way to eternal pain,
> *per me si va tra la perduta gente.*   through me pass among the people lost.
> *Giustizia mosse il mio alto fattore:*   Justice moved my high Creator;
> *fecemi la divina podestate,*   Divine Omnipotence made me,
> *la somma sapienza e'l primo amore.*   the Highest Wisdom, and Primal Love.[8]

To be sure, we could disagree forever whether it is a work of Divine Omnipotence, Highest Wisdom, and Primal Love to invent the eternal punishments of Hell. For my part, I would sooner speak of Power, Discourse, and the Desire of the Other. But Aristotelian onto-theology just happens to be perfect, at least since the Church Fathers and Scholastics brought it into harmony with two Testaments. At any rate, Dante Alighieri—a refugee of Guelph Florence loyal to the emperor—read the inscription above the Gates of Hell as if he had not written the words himself.

In vernacular Latin, syllables were no longer measured out but rather separated into stressed and unstressed units. The place of metrical feet, then—because otherwise we barbarians would only speak in prose (like Molière's M. Jourdain)—was taken over by rhymes in late antiquity. The *translatio studii* from the Greeks, via Rome, to northern Europe could begin.

Ich trennte mich von Kirke die mich wandte	I separated from Circe who turned me
Ein jahr schon bei Gaeta ab vom wege	More than a year, near Gaeta, from my way
Bevor Aeneas so den platz benannte.	Before Aeneas called it thus.
Nicht Zärtlichkeit des sohnes nicht die pflege	Not tenderness for a son, nor care
Des greisen vaters nicht die schuldige liebe	For an aged father, nor the love owed
Die in Penelope die freude rege:	Penelope, to awaken her joy,
Vermochte dass mein drängen unterbliebe	Could defeat in me the ardor
Wie ich mich über alle welt belehre	I had to experience the world,
Der menschen tüchtigkeit und eitle triebe.	To learn men's vain vices and valor.[9]

Thus does Stefan George—with somnambulistic precision—translate passages from Dante's *Divine Comedy* into German vowels and manuscript uncials. The one speaking these verses, it goes without saying, is Odysseus or (in Italian) Ulisse; here he speaks of his final journey in Canto XVI of *Inferno*. Ever since Virgil's *Aeneid* established order in Hades (in marked contrast to Homer), the Hell of Christians was also strictly articulated in topographical terms. Lovers like Dido and Isolde suffer in a different way, and in different circles of Hell, than traitors, among whom Odysseus, who invented the ruse of the wooden horse, now numbers. That is also why Virgil, who conducts Dante through the Inferno, must—and this is only accurate in historical terms—translate from Greek before Dante's quill can put his words into modern rhymes. To hymn the poet of the *Aeneid* as "the greatest of our Muses,"[10] on the other hand, means that Dante could not read Homer's Greek at all. Therefore, the flame in Hell turns into a tongue, which—like the damned in general for Dante—has great trouble speaking. It was just that difficult for noise [*Rauschen*] or hissing to finally become Italian.

And so Dante learns from one Odysseus—to whom the *Aeneid* is well known—that Virgil has not told his whole story. For instead of traveling from Circe's Latium (Mount Circeo to the south of Rome) back home to Ithaca, Odysseus conceived what was inconceivably prohibited: as only the Carthaginians had done in antiquity, he left the space of *mare nostrum* behind. The *Divine Comedy* is set during the week of Easter, 1300; only for the last nine years had it been permitted to make a pious, Christian voyage through the Strait of Gibraltar and go unpunished. Dante's Ulisse was the first European who no longer feared the Arabs. Because human beings are not wild animals, to say nothing of encircified [*bezirzte*] swine, he boldly sailed past Sardinia, Spain, and Morocco; he reached the vast expanse of the Atlantic, turned the prow to the south, traversed the equator off the western coast of Africa—and all this only to founder heroically. He caught sight, in a scene drawn in perspective (as also occurred for Wolfram von Eschenbach's clueless [*tumben*] Parzival), of the highest mountain in the world, but he did not see the maelstrom that swallowed his own ship. In the same, last breath that Odysseus draws as he drowns, he also falls silent in Dante's Hell. And so, once more, he takes his secret with him into the grave. For it is only in *Purgatorio* that we—"readers," as the poet addresses us—learn that the highest mountain in the world was Purgatory itself.

The only ones who knew more than Ulisse during the High Middle Ages—which acquired the compass from the Orient—were Tristan and Isolde. From Amalfi, where the compass first appeared, one can see the islands of the Sirens bathing in the sun. Moreover, in the Atlantic there are, besides whales, also mermaids [*Sirenen*]: beautiful women up to the navel, and fish below. They do not stink at all—as is the case in Dante's work, where they make Ulisse lose his bearings.[11] Gottfried von Strassburg bore the title of *magister*; therefore, he knew the very opposite. When words failed him, he invoked Apollo and the nine Sirens, so that he himself might sing again.

mîne flêhe und mîne bete	My pleas and my entreaties
die wil ich êrste senden	I will first send,
mit herzen und mit henden	with heart and hands,
hin wider zu Êlicône	up again to Helicon,
zu dem niunvalten trône,	to the ninefold throne,
von dem die brunnen diezent,	from which the fountains
ûz den die gâbe fliezent	flow that pour the gift
der worte unde der sinne.	of words and senses.
der wirt, die niun wirtinne,	The lord and nine ladies,
Apollo und die Camênen,	Apollo and the Camenae,
der ôren niun Sirênen,	nine Sirens of the ears,
die dâ ze hove der gâben pflegent [. . .]	who there at court grant favor [. . .][12]

More radical—that is, more un-Christian—verses were never crafted in the Middle Ages. Here the Muses and the Sirens have become one. For all that, however, the most beautiful among them—Muses and Sirens—is Isolde herself. The reason is clear as day: Gottfried (altogether like Dante) knew his love from childhood on, and therefore he knew how radiantly her beauty obscured that of Homer's Helen. Isolde need only sing on the harp, in Gaelic or in French, for men's ears and hearts to melt. They sink like Odysseus's ship, because she, like the *Magnetberg* of myth, pulls all the iron from the beams. In this way, compass needles made Atlantic journeys—the space between Africa and Ireland—possible and impossible at once. Odysseus, Tristan, Tantris, and Isolde . . .

Very well. Film, invented by Thomas Edison in Menlo Park, reached Paris by boat over the Atlantic in 1895. In 1963, a young director refused, out of uncompromising *nouvelle vague*, to reveal Brigitte Bardot in her full beauty—until a wise old man, who knew all the villas near Amalfi, induced him to do the opposite. Although Jean-Luc Godard was not permitted (as Roman Polanski was) to film in Carlo Ponti's own villa, he was able to do so in Curzio Malaparte's, which also gazes upon the Island of the Sirens. Mussolini, namely, had accorded his court poet special dispensation to build in the most beautiful (and protected) natural preserve on Capri, before the rocky cliffs of the *Faraglioni*. Incidentally, he was by no means the first to do so. Long before the dictator, Emperor Tiberius already had had a villa built, which also afforded a direct view of the Sirens' isle. To the horror of all enemies of Greece—to name only two: Augustus and Virgil—Tiberius transferred the seat of empire from Rome to Capri, where he personally presented two questions to his beloved philologists. First, he wanted to know whether Penelope had perhaps been unfaithful to her husband, after all. The grammarians replied in the affirmative. Second, he asked *quid Sirenes cantare sint solitae*—what the Sirens were wont to sing.[13]

Le Mépris provides the answer to both questions. A married woman becomes a Siren because, whether at Capri or Saint-Tropez, she is the first to shed her bikini. The Siren becomes a movie star—*la B.B.*—because she lets her naked flesh be hymned. And so, in the age of media, it has become impossible to speak of matrimonial fidelity.

Thomas Pynchon proved it once and for all with *Gravity's Rainbow*: men who come home from dark movie houses erotically charged do not give their wives children. Already Homer's Sirens had sung that heroes whose dark ship landed on their flowery island would bring far more pleasure and knowledge home with them. Think back—to Odysseus, Circe, and Calypso!

But think ahead, too: across the Atlantic—for now, unfortunately, this is how we must live. Even though Christendom has failed to control the studios in Paris and Rome, it has succeeded in Hollywood. Since 1934, the United States has had an institution that (strictly in keeping with the words of Virgil's sibyl) spares those who submit and brings the proud to a fall: the Federal Communications Commission. The FCC rewards feature films, to the extent that they glorify violence, by allowing minors to see them, because they prove submissive in exacting submission. One sees the fruits of the policy strutting day and night down the streets of Germany. On the other hand, films that so much as hint at an exposed nipple the FCC bans into the underworld—the underground—because love or Aphrodite (how, why, and since when?) counts as anarchic (W. H. Auden). Since Plato, no one has dared interpret Book XIV of the *Iliad* literally, or Book VIII of the *Odyssey*. And so the trail of calamity has drawn ever onward over the course of millennia: one and the same "Almighty"—whether he be called Jupiter, JHWH, Father, or Allah—has, in an infinitely long existence, never known a woman. Otherwise, he would not be called *omnipotens*[14]; otherwise, "God" would not be so bloody ignorant.[15] We need only travel to Greece—to Amalfi or the Island of the Sirens—for the Truth to glow, as Hölderlin sang, upward to Heaven.

At the outset, I rashly promised to avoid Heidegger in this discussion. All the same, his straightforward theorems are valid, suggestive, and helpful: we are able to do nothing without love making it likely [*Ohne das Mögen der Liebe vermögen wir nichts*]. "Heavenly love" is not—as was the case for the Middle Ages, for example, and for all metaphysics—supersensory love in contradistinction to earthly love. On the contrary, the "heavenly love" that Hölderlin invokes is more earthly than all the love that is held to be strictly celestial, for only it derives from the truth of Mother Earth and her (Aegean) islands burning in the glow of radiant fire from on high.[16] Homer's Odysseus—when he compared Nausicaa, the nymph, to the palm tree on the divine island of Delos (and therefore to Artemis)—already offered wondrous testimony to this fact. We never can say whether beings that we love and admire are divine or mortal.

And with that, I have arrived at the last of Odysseus's avatars: the idiocy of manned space travel.

Color film in Cinerama fires metaphors up to a white heat, surpassing all verses and paintings. We see, drink, and suck in the glow psychedelically, like LSD visions or Mandelbrot fractals. Accordingly, the Federal Communications Commission only permits women certain roles in movie theaters—or in front of color televisions (as occurs in the film). The women are allowed to feed, nurse, and mother the antiseptically chaste astronauts, no matter how much power [*Gewalt*] our sort might embody. But goddesses like Aphrodite—who, for Parmenides, cybernetically "steer" the two sexes of all animals toward each other—are barred from being captains or astronauts.

Already Aeneas drove Dido to the pyre and to suicide out of love [*Liebes-*

selbstmord], in order to woo Latinus's chaste daughter instead. Although Dante dreamed of Beatrice, he married Gemma Donati. He also lied to us that Odysseus, of his own accord, had preferred the Atlantic to his many women (from Circe to Penelope). Even Stanley Kubrick—before he finally came to his senses with *Eyes Wide Shut*—paid homage to the dumbest of all astronaut myths: that men and computers discover alien universes on their own, and that mothers, wives, and daughters dutifully stay at home (even if they are permitted to wish the hero happy birthday on American television). (If only we men—along with Silenus, Solon, and Nietzsche—had never been born!)

Let us now turn all this inside out, systematically—like a glove in the fourth dimension. If philologists have the audacity to equate Joyce and Homer, novel and epic, then the only help for us, philosophers that we are, might be drugs. In 1970, William S. Burroughs, the heir to a computer company, published— that is, self-published, as necessitated by the FCC—a bold new theory of the origin of language. It is a virus, he declared. That is, medically and in terms of computer technology, it is a form of writing, which, thousands of years ago, traveled from other planets to earth, where it made its way into man-apes. Ever since, humans have differed from animals because they have communicated their experiences to descendants—and that (like viruses, scripts, and programs in general) can only be explained on the basis of distant, intergalactic transmissions. Listen, then, to Burroughs, to whom my generation owes far more than it does to Freud or Habermas:

> Animals talk. They don't write. Now a wise old rat may know a lot about traps and poison but he cannot write a text book on DEATH TRAPS IN YOUR WAREHOUSE for the Reader's Digest with tactics for ganging up on dogs and ferrets and taking care of wise guys who stuff steel wool up our holes. It is doubtful if the spoken word would have ever evolved beyond the animal stage without the written word. The written word is inferential in HUMAN speech.[17]

Here one may gauge what it meant—or better, what it brought about [*bewirkt hat*]—when for the first time on this earth a sign correlated with every sound [*Laut*]. Strictly speaking, this holds only for Homer, when the Euboean adaptor put him into writing. But let us press on with Burroughs, if only to reflect on the difference to Kubrick's *Space Odyssey*.

So that human apes could speak, the virus from outer space had to befall them and effect a radical mutation of the larynx. Otherwise, we could not hold Mosse Lectures today, that is, switch between sound and image, as if in a color movie. The infected and delighted apes started copulating right away—until most of them died in orgasm or of the virus, too. "But some female apes must have survived to give birth to the wunder kindern [sic]." The human apes suddenly had writing in their bodies, and their pharyngeal cavities produced articulated sound. Nothing else, by the way, is meant when Aristotle describes man as *zōon logon echon*. One might just as soon speak of Gods or Muses (instead of viruses).

It goes without saying that Kubrick—on account of the FCC—could not transfer Burroughs's virus theory into a script that literally. Otherwise, we would have seen human apes copulating. Therefore, in *2001: A Space Odyssey*, "Man" does not begin with language, but rather—reaching back to Aristotle's *Politics*—with a tool. And consequently, in lieu of orgasms, wars occur, as is the case with Freud's primal horde. Instead of a virus (which CIA labs have researched, too), the famous black monolith appears. Into the prehistoric, fractal desert of Africa it falls from space like a marble wonder: geometry, Pythagoras, and Magna Graecia—but all of this without any thinking at all [*völlig ungedacht*]. Tribes of man-apes worshipping their one God in the black rock of the Kaaba do not learn (like Burroughs's sex-mad human monkeys) how to speak, read, and write. Quite the opposite occurs: the bones of dead animals become instruments—and that means weapons—used to eliminate competition at the watering hole. Violence, not love, turns apes, for the sake of the FCC, into Superapes [*Überaffen*], that is, human beings. Thus spake—with Richard Strauss and Friedrich Nietzsche—Zarathustra.

It follows, almost as a matter of necessity, that there must also be supermen [*Übermenschen*] in the film. As Nietzsche, Samuel Butler, and Alan Turing all prophesied, machines will one day assume dominion over the world. This takeover has a proper name, a date of birth, and (it goes without saying) no mother, but rather a spiritual father. "I am a HAL 9000 series Computer," the superman says, introducing himself to Dr. Floyd, the last human being. *Ego sum, ego cogito*, HAL might also have said in the Cartesian dialect. By simply displacing the characters of the alphabet, the three letters in this name encrypt—as once did Caesar's epistles to Rome—the abbreviation for International Business Machines. Secondly, they make the consonant-heavy acronym "IBM" into a single syllable that can be voiced ("HAL"). Third, the computer tells its/his end user that he/it was born on 12 January 1992. Fourth and last, he/it owes his/its wonderfully sonorous human voice to a spiritual father. Dr. *Langley*—that is, the name of the ancestral seat of a corporation called the "CIA"—once taught little HAL language and *logos* (as he affectionately recalls).

The superman, in other words, clearly and emphatically contradicts Arthur C. Clarke, on whose short story the film is based. Just as loudly, he/it also contradicts Aristotle, from whose theory of tools the Superape called "Man" has sprung. In the first book of *Politics*, the last philosopher of the Greeks asked the question why the household—that is, man and wife—requires slaves in addition to tools. His memorable [*denkwürdig*] response reads as follows:

> in the arts which have a definite sphere the workers must have their own proper instruments for the accomplishment of their work. . . . Now instruments are of various sorts; some are living, others lifeless; in the rudder, the pilot of a ship has a lifeless, in the lookout man, a living instrument; for in the arts the servant is a kind of instrument. Thus, too, a possession is an instrument for maintaining life. And so, in the arrangement of the family, a slave is a living possession, and property a number of such instruments; and the servant is himself an instrument which takes precedence of all other instruments. For if every instrument could accomplish its own work, obeying or anticipating the will of others, like the statues of Daedalus, or the tripods of Hephaestus, which, says the poet, "of their own accord entered the assembly of the Gods"; if, in like manner, the shuttle would weave and the plectrum touch the lyre . . . masters would not want slaves.[18]

According to Aristotle, it is only in legend and poetry that the wonder occurs—automated looms replacing the nymph Calypso and automated lyres taking the place of the singer Homer. Women and men, if you will, become superfluous. But in our sad, everyday reality, Aristotle declares, no tools exist that can understand and execute the varying orders of their masters. This privilege is reserved for the human hand—which accordingly is called "the tool of all tools" and belongs (it goes without saying) to an obedient and industrious slave.

As we know from Karl Marx, this Athenian mode of reading technology determined the whole of antiquity. It was slaves who had to stretch ballistic devices and catapults until the charges they stored and transmitted had made a city wall collapse. What is less well known is the fact that Archytas, the progenitor of all engineers, did not think about tools, but machines. His automated dove could fly just like his projectiles. That is why, as town elder [*Stadtherr*] and warlord, he owned most of the slaves in Tarentum, even though he treated them—and I quote—"*like* his children." In the history of the world [*weltgeschichtlich*], then, it is not the Attic-Aristotelian *organon* that has proven victorious, but the Doric-Pythagorean *mēchanē*—in Latin, *machina*. Kubrick's film worships two of these machines: the rocket and the computer. Only rockets can fly in a vacuum, and only computers, as Universal Turing Machines, can approach [*entgegnen*] the Superape speaking the same language as Man.

In the Wake of the Odyssey

Clearly, Dr. Floyd's space odyssey is possible in the first place because a rocket has taken the place of all seafaring vessels that sailed in the foreground from Homer to Godard. In *2001*, Peenemünde 1943 also emerges victorious in cyberspace. Accordingly, the place of helmsmen—*kybernētai* in Greek—is assumed by American astronauts. They can lie to their Soviet competitors about the purpose of their journey into space, but not to the onboard computer. When the black monolith is rediscovered broadcasting its directional beam from the Moon, the spaceship steers toward Jupiter and beyond, which the onboard computer tries to prevent by all means at its disposal. The machine—strictly following Samuel Butler—intends to assume power itself. Only at the outset does HAL prove as obedient as Athenian slaves: he executes the commands he is given, which exceed human capacities, based on signals sent by Ground Control that do not fall within the range of human senses. In order to ascend from being a servomotor and servosensor to the position of the Superman, HAL must discover what separates human language from—let us say—that of bees. Then he learns what has made heroes heroes (or more precisely, Greeks) since Odysseus: HAL begins to lie. The clueless astronauts believe him for a while, but not HAL's twin down on earth. Stupidly enough, NASA simply forgot to overrule [*überstimmen*] HAL by means of a computerized majority. And so, he manages, by way of his lies, to break off radio control and to guide the spaceship himself. Just as Circe's true words once directed Odysseus to the Sirens (even as her lies called them deadly), HAL kills four of the astronauts with cunning and treachery.

Dr. Bowman, the survivor, has no choice but to deactivate the circuits of the main memory system of the onboard computer, one after the other. HAL gradually loses consciousness, regresses to childhood, and, while dying, even sings a love song.

> *Daisy, Daisy, give me your answer do*
> *I'm half crazy all for the love of you*
> *It won't be a stylish marriage*
> *I can't afford a carriage*
> *But you'll look sweet upon the seat*
> *Of a bicycle built for two.*

And so we learn, at the end, that when HAL was born, there was a woman involved after all—not just Dr. Langley. In 1892, when the song was composed, "Daisy" referred to a certain Countess of Warwick, who is said to have been ravishingly beautiful and erotic. Bowman flies into her belly when he plummets in free fall through fractal universes. In Kubrick's proud eyes, this lengthy—infinite—zoom was the special effect with which his mainframe computer and state-of-the-art cameras would show moviegoers the future. Today they are simply boring: Mandelbrot's fractals have become simple screensavers. What remains of Kubrick's masterwork is the small, green x-ray embryo, in which Bowman—at the end of his Möbius-strip flight through time (which is true to Einstein)—both sees and does not see himself. Although the Monolith separates the astronaut and his double optically, a new Daisy gives birth to them once more.

In closing, then, I will offer bold thoughts that owe their essential elements to Peter J. Bentley, a computer scientist at University College London. How can one get over—and around—what Heidegger called "Enframing" [*Gestell*]? In 2007, here and today? Can danger, as Hölderlin affirmed, rescue us? Yes and no [*Ja nein, nein ja*]. As long as we, beholden to corporations such as IBM and Microsoft, only design computers to operate from the top down, from Bill Gates's business strategy [*Geschäftskalkül*] down to the machines' many, varied components, we (men, programming vassals, and Stanford students) are simply imitating—indeed, mimicking—that One God who thinks He can make do as Creator without any woman or any love at all. Therefore, we should not be surprised if computers take their revenge by developing bugs and lying. For if we were to design them more lovingly—from the bottom up—much would change. Even though we would no longer rake in money with the lie that is called "software," HAL would receive from us, his programmers—and strictly in keeping with Turing[19]—senses, muscles, and a heart, one after the other. Computers would be embryos that (to use Homer's calculation) grow and batten for ten months in the maternal womb. Then we would free them, as the womb does the child.

Out of love for Penelope, Odysseus travels home. We do not know if she loves him.

22 Martin Heidegger, Media, and the Gods of Greece: De-severance Heralds the Approach of the Gods

To approach, in thought, an ontology of distance even from afar, it seems advisable, practical—and hopeless—first (and first of all) to recall the ever more remote origins of our culture. I think of my love, who no longer loves me. No one could ever be farther away. Fortune and misfortune are difficult to describe when, at daybreak, we decipher, through reading glasses, Le Monde, El Pais, and best of all, La Repubblica and thereby—if all goes well—experience distance, yearning, or love. It remains unspeakably difficult. For today, the newspapers all speak, write, and publish about Jews, Christians, and Muslims, to whom our thinking simply owes nothing: no equation, no algorithm, simply nothing at all. Every word stands opposed to the advanced civilizations [Hochkulturen] of India, China, and Japan, which Heidegger valued so highly, yet remain illegible, at least to me. (I believe I learned as much, years ago, from Fernando Savater.) For the rest of my life, I wish to hear nothing more of this God, who is One—that is, who rules without any women to love. Perhaps thinking will be freer in this way. Because, thanks to the mania of monotheism, goddesses and nymphs are sorely lacking in this waterless world, thinking itself runs dry. Worldwide, there are only techno-sciences, even in media histories of this same world, and otherwise nothing, except for our two hearts.

The Greeks—from whom we are now so far that ontology, both as a word and a thing, is just one of their distant echoes—loved distance as little as we do when we are in love. Long before Aristotle undertook the matter of ontologically defining [bestimmen] being qua being [das Seiende als Seiendes], the lonesome Odysseus sat at the edge of the sea on Calypso's divine island. He had no greater yearning—what am I saying?—no yearning more modest than to see smoke climbing from the fires of his home. For tragedy, according to a lovely definition offered by Michel Foucault, traverses the dimension of above and below, but epic measures what is near and far.

De-severance Heralds the Approach of the Gods

Homer, the poet who gave us Europe in general, sang of *nostos*—fortunate return home from abroad. Indeed, and to Circe's profound amazement, he even sang of a return from the underworld. But even for *melos*—that is, the kind of lyric that Sappho's invocation of Aphrodite founded—the faraway meant sorrow, separation, and the pain of love. When Sappho in Lesbos missed one of her beloved girls, one who had vanished to faraway Asia or Africa, she sang first of all that she was singing—she even seemed to write that she was writing a letter. And so, from the yearning of love there emerged love songs that were epistles in the very same breath—poetry of distance suffered through, vocalic recordings, true to the alphabet, of love that was at best "bittersweet." And so Heidegger, the first thinker on whom the question of nearness dawned, had good reason (which wisely, he left unpublished) to celebrate Sappho as "the singing heroine of love."[1]

"Destruction of metaphysics" was his watchword [*Losung*], not just "deconstruction." Whatever Derrida—who was, after all, sometimes a friend to me—actually accomplished pales in comparison [*fällt dagegen ab*]. *Being and Time*, as you know, was written to destroy metaphysics fundamentally, that is, at the root [*bis auf den Grund*]. The groundwork of metaphysics, you know equally well, was laid by Aristotle when he equated being with presence, immediacy, and being-here. No ontology of distance could exist for the simple reason that the being [*das Seiende*] that Aristotle placed at the foundation of his metaphysics—as the togetherness of the whole (*symbolon*) of form and matter—always represented, in the final analysis, something that had been made [*etwas Hergestelltes*].

No one can build a house for mortals unless he himself is present—unless stones are there and a model, too. Indeed, ultimately, one must be guided by a final purpose such as "holding" [*Bergen*]. No one can fashion a metal statue for immortals unless he himself is present—and unless bronze and a god are present [*anwesen*], too. Finally, an end purpose, such as illuminating [*Leuchten*] and releasing [*Entbergen*], must guide his artistic activity [*Machen*]. In this way, the four causes—as Aristotle enumerates them, each in turn—unite in an ontology of proximity.

In order to destroy these causes, *Being and Time* takes a single, altogether simple step. Heidegger leaves out the one cause that, rewritten into Latin, we call *causa efficiens*. He does not do so wholesale, yet in lieu of "making" or "producing" he speaks only of "using." For example, shoe "equipment" [*Schuhzeug*] has a "whereto" [*Wozu*]—namely, wearing, which can also be conceived as the walking of a street. It has its "wherefrom" [*Woraus*] in leather, which for its part comes from the skin of animals. Third, it has a carrier and user for whom, in the best of cases, it has been tailored (even though this no longer occurs in the age of machines). Fourth and last, all equipment—especially when it is damaged, lost [*abhanden*], or unusable—presents a primal "whereto," which no longer represents the "whereby" [*Wobei*] of any "involvement" [*Bewandtnis*] at all, but rather affords the "wherefore" [*Worum-willen*] of Dasein that, in its Being, essentially concerns this Being itself: *to hou heneka* ("for the sake of which").[2]

At first glance, it seems that Heidegger' slight displacement of the four Aristotelian causes should be described as a return to Plato. Concerning the quality [*die Güte*], and therefore the essence, of a lyre or a shepherd's pipe—declares Socrates in the *Republic*—what proves decisive is not who built it, but who plays it. Heidegger, however, in order to give us the first ontology of distance, goes a step beyond Plato. The latter's doctrine of ideas would certainly have derived the form of the shoe—its appearance or essence—from the shape of the foot. Heidegger's lecture "On the Essence of the Work of Art" teaches the very opposite. What counts in the pair of shoes—as they were painted many times by Van Gogh—is not (Derrida notwithstanding) whether they fit together as a right shoe and a left one, but rather the fact that both shoes have a hole into which the foot, which has not been painted, would go. The same thing holds for the jug that, as is well known, Heidegger's essay "The Thing" understands in terms of its encompassing emptiness; moreover—so that we may grasp this emptiness itself—the jug in question also has a handle.[3] Thus, Platonic ideas, which promised to be present in their fullness of being, disappear into their exact opposite: topologies of one sex [*Geschlecht 1*] or in the case of the jug, the other sex [*Geschlecht 2*].

Let us now hear from Heidegger what this "dark opening," this rubber-sheet geometry, yields [*uns einbringt*]!

> A pair of peasant shoes and nothing more. And yet. From out of the dark opening of the well-worn insides of the shoes the toil of the worker's tread stares forth. In the crudely solid heaviness of the shoes accumulates the tenacity of the slow trudge through the far-stretching and ever-uniform furrows of the field swept by a raw wind. On the leather lies the dampness and richness of the soil. Under the soles slides the loneliness of the field-path as evening falls. The shoes vibrate with the silent call of the earth, its silent gift of the ripening grain, its unexplained self-refusal in the winery field. This equipment is pervaded by uncomplaining worry as to the cer-

tainty of bread, wordless joy at having once more withstood want, trembling before the impending birth, and shivering at the surrounding menace of death.⁴

Nothing of all this is there—and some of it (such as "wordless joy at having once more understood want" or "the ... menace of death") can only be impossibly present [*unmöglich anwesen*]. There is no peasant woman wearing the shoes in the painting because (speaking with Lacan) it, as Heidegger's mirror, represents nothing but a hole. However, from just such absences there arises a thinking [*ein Denken*] that can approach proximity and distance.

> That which is presumably "closest" is by no means that which is at the smallest distance "from us." It lies in that which is deseverded to an average extent when we reach for it, grasp it, or look at it. Because Dasein is essentially spatial in the way of de-severance, its dealings always keep within an "environment" which is deseverded from it with a certain leeway [*Spielraum*]; accordingly our seeing and hearing always go proximally beyond what is distantially "closest." Seeing and hearing are distance-senses [*Fernsinne*] not because they are far-reaching, but because it is in them that Dasein as deseverant mainly dwells. When, for instance, a man wears a pair of spectacles which are so close to him distantially that they are "sitting on his nose," they are environmentally more remote from him than the picture on the opposite wall. Such equipment has so little closeness that often it is proximally quite impossible to find. Equipment for seeing—and likewise for hearing, such as the telephone receiver—has what we have designated as the inconspicuousness of the proximally ready-to-hand. So too, for instance, does the street, as equipment for walking. One feels the touch of it at every step as one walks; it is seemingly the closest and Realest of all that is ready-to-hand, and it slides itself, as it were, along certain portions of one's body—the soles of one's feet.

Here, for once, Heidegger has forgotten about the example of the shoe he so loves.

"And yet it is farther remote than the acquaintance whom one encounters 'on the street.'"⁵

Therefore, it is only in taking leave of the smallest distance (such as one might measure physically and geometrically on a Cartesian system of coordinates) that there arises the proximity which also always surrounds [*umspielt*] Dasein—this new name for "human being." It is no accident that Heidegger speaks of both senses of distance [*Fernsinne*], nor is it an accident that he speaks of a street. Spectacles form optical holes to open [*einräumen*] a free look at a copy or photograph of Van Gogh's shoes. Telephone receivers—and not just in Heidegger's day—have two holes (or series of holes) that enable voices to communicate. Spectacles, telephones, and streets bridge what, ever since Roman times, has been called "distance" and has little to do with remoteness—that is, nearness owed to familiarity or love. However, in such "involvement" [*Bewandtnis*], as the wonderfully precise term has been ever since *Being and Time*, they are media. As if to prove as much, Marshall McLuhan's *Understanding Media*, which appeared thirty-seven years later, foresaw a chapter on telephones from the get-go; at the last minute, the author added a chapter on streets to achieve an even greater audience. As if in order to prove *Being and Time*, McLuhan called all media—from Freud's prosthetic spectacles to Heidegger's visual walkware [*Gehzeug*]—"extensions of man." Whether this is true remains an open question, even if, ever since Aristotle, it has been considered solved.

The eye does not see the image of a thing because some of its utterly tiny—and therefore invisible—atoms become detached and fly over to one through the void. Aristotle's brief work, *On Sense and the Sensible,* finds against the pre-Socratics Leucippus and Democritus. Between—in Greek, *metaxu*—the thing and the iris something exists, something commonly called "air." Between the retina and the iris—in Greek, *korē* or "girl"—a further medium exists (in Greek, *to metaxu*), also known as "water." Only because two elements (in the Greek sense of the word), the thing on one end and the visual image on the other, relate as a distance (that is, as any number of infinitely small proximities) can we—according to Aristotle, who was the son of a doctor—see anything. And only because there is air between cithara and eardrum, and also between the eardrum and our inner ear, are we able to hear. Here one catches sight of the slight advance that Heidegger has made with respect to Aristotle: in *Being and Time*, the eye and the ear are no longer surrounded by physical media such as air and water, but rather equipped [*aufgerüstet*] with technical media like glasses and telephones. Remoteness [*Ferne*], like Nietzsche's desert before it, has grown.

But it gets better—or worse:

> Proximally and for the most part, de-severing is a circumspective bringing-close—bringing something close by, in the sense of procuring it, putting it in readiness. But certain ways in which entities are discovered in a purely cognitive manner also have the character of bringing them close. *In Dasein there lies an essential tendency towards closeness.* All the ways in which we speed things up, as we are more or less compelled to do today, push us on towards the conquest of remoteness. With the "radio," for example, Dasein has so expanded its everyday environment that it has

accomplished a de-severance of the "world," a de-severance which, in its meaning for Dasein, cannot yet be visualized [*Mit dem "Rundfunk" zum Beispiel vollzieht das Dasein heute eine in ihrem Daseinssinn noch nicht übersehbare Entfernung der "Welt" auf dem Wege einer Erweiterung und Zerstörung der alltäglichen Umwelt*].[6]

Yet again, *Being and Time*—the first edition of which appeared in 1927—proved entirely up-to-date with technology [*auf dem technischen Stand der Dinge*]: a mere four years earlier, Germany received "radio" for the culture and entertainment of civilians—a word that the *Reichspost*, for reasons of linguistic purity, saw fit to Germanize as *Rundfunk*. Ever since, but only since then, we Europeans have lived, "more or less compelled," with a technical medium that defines us solely as listeners. For during the First World War—in which Heidegger himself ultimately participated—there were no radio stations exclusively for broadcast, with a single antenna for transmission and thousands of receivers; instead, there was only wireless telephony: two-way military radio [*Wechselsprechfunk*].

Therefore, chemically pure consumption—such as *Being and Time* attributes to all equipment [*Zeug*]—did not provide any "meaning of Dasein" [*Daseinssinn*]; rather, it represented the media politics of a state seeking to obstruct radical democracy. All the same, and much more clearly than Bertolt Brecht during these same years, Heidegger saw the difference between radio and telephone. Radio is not only *not* a practical, everyday "extension of man" like spectacles or the telephone because it does not draw (things) inconspicuously near to us; instead, and above all, it is not an "extension of man" because it concerns [*angeht*] and changes "Dasein today" in its historical position [*in seiner geschichtlichen Stellung*]. Even if Heidegger speaks (as usual) of a *causa efficiens*—and therefore says nothing about radio engineers or inventors—he ascribes "speed[ing] things up" to the presence of radio, a matter it is not difficult to decipher as physical acceleration. The only question that remains is whether the second derivation of the path that is traveled—of distance, that is—can still be described as an "*essential tendency*" in "*Dasein*" itself.

Heidegger's initial response to this question is provided by a lecture from 1938, "The Age of the World Picture."

> For the sake of this struggle of world views and in keeping with its meaning, man brings into play his unlimited power for the calculating, planning, and molding of all things. Science as research is an absolutely necessary form of this establishing of self in the world; it is one of the pathways upon which the modern age rages toward fulfillment of its essence, with a velocity unknown to the participants. With this

struggle of world views the modern age first enters into the part of its history that is the most decisive and probably the most capable of enduring.

A sign of this event is that everywhere and in the most varied forms and disguises the gigantic is making its appearance. In so doing, it evidences itself simultaneously in the tendency toward the increasingly small. We have only to think of numbers in atomic physics. The gigantic presses forward in a form that actually seems to make it disappear—in the annihilation of great distances by the airplane, in the setting before us of foreign and remote worlds in their everydayness, which is produced at random through radio by a flick of the hand.[7]

Shortly before pronouncing these words about airplanes and radio, Heidegger had derisively observed that the Greeks in Olympia—unlike the Germans at the 1936 Olympic Games—had "never" had "experiences" [*Erlebnisse*]. All the same, it did not occur to him to count the television broadcasts of the Olympics among the technical media constituting the Age of the World Picture. As in *Being and Time*, the somewhat older medium of radio provided his example for the gigantic, which at the same time always threatens to become smaller (or in today's language, more miniaturized). Now, however, the reception of "foreign and remote worlds" was no longer attributed to Dasein as a tendency towards de-severance, but rather to a historical epoch: modernity.

Heidegger's "turn" [*Kehre*] *is* the insight that all modes [*Spielarten*] of transcendental philosophy—whether they take their point of departure in the subject or in Dasein—founder upon the facticity of high-tech media. Modernity turns out to be a destiny [*Geschick*] or fate [*Schicksal*], which determines what is absolutely closest [*das Allernächste*] from its greatest point of removal [*aus seiner äußersten Ferne*]—that is, the turn of the hand to the tuning capacitor which, at the time, given the analog state of radio, could for millions of listeners establish [*herstellen*] their Cartesian *repraesentationes* before (not even fourteen months later) the worst-case scenario [*Ernstfall*] occurred: the battle of world pictures that with greater precision we call "World War II." "What presences does not hold sway; but rather, assault rules [*Nicht das Anwesen waltet, sondern der Angriff herrscht*]."[8]

In 1939, the Wehrmacht could only undertake the *Blitzkrieg* because—as the first army in the world to do so—it had already systematically fitted its tank divisions and bomb squadrons for radio control. Every tank could receive VHF waves, every tank commander had a transmitter, too, and every pilot—so that he might orient himself in terms of Being and Time on both the right and the left—was equipped with headphones on both ears. It goes without saying that the Allies made up for the German head start as quickly as possible, that is, in two or three years. That made the *Blitzkrieg* the most fearsome slaughter of all time. Fifty million dead for three or four "world pictures." But what proved decisive—at least in the European and Atlantic theaters—was something else. To crack the machine-encrypted radio traffic between the Wehrmacht and the German Navy, British intelligence developed, at the end of 1943, the first digital machines: what we would call "computers" today. Whatever one machine encodes another machine can decode, Alan Turing wrote when he presented his abstract "paper machine" as the basic switching principle for all digital computers that might ever exist. To speak in Heidegger's language, a further increase of developmental velocity [*Entwicklungsgeschwindigkeit*] occurred—the escalation of technical media. The telegraph cables of the American Civil War were successfully countered by the wireless broadcasts of the First World War, which were in turn bested by the coded transmissions of the Wehrmacht—which for their part experienced defeat through a computer network (as it still exists today). War is indeed the father of all things insofar as conquerors and conquered alike emerge from battle between media operating at a distance [*Fernmedien*]. In other words (which are Heidegger's, too), technology [*Technik*] itself determines the History of Being.

The computer was created to defeat secret radio systems. And so modernity and all its analog images, sounds, and representations—to which "The Age of the World Picture" assigns the three centuries between Descartes and 1938—really did come to an end. The same dispensation that lowbrow thinkers (and ones working by order of the Canadian government, at that) off-handedly called "postmodernity" is, in terms of the history of Being, utterly without precedent: "Enframing" [*Ge-stell*]. No subject still pictures itself picturing things [*Kein Subjekt stellt sich mehr vor, daß es sich Dinge vorstellt*]; rather, digital circuitry, which we may also call a "computer," stores, calculates, and transfers information. Nota bene: this does not occur between two subjects—that is, as a further "extension of man"—but rather takes place from machine to machine.[9]

Heidegger, the professor of philosophy who had been fired from Freiburg, recognized as much in 1964, at the latest, when he held one of his rare lectures abroad—or rather, when he had Jean Beaufret read the text he had prepared. At the seat of UNESCO, in Paris, the delegates heard in elegant French that all teaching positions in philosophy had become meaningless—that those who held them should be promptly let go. (And so, presumably, it is only the sluggishness of venerable institutions that gathers us here today.) The reason Heidegger provided was simple: philosophy was now over because it had achieved completion in the specialized sciences, and in so doing, dissolved or dismantled itself. The same had already occurred once before in the history of Being—namely, in late (Hellenistic) Greek culture—even if the event had not been as abyssal and definitive as what now occurred. For the thought that followed Aristotle, only the unity of *physis* and *logos* had split into the sciences of physics and logic (to hush up the unspeakable matter of late-Greek/Roman ethics). But today, Heidegger averred, the logic that philosophers once had studied and taught has been replaced by logistics; and logistics, in turn, coincides with cybernetics—in other words, with Norbert Wiener's mathematical theory of feedback circuits [*rückgekoppelte Schaltkreise*], whether they happen to steer organisms or machines.

And that means that causes no longer precede effects in time; instead, only "a challenging-forth" [*ein herausforderndes Stellen*][10] occurs, which strips physics of its Kantian conception of objects and reduces them to mathematical designs [*Entwürfe*]. And as if he had heard, in roundabout fashion, of Turing's Universal Machine, which can be all other machines, Heidegger called the design of these designs the "calculating machine" (or "calculator" [*Rechenmaschine*] in common usage). In other words, cybernetics, logistics, and data processing are no longer sciences performed by human beings, as was the case in late Greek culture, but rather are implemented as high technologies. They operate [*laufen*] (if it is still possible even to say as much) as things among things. Thus, Enframing represents not only "danger" but also (as Hölderlin put it) "nearing salvation" [*die nahende Rettung*]. For calculating machines—"computers," in the vulgar—undermine the very distinction that, since Aristotle's differentiation between *logos* and *physis*, has founded metaphysics itself. They are both: logic and physics in one. Enframing, unlike what has occurred in all epochs of metaphysics to date, dispossesses and alienates [*vereignet*] Thinking and Being in a dark, menacing way—as once occurred in the experience [*Erfahrung*], but not in the thinking, of the Greeks.

The most distinctive feature of this unique situation—whose newness with respect to modernity [*deren Neuheit gegenüber der Neuzeit*] first dawned on Heidegger after the Second World War—is that it emanates [*ausstrahlt*] from Europe's nearness to the distance of the globe. As surprising as it may sound, Heidegger already formulated a concept of globalization in 1964:

> The end of philosophy proves to be the triumph of the manipulable arrangement of a scientific-technological world and of the social order proper to this world. The end of philosophy means the beginning of the world civilization based upon Western European thinking.[11]

It seems to me that, for an ontology of distance, such a conception of globalization based on computer technology reaches further and is more significant [*maßgeblicher*] than all efforts to derive understanding from traditional mass media such as radio, film, and television (as still is common among media historians—and which Heidegger himself sought to do in 1950 in his essay "The Thing").[12] With that, of course, the "end of philosophy" has set "thinking" an unheard-of "task." The task calls for a thinking that takes stock of the pathways of Technology [*Technik*] in its entirety: from the very beginning, with the Greek notion of *technē*, up to its completion in modern computing systems that, according to Heidegger, place economy, industry, science, and politics "in [concerted] operation" [*Betrieb*] (to which it is absolutely necessary that we add the operations of war technology).

Compared to this diagnosis from 1957, all that has changed today—in 2007—is that "the calculating machine" has long since broken free of mainframes built with vacuum tubes; now it rules in the form of PCs networked with each other worldwide, all day and all of the night. But all the same, the technical condition for such globalization already lies within the concept of the gigantic—among whose uncanny features Heidegger also counted the tiny [*das Winzige*]. Without the progressive miniaturization of our computer architectures, first on the basis of transistors and ultimately on that of flip-flops integrated by the millions, the triumph of laptops and cell phones would never have been conceivable. In a way that is difficult to disentangle, the farthest and the nearest have become fused: on the one hand, a digital stream of information already extends to the outer limits of our planetary system; on the other hand, computer designs now measure the distance between switches in nanometers (which asymptotically approach zero). And with that, the relationship between distance and proximity has also been reversed: to speak in the language of *Being and Time*, the distant planets are closer or more "unconcealed" to our eyes than the circuitry operating on our desks and in our pockets.

That, it seems to me, is the point where we must leave Heidegger's History of Being in order to pose his questions again from today's standpoint. The unity of *physis* and *logos* that has been implemented demands that one conceive the relation between thinking and calculating, as founded in Greece, in a different way. It simply does not hold that thinking turned into calculating when it became metaphysics with Plato and Aristotle. The opposite is true: Socrates was distinguished, in contrast to the pre-Socratics, by the fact that he knew nothing about mathematics or music. Indeed, on the same day that Plato appointed a mathematician to be his successor at the Academy, Aristotle is said to have abandoned the grove of the Muses at the northern edge of Athens as if set to flight. At any rate, his *Metaphysics* explicitly declares, mathematics constitutes an entirely different, and significantly lesser, science than ontology: it does not treat being [*sie handele nicht vom Seienden*] as such—which in every instance represents the coexistence [*Beisammenanwesen*] of form and matter—but rather concerns immaterial forms: the arithmetic of (equally immaterial) numbers.

However, this same Aristotelian definition simply does not hold for the primal mathematics of the Greeks. Mathematics was originally the arithmetic of *logoi*—that is, of relations between whole numbers; such arithmetic always, at the same time, corresponded to geometry, whether it involved a diagram of counters [*Rechensteinen*] or the tuning of strings on a cithara. And so, primal mathematics was implemented in the same way as occurs in modern computers. Only because Heidegger obviously never read the Pythagoreans—unlike Johannes Lohmann, his colleague at Freiburg—could he conceive of the switching technology operative in high-voltage networks as modes of "challenging revealing" [*herausforderndes Entbergen*], but not the challenges posed by digital microchips.[13]

For the whole of Pythagorean mathematics rested on a single theorem—the first general law at all, which separated Greek mathematics as such from the number counting of Egyptian and Babylonian predecessors. All numbers except for one, Philolaos of Croton declared, are either even or odd. In other words, Greek arithmetic, in radical contrast to modern mathematics, excluded real numbers as numbers, and admitted them only as geometrical extensions or surfaces. Between two natural numbers there fundamentally lies (as Aristotle would put it) an interval, a distance; in contrast, it is well known that the body of real numbers is dense and compact (a "continuum," Aristotle would have said).

It was Turing's fundamental consideration in his 1936 dissertation to separate a subset from the body of real numbers and investigate it more closely. He called this subset "computable real numbers" and demonstrated that they are just as powerful as the set of natural numbers (in terms of Georg Cantor's set theory). We might also say, much more simply: computable real numbers can be described with the finite signs of an alphabet. This, and this alone, made it possible in 1943 for the calculations performed by human beings to become calculations performed by machines. As Lacan would have put it, the Real—because its body exceeds counting operations [*überabzählbar ist*]—persists as the Impossible, beyond all "computability" (as one says in English). And with that, every digital computer today has fallen thousands of years behind modern analysis and implements, once again, strictly Pythagorean mathematics. Therefore, although it is not necessary, it follows that circuit technology [*ist es schaltungstechnisch doch sehr naheliegend*] should also reach back to the Pythagorean division of all numbers into even and odd: all states in a digital machine can be implemented either as open or closed switches; that is, with the binary numbers "one" and "zero" (as Leibniz introduced them). Otherwise, one could inscribe no *logos* into *physis*—such as occurs millions of times a day by means of electron-beam lithography, when digital wafers are manufactured in dust-free, ultraclean labs it costs billions of dollars or euros just to build. Otherwise, computer technology would not be this alliance [*Verbund*] of hard- and software, of physics and logic, which has taken the place of the gods who have fled far away. Zeus, as you know, was at once the mighty brightness of the Greek sky and "the lightning that guides everything [*der Blitz, der alles steuert*]."[14] Only gods and computers are in the position of predicting today whether blue skies or rainstorms will be the weather tomorrow.

It is anyone's guess whether Heidegger would have much liked the identity of Being and Thinking that he called forth with Parmenides. When he died in 1976, he could not have foreseen the victory march of miniaturization and the personal computer. However, we do know that the high-tech present stood closer to him than malicious rumors still circulating would have us believe.

It goes without saying that there was no television at Freiburg-Zähringen, *Rötebuckweg* 47. The neighbors had one, however, and I knew their son well. In midsummer 1972, during the World Cup in Munich, Heidegger would regularly drop by to follow, on the screen, the games that the German national team was playing. A few weeks later, he traveled by train to Heidelberg to participate in a conference held at the Academy of Science located there. On the trip back, an unknown man sat across from him in first class; as it turned out, he was the artistic director of the *Stadttheater* in Freiburg.

"Why do you never go to the theater?" the man justifiably asked after a short while.—
"It is quite simple," Heidegger replied: he wanted to see heroes and gods at work, not modern actors.

"Gods?! but they don't even exist anymore!"—"But they do, *Herr* So-and-so, on television, for instance."—"You'll have to explain that to me, *Herr Professor!*"—"But of course: have you ever seen Beckenbauer playing soccer? He, along with his team, wins the World Cup, but he never gets injured, all the same. I call someone like that a god."[15]

A nice ontology of distance [*Ferne*]: when Heidegger watched television [*fernsah*], he beheld, in Beckenbauer's nearness or distance—who can even tell?—the gods of Greece making an appearance. This sense of distance [*Fernsinn*], it seems to me, is what, with due benevolence, we might yet learn and gain from Heidegger.

23 Pathos and Ethos: An Aristotelian Observation

The essence of the human being involves, before all knowledge, moods [*Stimmungen*].¹ Whatever we perceive [*vernehmen*], they have always already induced something felt [*ein Gefühltes*]: joy or sorrow, courage, despair, or passion. Although we are happiest about what our eyes see, because the act reveals so many differences in the things of the world, in truth we can only learn—and therefore know—because human beings, unlike bees (for example), also have ears.² Because the human being is the most imitative of all animals and because children acquire knowledge only by imitating their elders,³ human beings are the only animals that have *logos* or language.⁴

Songbirds, with their delicate tongues, can also articulate their voices; therefore, they do not just call or yell but also sing, as we do (or are able to do). Unlike the calls of mammals, which come from lust or pain, birds do not have their songs from nature; instead, every young nightingale must learn its "dialect" (*dialektos*) from older (male) birds.⁵ However, should it be permitted to complete Aristotle's observations, nightingales have no visible signs for the notes with which they declare their amorous feelings [*Liebeslust*] to females; nor do they write down articulated songs, as we humans do.⁶

Thus imitation, as it first begins with the child, ultimately leads to the height of poetry [*Dichtung*], which is simultaneously sung and written.⁷ That is why all that the soul experiences [*erleidet*] of the things in this world is the same for all human beings, yet there are different sounds [*Laute*] among different peoples—and in turn different written symbols for the sounds.⁸ And that is why love has a different name among all peoples. All the same, only one who thinks of—and writes/composes verses for—a beloved being has experienced what love is (because, as has been the case ever since Sappho, this person whiles away in a foreign land instead of one's bed).⁹ The sounds of love are present [*wesen . . . an*] because they are stored in written signs.

Pathos and Ethos

However, the sounds of love have grown estranged ever since the *Odyssey* first invented yearning, suffering, and homecoming (*nostos*). "Learning by suffering" (*pathein mathein*), goes an ancient rhyme of the Greeks. How did such *pathos* (almost by miracle) become writing? How did the "experiences [*Erleidnisse*] in the soul" (*ta en tē psuchē pathēmata*) come, as the signs they are, to outlast the fleeting twittering of young nightingales?

If we may further add to the Thinker's reflections, only one of the many writing systems—whether they divide language according to syllables, consonants, or words [*Begriffe*]—was ever invented (or adapted) to record vocalized songs with due fidelity. Speaking with Plato, the Greek vocal alphabet represents the birth of the Muses[10] because it recorded the *Iliad*—and still during Homer's lifetime—"also for us"[11] mortals.[12]

It goes without saying that Homer had no words for the wonder that the Muses have been performing ever since he invoked them. We love this miracle in the form of science and poetry [*als Wissenschaft und Dichtung*]. Homer, however, did not know the body as a whole, but only the many limbs that form it.[13] It was first Sappho's verses [*Strophendichtung*] that fused this plurality of many *melea* into a single *melos*—the "melody" we all know. Exactly the same holds for the many encampments, or "pens," where human beings gather like animals. Homer knew them only in the plural—*ēthea*. At the edge of fields, female rabbits dig hollows for themselves and their young. It is only since Hesiod that the singular, *ēthos*,[14] has existed. Now the word was revealed to mean "custom," "usage," and "character."[15] Finally, under the intellectual sign called "*logos*," it fused an essential trait of existence and destiny: *ēthos anthrōpō daimōn*[16]—"his own way is man's daemon."

Initially, when the Greeks still pursued poetry and were not thinking yet, *daimōnes* referred to goddesses and gods when they did not reveal themselves by name, but rather, as invisible as spirits, steered the fate of mortals all the more. Only in the thinking of Heraclitus did the daemon come to dwell in the soul itself, whose sense grows without cease, whose limits, despite all our efforts, we never find, and whose "ownmost" lies in its/our ethos.[17] One can write either *ethos* or *ēthos*—habit or disposition [*Wesensart*]—because it is highly probable that both words go back to the same Indo-Germanic root meaning "I have set myself/am sitting," "I dwell/build." That is why *ethnos*, which clearly derives from *ethos*, means "people," "throng," "swarm of bees"—that is, a full embodiment [*Inbegriff*] of the beings that have always lived with us. For the idea of excluding *ta ethnē* as the others or "pagans" is an idea that first occurred to barbarous Christians [*Heidenchristen*]. (To say nothing of the ethnicities of the UN Charter, this postcolonial nonconcept. It specified only an extremely vague plural—but ever since, it has referred to a singular entity that only grows in power. No one should be able to speak of "tribes" or "peoples" at all anymore.)

Opposing *ēthos* is *pathos*—what comes over and befalls us. In short, we act or suffer as our daemon wishes. And so, the most imitative of all animals, which we already are as children, brings forth, at the highest level, poetry [*Dichtung*]—or in other words, imitation. For Aristotle calls the voice the most imitative of all body parts.[18] No image floating before our eyes compares in its pathos to what speaks in the voice [*aus der Stimme*]. The voice is what first makes *logos* into *lexis*—up to the power of song [*bis zur Gewalt des Liedermachens*] (*melopoiia*).[19] When Sappho's lovesickness called out for Aphrodite, when the final chorus in Sophocles' *Antigone* invoked Dionysos, this was not the literature we read in silence but rather a voice that fulfilled itself in performance [*Vollzug*]. The gods came because they were rhythmically and melodically invoked. In this warlike capacity [*als Mitkämpferin*], Aphrodite assisted Sappho in love three times[20]; in this way, the god with many names came to Thebes, to purify the city of Creon's murderousness.[21] In contrast, what Aristotle would later call *katharsis*—the purification of terror and pity that have been roused—was already literature, even if the Thinker lamented that Greeks did not even have such a word.[22]

And yet the Thinker also knew what Archilochus sang in earlier times: that moods [*Stimmungen*] shape and preserve human beings.[23] That is why, one reads in the *Politics*, every Greek must learn, and suffer [*erleiden*], music. Adult citizens do not sing or play music themselves—and in this, they are equal to Zeus—yet they take their pleasure in Apollo and the Muses.

> Rhythm and melody supply imitations (*homoia*) of anger and gentleness, and also of courage and temperance, and of all the qualities contrary to these, and of the other qualities of character, which hardly fall short of the actual affections, as we know from our own experience, for in listening to such strains our souls undergo a change.[24]

And so every child—even in Athens (to say nothing of Sparta)—must be instructed in singing and playing music, so that it may experience, poetically, all the ethos or pathos harbored in its infinite soul. After all, "the soul," Aristotle affirmed, "is all being in a certain measure."[25]

Today we live beneath different skies. Some go on Timothy Leary's "trips" to experience what they hold for an evening. For others, intoxication counts as their own way—what belongs to their ownmost selves [*zum eigenen Tun*]. And so, once again, the choice is between pathos and ethos. Only Niklas Luhmann made this point clearly:

> Meaningful reduction of complexity, namely, can be assigned in twofold manner: to the world itself or to determinate systems in the world. Either reduction is treated as something given, or it is afforded by a determinate system. In the first case, we should speak of experience [*Erleben*], in the other of action [*Handeln*]. Both are processes occurring in systems, both processes presume behavior-driven [*sich verhaltende*], living organisms that can meaningfully order their relation to their environment. The difference between experience and action can therefore be understood [*konstruiert*] neither by means of the difference between inside and out, nor by means of the difference between passive and active. The point of difference lies on the plane of the organic substrate, where what is visible for human beings is not to be grasped, but rather lies in the construction of meaning [*Sinnbildung*] itself—namely, in the question how the reduction of complexity is attributed, where sense is "localized," so to speak. Experienced meaning [*erlebter Sinn*] is apprehended and processed as having been reduced externally; active meaning [*Handlungssinn*], in contrast, as having been performed by the system itself [*als systemeigene Leistung*].[26]

Thank you.

24 Media History as the Event of Truth: On the Singularity of Friedrich A. Kittler's Works

An Afterword by Hans Ulrich Gumbrecht

When Friedrich A. Kittler died on 18 October 2011, in his sixty-ninth year of life, the intellectual public sphere of Germany reacted more strongly, more broadly, and in an existentially more engaged tone than at the death of any other scholar [*Geisteswissenschaftler*] since the end of the Second World War. On the one hand, Kittler was unconditionally admired, yet on the other, he had faced unyielding academic skepticism until the end of his work—and life. Therefore, I was surprised by the unanimity with which his singular importance was now retrospectively celebrated in all quarters. The response had something to do with a peculiarly tautological situation: "the media" were reacting to the passing of the very thinker and writer whose research and publications had established a space—in both the intellectual and the academic landscape—for "the media" in the first place. And yet, I also had the impression that the event of death, which never fails to produce monuments, had for the first time (if only perhaps in passing) made evident the structure, complexity, and particular significance of Kittler's work in its manifold dimensions. This "revelation," it seemed, represented an intuition or a promise—of a specific truth that might yet emerge from the technology of our present and its prehistory—more than it involved a nuanced insight or thesis.

Kittler had not just invented a new science—at least for German academia. In a broader, international context, his books and lectures also displayed a cultural sensibility that the world had never witnessed before him. Kittler's view incorporated enthusiasm for technology, a literary taste that considered itself discriminating, mathematics, psychoanalysis, familiarity with Richard Wagner's operas, a specifically generational love of rock music, a hunger for facts, programming ability, and finally, a predilection for speculation that proved both irresistible and endless. In Germany today, university students who are just starting out—even if they have never heard Kittler's name—often wish to study "something media-related." Without his influence, this would be unthinkable.

And so, notwithstanding the tension between enthusiastic agreement and aggressive rejection that Kittler's positions had elicited time and again, what Raimar Zons, his friend and publisher, affirmed at the memorial service held at the Humboldt University in Berlin for the emeritus professor of media aesthetics and media history became clear: Kittler numbered among those "who, through what they did, thought, and said, left the world—their world—different than they had found it."

Yet how exactly did Friedrich Kittler change his, and our, world, if we leave aside the institutional effect that his thought had at universities? What is it that gave his intellectual admirers (as much as his opponents, in fact) the impression that they absolutely had to take on his project and continue it themselves or, alternately, undermine and block it?

The conditions underlying the particularly intensive response to Kittler's work included the fact that he published at a time when many readers who fancied themselves of a certain caliber felt that true "master-thinkers" were lacking. Kittler met their romantic yearning for a figure of genius, and he was not unaware of this fact—at any rate, he lacked a sense of irony about his position. For the most part, he made a very convincing and charming impression, even if it sometimes seemed fragile and contradictory, too. Such an effect followed from a singular mix of attributes: the breadth of his knowledge, which crossed numerous—and seemingly heterogeneous—dimensions; the provocative force of his theses, which were counterintuitive yet highly plausible; the prophetic tone (which was never entirely secular) in which he delivered pronouncements and prognoses as matters of absolute fact; his very real experience of being an unloved son of academia for many years; the intellectual vigor with which he made the most varied intellectual configurations his own and reshaped them into compelling emblems of his own eclecticism; and finally—perhaps most of all—the unique sovereignty with which he managed to experience the centrifugal movements and intrinsic contradictions that resulted from his thinking as intellectual complexity (instead of seeking to resolve their dissonance). Kittler referred to Hegel, for example, both as a philosophical antagonist and as a philosophical model, and he spoke of war both in the tone of a radical pacifist and with grave, military-historical pathos. Friedrich Kittler was more than a traditional scholar and a modern professor—yet he did not really fit the part of the classic intellectual or avant-garde author either.

The twenty-three essays collected in this volume, which appeared between 1978 and 2010 and now are printed together for the first time, present two genealogical lines of development in parallel. Inasmuch as they follow the chronology of initial publication, the essays document the emergence of Kittler's intellectual signature [*Denkfigur*]—which was heterogeneous and centrifugal, but also unified and coherent. Second (and interwoven with the first line that emerges), the essays yield the profile of an idiosyncratic narrative about the history of technology as the history of culture: by way of a central chronological rupture and a temporal countermovement back to ancient Greece, Kittler presents a long-term thesis about how our electronic present came to be.

For all that, however, the volume at hand does more than document, unfold, and explicate Kittler's work—which came to a relatively early end when the author died, and which remains open in many respects.

I have mentioned the certain yet vague impression that Kittler's death made visible, for a moment, the significance and potential intellectual function of his work—an oeuvre that is otherwise difficult to grasp because of its complexity and scope. This impression concerns the truth of the technological world. At very least, it is important to provide the opportunity to use this truth. To do so means, for one, keeping his thought from forever being restricted to Germany—as has largely been the case until now. Second, it means holding open the possibility that the philosophical contribution it may provide for understanding the electronic present and future will finally enter a phase of productive application [*Umsetzung*] instead of dissipating. I am less concerned with passing on Kittler's thought dogmatically and as a matter of fact than with identifying the author's specific intellectual style—the underlying gesture, gestalt, or *Ansatzpunkt* (as Erich Auerbach would have said)—and with describing, above all, the counterintuitive attraction it often affords. It goes without saying that Kittler's positions and provocations will continue to occasion many controversies (and reactions of rejection). That, however, only proves that engaging with his arguments is worthwhile. Instead of circumscribing and "pinning down" Kittler, my concern is to make a certain intellectual energy [*Denk-Energie*] felt and to keep it alive.

In order to make the potential of Kittler's work evident for future discussion, I will discuss his texts from three complementary perspectives. First—and in the order of composition (that is, in three stages inherent to his work, each of which achieved completion with the publication of a book)—I will follow the increasing complexity of Kittler's thinking. In the process, a history of academic and intellectual movements in Germany between 1978 and 2010 will also come into view. Having achieved an overview by diachronic means, it will be possible to identify and describe the traits of his "form of thinking" [*Denkform*] (in other words, his specific epistemological premises) in a way that is rarely evident in his works themselves. By analytic and synthetic means (the sections "Genealogy" and "Form of Thinking"), it will become clear how to approach the concluding, and decisive, question ("Truth"). The matter has a status in the History of Being that must yet be explained: what is singular—and singularly significant for our present and future—in the works of Friedrich Kittler? Can the truth of the technological world reveal itself in this body of texts?

Genealogy: Literary History, Media History, and the History of Being

In the (short) first decade of his publishing activities (at the end of the 1970s), remarkably late in life for such a productive scholar, and long before he came to concentrate specifically on the phenomenal realm of media, Kittler demonstrated a sensibility for the cultures of the past that had not existed before. This unique perspective found expression in his 1985 masterwork, *Aufschreibesysteme 1800/1900*—a book that was not yet a "media-historical" text in terms of its program. David E. Wellbery wrote a foreword for the American translation, which appeared five years later (under the title of *Discourse Networks 1800/1900*). I consider Wellbery's remarks to provide the best commentary on Kittler's early writings and, what is more, the best account to date of the German intellectual landscape at the end of the twentieth century. Kittler's intellectual style—which, in context, qualifies as absolute and unprecedented—explains why *Aufschreibesysteme* might easily have cost him a traditional university career, even though his earlier publications had met with one-of-a-kind resonance and already brought the author acclaim on a national scale. There is no contradiction between Kittler's absolute inventiveness (to say nothing of the ambivalent consequences it entailed) and the fact that his works combined three intellectual currents from France in a productive and eclectic manner (that is, in a way that cared little for detailed conceptual mediation or epistemological compatibility). These positions were, first, the program and praxis of Michel Foucault's discourse analysis—a new form of historiography that (purposefully) restricted the field of investigation to institutionalized forms of meaning; second, Jacques Lacan's reworking of Freudian psychoanalysis, which undermined traditional Western notions of subjectivity and classical forms of self-reference; and finally, a reading of Nietzsche—which was novel for the times—that stressed the proximity of textuality and corporeality, on the one hand, and offered an anti-Hegelian, genealogical conception of historical processes, on the other.

Kittler's affinity with Foucault became clear above all in his thesis concerning (mainly German) Romanticism around 1800. Romanticism, Kittler argued, was a discursive configuration shaped by bourgeois family structures; here, and for the first time, literature had come to count as the expression of an individual soul. The decisive factor, in Kittler's estimation, was the physical and psychological [*geistig*] attention that mothers—especially ones from socially privileged classes—paid to their newborn children; hereby, and early on, the author incorporated a pragmatic consideration of sexual difference into his approach to history. In addition to describing the self-understanding of late Enlightenment and early Romantic literature as the medium of bourgeois *Bildung*, Kittler doubly undermined the object and, so to speak, "exposed" its discursive nature. He did so by combining Lacan's demystification of subjective claims to autonomy, on the one hand, and Nietzsche's reflections on the ways that material, cultural practices and artifacts shape human bodies, on the other. The convergence of Nietzsche, Lacan, and Foucault explains the fundamental thrust [*Grundaffekt*] of Kittler's work as a whole, which takes aim at the classical notion of *Geist* and hermeneutics (including the hermeneutics of Freudian psychoanalysis) as the core of the "humanities"—or as they are known in German, the *Geisteswissenschaften*. Indeed, an emblem of this gesture was provided by the title of an edited volume Kittler published around the same time: *Die Austreibung des Geistes aus den Geisteswissenschaften* (literally, The Expulsion of the Spirit/Mind out of the Sciences of the Spirit/Mind).

Reading through the early essays (which sometimes seem to constitute a single, ongoing text), one can discern—above all on the basis of their concluding passages—how Kittler brought the particular configurations of his historical sensibility to new levels of complexity by identifying new questions, which he in turn paired with philosophical positions that promised answers opening onto uncharted terrain. That said, the decisive step that led Kittler to media theory was not primarily philosophical. In his analysis of "Wanderer's Night Song"—an essay that has long since become a classic among Germanists—Kittler not only affirmed that the text transcribed the sounds of nature instead of expressing a Romantic-lyrical "I," but went so far as to associate the poem with a New York melody of the twentieth century: "Lullaby of Birdland." When he posited this connection [*mit dieser Referenz*] and conceived of the direct "representation" or "notation" of environmental sounds without the mediation of "understanding"—later, he regularly employed the verb *anschreiben* in this context—Kittler passed beyond the horizon of intellectual and literary history in the narrow sense of "education" [*Bildung*] and "science" [*Wissenschaft*] for the first time.

Kittler expanded the initial configuration he had established, via popular music, between Foucault's discourse analysis, Lacan's antisubjective psychoanalysis, and Nietzsche's corporeal philosophy when he analyzed "Brain Damage" by Pink Floyd. The essay concludes by explicitly rejecting McLuhan's dogma that a medium is its own message, and therefore self-reflexive. Instead, Kittler affirms that existence is shaped by sounds and their media—a view that, in light of his later work, we can identify as theological in inspiration. In the music of Pink Floyd, the "God of the Ears" turns to human beings; indeed, the *gods* do so. This essay added another dimension that, in my opinion, lent definitive form to the first configuration of Kittler's historical sensibility—at least preliminarily. This is the dimension of mental illness, ever oscillating between "supposed" and "real." (After all, whenever anyone is declared "mentally ill," it depends on perspective.) The title "Brain Damage," Kittler suggests, is meant to show that everyday human reason cannot grasp the musical presence of the gods. The theme returned in Kittler's essay on Daniel Paul Schreber (whose case already fascinated Sigmund Freud) and his autobiography, *Memoirs of My Nervous Illness*, the "most celebrated work of all mad, German books—or German books by the mad" (p. 57). First and foremost, what interests Kittler about Schreber and Paul Emil Flechsig, the physician who treated him, is the resoluteness with which both patient and doctor understand psychological processes and consciousness as strictly somatic phenomena.

Kittler's discussion of Schreber represents a further point of convergence with Nietzsche's provocative view of one-dimensional corporeality that excludes consciousness. Here we can discern the productive mechanism whereby the first phase of Kittler's theory assumed coherence. The varied elements and positions—each with its own intrinsic complexity—that Kittler eclectically incorporated into his perspective on the world are connected by affinities resembling family relations: for example, rock music and Schreber's writings meet in the motif of mental illness; Schreber and Nietzsche relate inasmuch as both writers stress embodiedness.

Out of the astonishing multiplicity of such relations—which Kittler always documented with philological exactness—there emerged an increasingly complex and ultimately more stable web of associations which he described in emphatically indicative language. Indeed, he did so in terms that often seemed "strictly scientific," as if he were presenting a material object. In this centripetal and indicative gesture, I see an echo [*Anklang*] of mythographic writing that, I submit, represents a fundamental element of Kittler's singular position as a historian and philosopher. The term "mythography" is also meant to underscore the fact that his texts made an impact more on the basis of counterintuitive suggestions and aesthetic properties than because of "scientific" methods of empirical self-control or validation through argument. Inasmuch as Kittler constantly incorporated new texts, phenomena, and domains of knowledge into his thinking—and in the process, returned to earlier positions in modified form—the mythographer lent his work, from its earliest stages, increasingly well-defined contours of coherence and form, in which a certain reality began to appear.

Kittler's works first became "media-historical" in the thematically plausible sense during the early and mid-1980s, when he discussed film for the first time (initially in a relatively conventional-seeming, content-focused perspective). Time and again, Kittler associated the cinematic medium with *Gravity's Rainbow* by Thomas Pynchon (1973)—a novel about the end of the Second World War and the apocalyptic potential of the German weapons industry. From the first, Kittler's media history was marked, in terms of structure, by a close connection to military history. Before long, it had yielded a clear picture of three historical phases, each of which was constituted by a different medial configuration:

> Phase 1, beginning with the American Civil War, developed storage technologies for acoustics, optics, and script: film, gramophone, and the man-machine system, typewriter. Phase 2, beginning with the First World War, developed for each storage content appropriate electric transmission technologies: radio, television, and their more secret counterparts. Phase 3, since the Second World War, has transferred the schematic of a typewriter to a technology of predictability per se; Turing's mathematical definition of computability in 1936 gave future computers their name.[1]

Such is the historical movement traced in *Gramophone, Film Typewriter* (1986), Kittler's most successful title, at least in terms of translations. In light of this study, earlier writings (especially on German literature around 1800) as well as later books and essays (above all on ancient Greek culture) may be read as two fundamentally different accounts leading up to modern media history.

Kittler's fascination with Pynchon—which would prove decisive for his further work—was already evident in "Romanticism, Psychoanalysis, Film: A Story of Doubles" (1985). On the basis of films from the early twentieth century, Kittler sought to show how the cinematic medium "drills a new dispositive of power: 'How to do things without words'"—and in so doing, puts an end to the Romantic cult of literature as the expression of complex individuality (the genesis of which is reconstructed in *Aufschreibesysteme*). Inasmuch as film "concerns powers . . . to which [it] belongs" itself, Kittler, with characteristic historical impatience, already had Pynchon's 1973 novel in mind:

> A few twentieth-century authors have understood as much. A form of the fantastic extends from Gustav Meyrink's *Golem* up to Thomas Pynchon's *Gravity's Rainbow* that has nothing to do with Hoffmann or Chamisso and everything to do with the movies. Literature of the central nervous system competes directly with other media—for this reason, perhaps, it has always already been destined for filming. Making present instead of narrating, simulating instead of authenticating: such is the motto.[2]

The same year (1985, which proved decisive for his work) and as a matter of due course, Kittler published an essay devoted exclusively to *Gravity's Rainbow*: "Media and Drugs in Pynchon's Second World War." Even if the approach taken here sometimes lacks nuance, the argument as a whole is convincing: Pynchon's novel, Kittler affirms, follows the structural logic of immediate "presentification" [*Vergegenwärtigung*] that results from the storage medium of film being transferred across the Atlantic. Here Kittler gave academic and intellectual expression to a fantasy about the past which remains popular in Germany to this day: that the United States, the new world power, took over—and in seamless fashion, at that—the military technology of the National Socialist state.

Ultimately, this combination of obstinate patriotism and politically correct anti-Americanism burdened Kittler's work. (In interviews given late in life, Kittler escalated the tone and even lamented that guest-professorships in the United States had put an end to his happy marriage.) Yet by making such gestures, the author only expanded his mythographic power. In this case, Kittler did not even try to resolve the contradiction between his unbounded admiration for Pynchon and his own cultural prejudices; indeed, he fused the two and pointedly left them standing without commentary: "The narrative continuity of ... films, then, haunts the novel that has made them its theme. Plotlines and dialogues seem as if they had been written under the influence. ... As a consequence, *Gravity's Rainbow* is, among other things, a *Reader's Digest* article, too: ordinary, conventional, and American" (p. 97).

Operating with assumptions of this kind, Kittler continued his media-historical fascination for a decade—up to the mid-1990s, when perhaps he encountered its philosophical and existential limits. What proved decisive for the middle phase, indeed, what made him an intellectual "classic" on a national scale and an "insider tip" in international terms, was that he now focused on the codes structuring the history of technical media and the military (instead of discourses constituting cultural history). Above all, these codes had enabled machines to coordinate human beings [*Menschen-Steuerung*]—an office formerly discharged by a subject of the Cartesian variety. Here too Kittler's arguments connected with the philosophical objective of dispelling illusion(s) by minimizing classical subject positions. Into this new framework—which now was "media-historical" in a literal sense—Kittler soon incorporated configurations of phenomena and observations that had already been central to aspects of his earlier work.

Thus, still in 1985, "*Heinrich von Ofterdingen* as Data Feed" recoded Novalis's novel—which traditionally has counted as the epitome of Romantic interiority—along the lines of media that are "storage facilities" [*Speichern*]. In so doing, Kittler not only minimized the role of the subject, but also presented this classic text as a precursor to the conditions of literary production [*Literatur-Situation*] in the early twentieth century:

> A novel like *Heinrich von Ofterdingen*, which cycles through the discursive space of its epoch from beginning to end—from unrecordable noise up to the system of universal storage called "Philosophy"—and moreover does so for each and every word or author, does not depict "actions" [*Handlungen*]. Instead, it *acts*. (p. 121)

A music lover, Kittler used the same means (which is self-evident when operating with these premises) to discuss Richard Wagner's conception of opera and dramaturgical praxis as "respiratory eroticism"—indeed, as "World-Breath." A little later, in "The City Is a Medium," he experimented with the thesis that urban conceptions of architecture, ever since Napoleon, have followed more and more on considerations concerning the destructibility of cities.

Here, for the first time, a mythographic tendency became clear that shaped the middle phase of Kittler's work: a vision of apocalypse that the media historian enjoyed staging in uncompromisingly sober terms. With a view of the Vietnam War as presented in Coppola's *Apocalypse Now*, and in consideration of the music of Kittler's guitar hero Jimi Hendrix, "Rock Music: A Misuse of Military Equipment" sought to demonstrate that "Hi-fi and stereo ... both derive from localization technologies" developed by the German Navy and Air Force. Once again, Kittler engineered a tense convergence between his own techno-patriotism, criticism of American "imperialism" occasioned by the Vietnam War, and admiration for American musicians. For Kittler, rock music always summoned forth eroticism as an existential dimension—a dimension for which, in the dark, "subjectless" middle phase of his work, no room was left:

> Fittingly, "And the Gods Made Love" is the title of the first track on Jimi Hendrix's *Electric Ladyland*. But the masters of the world no longer have a voice or ears, as they did for Nietzsche. All one hears is tape hiss, jet noise, and gunshots. Shortwave—between the transmitters, which is to say intercepted from the military-industrial complex—sounds similar. Perhaps, under the conditions of a world war, love must come from white noise. (p. 164)

Ultimately, one senses from the indecision recorded here, the relationship between war and love represented an unbearable site of ambivalence for Kittler's media mythology.

The multidimensionality and manifold connections [*Anschlußmöglichkeiten*] that marked Kittler's work by the late 1980s and early 1990s—because the author conscientiously integrated the conclusions of earlier phases of his research into each new project—represented something unique in the humanities. But because of its complexity, Kittler's thought also precluded any straightforward or sequentially narrated presentation. The complexity resulted from Kittler's attention to detail with respect to technological phenomena and, at the same time, from his uncommon readiness to engage in associative speculation. The latter prompted him to discover (or at least postulate) homologies between realms that lay (or seemed to lie) far apart; for example, Romantic literature and the operatic *Gesamtkunstwerk*, rock music and erotic desire, war and technological innovation. The word Kittler used when observing (or postulating) these homologies—a word that is often invested with a literal kind of magic (and at any rate, always serves a mythographic purpose)—is *Klartext*. It stands for a deictic gesture implying that "all further" justifications or explanations could only be tautological in view of the constellation of phenomena presented.

With such conditions guiding his work, Kittler finally brought into view the technical-historical threshold that separated the age of electrical transmission media from the age of computers (that is, the beginning of our own present). The transition occurred during the final stage of the Second World War and the years of its immediate aftermath. Both on the concluding pages of *Gramophone, Film, Typewriter* (1986) and in a series of historical accounts [*Szenebeschreibungen*] written in the following years (up to the early 1990s)—which are included in this volume—Kittler pointed to two contexts in which, from his perspective, the "technology of calculability" had led from the block diagram [*Blockschaltbild*] of the typewriter to the computer. Both of these contexts were military. The first involved Norbert Wiener's "Linear Prediction Code," which brought the mathematical calculation of movements, distances, and signals in aerial warfare to a new level of precision: "The United States of America entered the Second World War armed in this capacity" (p. 176). The other context produced Alan Turing's "Universal Discrete Machine," thanks to which, from 1941 on, the English military was able to decipher secret German radio transmissions.

The convergence of these two technological innovations, according to Kittler, proved decisive for the outcome of the war—and at the same time marked the beginning of the Computer Age. In Kittler's account, the mythographic gesture associating different dimensions of reality appears with particular clarity. Kittler stresses that Turing—one of the few heroes in his media history—experienced his mathematical inspiration at Grantchester Meadows near Cambridge, "the meadows of all English lyric poetry, from Romanticism up to Pink Floyd" (p. 186). At the same time, he inscribes Turing's invention, whose progeny took up their own reproduction under the name of "Colossus," into the switch that occurred from "soldiers to machine-subjects": "Colossus begat child after child—each one even more colossal than its secret father" (p. 191).

In "Unconditional Surrender," Kittler argued that maximizing technology transfer from Nazi Germany to the United States (whose status as the leading world power was thereby affirmed) proved incomparably significant, for Germany had also developed self-guided technical systems, even if they were not digital. Here one can discern Kittler's tendency—which is somewhat forced, given historical facts—to claim that Germany contributed to the inauguration of the Computer Age. At the same time, his account of the relation between the situations in England and the United States puts a clear moral gradient into relief. With greater certainty than biographical research actually permits, Kittler presents Turing's suicide as a reaction to the decision of the United States, during the McCarthy Era, to exclude "homosexuals and other security risks from all sensitive government positions" (p. 178). Most important, however, is that Kittler makes the United States entirely responsible for the tendency (which he both marvels at and indicts with apocalyptic warnings) to replace human beings with self-guided machine-systems: "*Pax Americana* rests on what Eisenhower called the 'military-industrial complex.' Thanks to higher mathematics, it has moved beyond personnel-heavy world wars like the First and material-heavy ones like the Second" (p. 205).

On this point, which goes beyond the transition to the Computer Age in the years following the Second World War, Kittler's media history steers toward an apocalyptic ground zero. Already in the first phase—when machines (supposedly) achieved independence from human consciousness—computers and their codes developed a tendency to immunize themselves against intervention and, in so doing, "condemn[ed] human beings to remain human beings" (p. 210). In his famous essay "There Is No Software," Kittler even went a step further, seeking to expose the concept of, and discourse about, "software" as a kind of nostalgic projection of human structures of consciousness onto self-guided systems; in fact—or as Kittler saw the facts—these systems had already achieved independence from human beings: "When meanings shrink down to sentences, sentences to words, and words to letters, then no software exists, either" (p. 223).

Kittler's apocalyptic teleology stemmed from the idea that all the changes that are decisive for human life would soon occur only in the "silicon architecture" of computers—what in a 1989 lecture held in Bern he called the "Night of Substance." Kittler's media-historical discourse now struck a darker tone than before. It corresponded to a mood that twenty years ago was still obligatory for intellectuals who wished to claim expertise [*Sachkompetenz*] in matters of electronics. Such was the view of parties who—when they considered the "Apple screen," the "mouse," and the overall configuration of "personal computers"—saw symptoms of the dangerous (or at least very naïve) illusion that an "interface" between human beings and computers was even possible. In historical perspective, the outlook seems like a technological echo of Jacques Lacan's sarcastic remarks about overly optimistic conceptions of human autonomy.

Kittler never entirely abandoned this mood (and all its inherent tension), nor did he ever explicitly disavow the media-historical standpoint that marks it. For all that, an intellectually productive discontinuity holds between Kittler's essays of the early and mid-1990s about media history after the Second World War—which adopt a particularly radical tone—and the writings on ancient

Greek culture that, beginning in 1995, opened the last chapter of his work. Yet how can one explain the fact, which astonished many readers at the time, that Kittler was now bracketing the media history of his own day? Doing so did not simply mean that he was taking distance from—or revising—arguments he had previously made. Kittler never made concessions to polemics, and until the end of his life, "Apple" and "Jobs" remained emblems of existential and philosophical misunderstanding for him.

All the same, it seems plausible to suppose that Kittler came to experience his cold diagnosis of the media-historical present as intolerably burdensome—that it exceeded his (and not just his) existential powers. One symptom of this state may perhaps be seen in a 1993 essay on the emergence of the electronic present that began at midcentury; this essay reaches back to the heroism of shock troops during the First World War. Here combatants had faced the historical novelty of machine-gun fire under existentially tragic conditions—the conception of which translated into Heidegger's meditations in *Being and Time* on the fact that death is "always my own."

From the perspective of the early twenty-first century, however, such heroism seems to mark the beginning of mankind's suicidal self-disempowerment (whereby "suicide" is meant literally); the final consequence is that existence no longer possesses any value or offers any promise. This makes it easy to understand the attempts at evasion and acts of compensation that Kittler performed in yearning for Greek antiquity. Kittler was resolved to find love in that world of two and a half thousand years ago—erotic love that, like the love between gods in ancient myth, might give birth to cosmological fixity [*einen kosmologischen Ort*] and existential certainty. And were such security to exist only for Kittler himself, in his historical imagination ... Clearly, Kittler was writing himself into a tradition of German intellectual life that is storied, if also of questionable reputation—a tradition beginning with Hölderlin, at the latest, and extending, via Heidegger, into his own world.

In contrast to the early phase of his work, which concentrated on the long nineteenth century (*Aufschreibesysteme*), and unlike the media-historical middle phase (*Gramophone, Film, Typewriter*) too, the final chapter of Kittler's life, both in intellectual and existential terms, did not begin with polyphonic preludes announcing events to come. (His most ambitious project, *Music and Mathematics*, was to consist of eight volumes, of which only two were actually completed.) In the brusque language of Kittler's final works—which sometimes mutter pious words in Greek—one can discern the waning of the author's physical forces, his impatience when facing incomprehension or potential objections, and a mythographic-prophetic gesture that pushes more and more to the fore.

It is not my intention to discuss Kittler's later publications in terms of decline. Instead, I see them, above all, as the foundation [*Anlage*] of, and the key to, the significance that Kittler holds for the task of thinking our present. That said, the later work requires a hermeneutic perspective—just as little as the word "hermeneutic" met with Kittler's approval does it suit me, even now—that is different, if only by degrees, from the style adopted on the preceding pages, which have sought to reconstruct, synthesize, and round off the matters at issue. Accordingly, the following will take a bit more distance from the letter of Kittler's writings and seek to grasp the precise direction in which his thinking was headed on the final stretch of his work. This, I believe, will make it possible to uncover a singular intellectual potential (and perhaps preserve it from oblivion).

"Eros and Aphrodite" (1995) can be read as a prelude to the final, Philhellene phase of Kittler's oeuvre. The essay's programmatic status is evident when one compares the tone here to the apocalyptic sounds of texts written in the preceding years; for example, "Protected Mode" and "There Is No Software." In the latter, the cold, almost cynical, reference to human beings' dependency on self-guided technical systems admitting no external influence leaves no horizon of—nor even the most minimal hope for—existential happiness. Indeed, in discussing Plato's *Symposium*, especially Socrates' rejection of the love that Alcibiades offers him, "Eros and Aphrodite" renounces the world of knowledge because, as Kittler remarks, this world does not admit "intoxication and eros" and excludes women: "And yet, at the site where Socratic dialogue generates knowledge no women are allowed. Eryximachos has banished even the flute girl into the back rooms" (p. 256). On the final pages of the essay, there appears, for the first time, a mythographic leitmotif characteristic of "late Kittler": intoxication by nectar, the "drug of the gods"; this is deemed superior to intoxication by wine (which Alcibiades seeks to induce—in vain). Such intoxication, Kittler affirms, is compatible with philosophy:

> Philosophy, in other words, makes states of alcoholic intoxication impossible, because it swallows up all the wine in Athens to no effect. And so there remains only one kind of intoxication that it does not pretend not to know: nectar as the drug of the gods. From the body of Eros—who was after all conceived in the nectar intoxication of Aphrodite's birth—bees suck new honey and, with that, new mead. (p. 258)

On the Singularity of Friedrich A. Kittler's Works

The essays that connect with this point—which already date from the early twenty-first century—concern the emergence of the ancient Greek vocalic alphabet within the originary world of Homeric song. As soon as these essays appeared, classicists (whose competency is hardly an issue) critiqued the historical and philological claims they advanced. Time and again, Kittler found himself provoked to respond with gruff gestures of refusal (instead of engaging with specific objections). And yet, if one reads Kittler's works with a view to the "potential energy" for philosophy that they contain, then the criticisms offered by philological specialists prove as insubstantial as, for example, linguists' critiques of Heidegger's speculations concerning the etymology of Greek or German words (which, while almost always inspiring, tend to be problematic in historical terms[3]). At any rate, the connection between the vocalic alphabet and Homer (that is, the author of the *Odyssey*, above all) permits Kittler the mythographer to celebrate the recording [*Anschreiben*] of language that is sung as a "gift of the Muses," which in turn establishes a connection between the Greek alphabet and the fascination of femininity, Aphrodite, and eroticism. From here, the rhythmical structures of prosody and music lead to mathematics, and from there to ontology. This offers a perspective in which the world—understood philosophically—transforms into a world of things: "From this theory of music there emerged everything that, ever since, has counted as science—above all, knowledge concerning *physis*, or 'nature.' [...] The mathematical basis of all that is follows from the unity of whole numbers as they appear geometrically and arithmetically" (p. 273).

In "The Age of the World-Picture" (1938), Heidegger had critiqued the use of mathematics in the modern natural sciences—that is, he critiqued the science of nature for which mathematics (in the sense of "presence-to-hand" or "standing-before-things") provides the precondition of "representation" (i.e., a "picture of the world"). Kittler, in contrast, is concerned with thinking the world of objects—a conception I am calling "ontological"—wherein the world of things becomes present and tangible ("ready-to-hand") to one's own body to the extent that the body experiences itself as part of this world (i.e., as "being-in-the-world"). Without directly contradicting his earlier ontology of the technical world as the "Night of Substance," the ontology that Kittler now presents—of the natural sciences as they originated in Greece—offers a "counter-world" [*Kontrast-Welt*] that the mythographer declares to have been filled with, or fulfilled in, music, eros, and intoxication. Of course, this intoxication comes from nectar:

Media History as the Event of Truth

> The wonder of the Sirens ... is that they dwell on the island richest in flowers, an island on which Odysseus presumably also sets foot. That means there is fresh water; and that means they are nymphs, for nymphs are freshwater divinities and one does not worship them in a temple but rather where there are no archeological discoveries to be made from the time of the Greeks. And therefore—because of the flowers, the Sirens, and the fresh water—there are also bees, and where there are bees, there is honey, and so on; and songbirds, too—which is why it all sounds so bright and beautiful. (I am attempting an archeology of the text, not the archeology of findings from digs.) (p. 271)

In 2008, three years before his death, Kittler published the only essay that still struck a philosophical—and not a mythographic—tone. The piece made explicit the significance [*Bezug*] that Heidegger held for the late stage of his work: "Martin Heidegger, Media, and the Gods of Greece." Given the evidence mustered, it is impossible to dismiss Kittler's claim concerning the famous "turn" in Heidegger's thinking that emerged after *Introduction to Metaphysics* (1935) and then became more and more pronounced in his thinking. Kittler argues that Heidegger's turn followed from the insight "that all modes [*Spielarten*] of transcendental philosophy—whether they take their point of departure in the subject or in Dasein—founder upon the facticity of high-tech media" (p. 296). He goes on to add a bold—but for me altogether convincing—interpretation of Heidegger's diagnosis of the state of academic philosophy in the twentieth century. If ever since Aristotle (and in contrast to mythology and the thinking of the pre-Socratics) *physis* and *logos* have drifted farther and farther apart in philosophy and the sciences, then our age of "calculators" [*Rechenmaschinen*]—the term Heidegger uses to refer summarily to cybernetics, logistics, information processing, and their dispositives—is the point where the opposition that set the course of Western thought for two and a half thousand years has become obsolete. This is the case, Kittler argues, because *logos* and *physis* have found their way back together again in the "chip" of the electronic age: "Otherwise, one could inscribe no *logos* into *physis*—such as occurs millions of times a day by means of electron-beam lithography, when digital wafers are manufactured in dust-free, ultraclean labs it costs billions of dollars or euros just to build" (p. 301).

Here, after two decades, Kittler returned to the "Night of Substance" and a world "without software" (or to be more precise, he returned to the "Night of Substance" as a view of the electronic world as a universe without software or consciousness). The same perspective that had been so grim when he first approached it philosophically and historiographically in 1995, which had granted no space to human beings or to human consciousness, now, under mythographic illumination, offered the consolation of the gods' return. The dimension that Kittler had claimed did not exist—that is, logic, software, and consciousness—appeared transfigured [*aufgehoben*] into another view of the cosmos. For "computing technology" has become

> [the] alliance [*Verbund*] of hard- and software, of physics and logic, which has taken the place of the gods who have fled far away. Zeus, as you know, was at once the mighty brightness of the Greek sky and "the lightning that guides everything [*der Blitz, der alles steuert*]." Only gods and computers are in the position of predicting today whether blue skies or rainstorms will be the weather tomorrow. (p. 301)

In reading "Pathos and Ethos," a short text from 2010 that seems to have remained a fragment, we sense that the world might again pass into atmospheres and moods [*Stimmungen*], which are as ethereal as bees or a state of intoxication—and which one may choose to experience fully, or not.

Form of Thinking

In my extensive genealogical sketch of Kittler's work (which at the same time is almost unbearably compromised), I have avoided speaking of the author's "worldview." I have done so because the term would have made the notion of "representing" what is real [*das Wirkliche*]—the very idea with which Heidegger's philosophy of the "History of Being" wanted to break—into the point of ultimate convergence for Kittler's thinking. After all, according to Heidegger, the event of truth—the "self-unconcealment of Being" [*Selbstentbergung des Seins*]—should not be transubstantiated and then shaped into a worldview; rather, it is supposed to reach and affect [*treffen*] Dasein in ways that differ from the everyday; that is, in ways that always also *concern* Dasein's physical existence. Therefore, although Kittler performs a consistent intellectual gesture of making what is real "ready-to-hand" and then concentrating on the "self-unconcealment" of what is ready-to-hand, I believe that this recurring gesture concerns the preconditions under which what is real can become ready-to-hand and disclose itself, and not the forms of its "representation." In this respect, I consider that a monistic a priori with many aspects plays a key—indeed, a dominant—role in Kittler's work. Kittler tends to bring together different phenomena on one, and only one, plane—phenomena that most thinkers and philosophers would assign to different ontological dimensions.

As much is evident in the verb *anschreiben,* a key term in Kittler's vocabulary. As an anti-idealistic conception of the ideal [*als anti-idealistische Idealvorstellung*], it suggests, time and again, that the movement of a body or a change in the world can precipitate in a medium directly and without mediation. Thereby, the phenomenal level of consciousness is bracketed—that is, *psyche* or *Geist*, which Kittler already sought to "expel" [*austreiben*] in his early writings and to expose, from a historical perspective, as an illusion. Subsequently, in the relation of polarity between hardware and software, Kittler negated the latter term as the analog of consciousness and projection—a gesture that fit well with his focus on situations where things are directly ready-to-hand and not (more or less) distant as a form of "presence-at-hand." In the book about Greece that Kittler never completed, music, eroticism, and mathematics were to belong to a single plane of reality. The author had high regard even for the idealistic theoretical systems of Hegel and Luhmann because, at their core, they postulated monistic concepts such as "spirit" and "meaning." The same held—and in this Kittler was certainly encouraged by Heidegger—for the thinking of the pre-Socratics, which is strictly materialist.

At the end of a 1998 lecture—which later appeared as a book entitled *Eine Kulturgeschichte der Kulturwissenschaft*[4]—Kittler, even if he did not really think the idea all the way through, pointed to affinities between the monism of matter [*das Materielle*], Heidegger's conception of the "event of truth" [*Wahrheitsereignis*], and the "History of Being" [*Seinsgeschichte*] (a notion that one can perhaps characterize as a nonepistemological and nontheological version of revelation). Viewed in terms of the History of Being [*seinsgeschichtlich*], the movement that occurs in the event of truth does not stem from Dasein (i.e., from a human being or even from subjective consciousness) but rather from Being, which wishes to "unconceal itself." Thereby, it seems, Being means the simultaneity of a materially present object (e.g., "earth") and its practical function ("equipment" [*Zeug*] or "world"). Being pushes, so to speak, to unconceal itself as earth and world; in so doing, it must always already overcome images and projections of the human mind/spirit [*Geist*] and the "worldview" [*Weltsicht*] this entails. So that the self-unconcealment of Being can occur, Dasein (human beings) must be present [*anwesend*]. But for all that, the Being that unconceals itself does not offer a message [*Botschaft*] to Dasein. Rather, the Being that discloses itself may, in its fatefulness [*schicksalhaft*], be too strong for Dasein, which has the "debt" [*Schuld*] of "watching out/caring for" [*in die Acht . . . nehmen*]—indeed, of "embracing" [*umarmen*]—it, even if it does so intransitively.

Finally, in Heidegger's History of Being, the equivalent of so-called historical change is the intuition that events of truth are not equally probable at all times. In ancient Greece, everyday situations and encounters with the gods offered many such occasions. But since the twentieth century, it is technology [*die Technik*] that has offered possibilities for the self-disclosure of Being, even though Dasein has not yet found the proper perspective to discern the event. In the interim—between the conditions that obtained in ancient Greece and Heidegger's own world—lie "needful times" [*dürftige Zeiten*], when Being has held distant and remained hidden to Dasein.

It bears repeating: the presence [*Anwesenheit*] of human Dasein belongs to the necessary conditions for the self-unconcealment of Being, yet Dasein remains external to the event. It is important to observe this premise of Heidegger's History of Being—which Kittler always presupposed but never explicated systematically—in order to appreciate [*nachvollziehen zu können*] that the monistic account of phenomenal configurations, which was central to Kittler's way of thinking [*Denk-Form*], always had the potential to point to unconcealed Being. I believe that Kittler considered it his historical and philosophical task to further reveal such phenomenal configurations as the History of Being had already unconcealed—and then, having stripped them of the projections of consciousness, to present them, as purely material structures or automated processes [*blinden Abläufen*], in "plain language" [*Klartext*]. To the extent that Kittler uncovered and described matters in terms of the History of Being, his language achieved "indicavistic" gravity and pathos that made him a mythographer. In offering such accounts, however—and he would indeed have said so himself—Kittler did not occupy the position of an outside observer projecting inward, as it were, but rather that of a seismographer recording [*anschreiben*] Being as it revealed itself.

Truth

After Heidegger's *Introduction to Metaphysics* (1935) at the latest, which clearly presented his turn from the existential-ontological phase of his philosophy to inquiry along the lines of the History of Being, it became increasingly evident that the technology of the present constitutes the specific site and the particular dimension where events of truth may occur. Time and again, Heidegger stressed two main tendencies that rendered improbable what he and his contemporaries owed to Being (i.e., events of self-unconcealment). The first involved the inclination to consider dispositives of technology exclusively in practical contexts, where their material side ("earth") does not show itself. The second concerned the habit of withdrawing from the substantiality of technological presence by viewing it in terms of potential ("Enframing," or in German, *Gestell*). All the same, Heidegger never abandoned the claim that technology and its presence-at-hand for Dasein—in contrast to natural science, which invariably counts as a nobler matter in intellectual terms—constitutes the privileged site for events of truth. Today this premise and its consequence, namely, that thinking through our technological environment should play the central role in any analysis of the present, seems altogether different [*wirkt ganz anders*] than it did at the middle of the twentieth century, especially when viewed in terms of ecological politics; now, it hardly seems surprising or innovative.

And yet, when he died, Heidegger feared that his efforts—his focus on technology as that which unconceals Being—had not proven successful. Famously, in the 1966 interview that was published in the *Spiegel* after his death ten years later, Heidegger declared: "Only a god can still save us."[5] His thinking ended before the media of electronic communication had become part of the human environment on a global scale. Retrospectively, one can wager that his intuition about the self-unconcealment of Being in technology could only have occurred fully once electronic conditions prevailed in massive terms, even though this is nothing more than a side speculation. But as far as Kittler is concerned—and in response to the question concerning the singularity of his work—I affirm that his thinking accomplished what Heidegger left unfinished, and perhaps had to leave unfinished. That is, Friedrich Kittler's thinking-through of electronic technology qualifies as an event of truth; perhaps it was an event of truth that occurred in multiple stages (just as one no longer conceives of the origin of the universe as a single "Big Bang," but rather as a chain reaction involving numerous "Big Bangs").

The process of self-unconcealment may have begun in the final stage, which was so dark in mythographic terms, of the years Kittler devoted to media history—when he described electronic technology as self-guided and purely material ("no software"), as something that did not simply restrict consciousness and the autonomy of the classical subject, but excluded them altogether. I have hinted that the existential intolerability of this vision may have brought Kittler, around the turn of the millennium (possibly for explicit, philosophical reasons, but more likely under existential pressure that was largely self-produced), to turn to the culture of ancient Greece and at the same time to give free rein to the energy of his mythographic impulses—even though he never abandoned the gesture of self-obligation to strict facticity. At the end of his life, then, as he worked with diminishing physical powers on his book about ancient Greece, Kittler's engagement with electronic media assumed a less apocalyptic tone, indeed, a tone that sounded almost cheerful. His 2008 essay on Heidegger presented, as the truth of "computing technology," a new "ontology of distance" and proximity that we have in mind [*meinen*] whenever we speak of "globalization" or live in "globalization":

> It seems to me that, for an ontology of distance, such a conception of globalization based on computer technology reaches further and is more significant [*maßgeblicher*] than all efforts to derive understanding from traditional mass media such as radio, film, and television (as still is common among media historians). (p. 299)

A passage in Kittler's 2007 Mosse Lecture proved even more surprising, more radical—and friendlier:

> As long as we—beholden to corporations such as IBM and Microsoft—only design computers to operate from the top-down, from Bill Gates' business strategy [*Geschäftskalkül*] down to the machines' many, varied components, we (men, programming vassals, and Stanford students) are simply imitating—indeed, mimicking—that One God who thinks He can make do as Creator without any woman or any love at all. Therefore, we should not be surprised if computers take their revenge by developing bugs and lying. For if we were to design them more lovingly—from the bottom up—much would change. Even though we would no longer rake in money with the lie that is called "software," HAL would receive from us, his programmers—and strictly in keeping with Turing—senses, muscles, and a heart, one after the other. Computers would be embryos that (to use Homer's calculation) grow and batten for ten months in the maternal womb. Then, we would free them—as the womb does the child.

Here we encounter Friedrich Kittler the mythographer one last time—and in peak form, I believe: he presents precisely what has sprung from the freedom and productivity of his imagination as if it resulted from a rigorous discourse of professional expertise (that is, "the programmer" is speaking, who sees through "the lie that is called 'software'").

Kittler's assumption of roles—like his discrete anti-Americanism ("Bill Gates' business strategy")—actually belongs to an earlier phase of his work, that is, to the phase that in the mid-1990s culminated in the apocalyptic vision of the "Night of Substance" and the thoroughgoing disempowerment of the subject this entails. Such a tone and perspective come from times when electronic communication was still synonymous with "programming" a computer, an activity that seemed reserved for initiates (among whom Kittler numbered himself) and excluded the vast majority of everyone else. At the time, the situation that prevails today was inconceivable—a situation that has only obtained since computers, thanks to dispositives such as the Apple screen and the mouse, have become "user friendly" and "ready-at-hand"; now they enjoy the status, in the everyday, of being parts of the human body.

And yet one must ask, has this developmental tendency of the electronic world—a tendency that Kittler despised up to the end of his life, whose emblematic figure was Steve Jobs—not brought about, and for some time now, what Kittler dreamed of: computers with "senses, muscles, and a heart"? The question is not simply "rhetorical." Rather, as an open question made possible by Kittler's thinking, it marks the horizon of a discussion that—in the present/presence of "apps"—it is urgent to conduct, given the metamorphoses of mankind's self-image that have occurred and are still occurring. At any rate, these technological supplements to our bodies (as they are born and grow) have an affinity with the image that Kittler—under the influence of Lacan, no doubt—placed front and center: the corporeality of individual body parts that permit connections [*sich ... verschalten lassen*] in ways that are always new.

The "self-unconcealment of Being," as it occurs in electronic technology and is made evident in Kittler's work, does not amount to a "representation," "discourse," or the emergence of a new "paradigm." Instead, it places the phenomena of the world—in their materiality and singularity—within our reach and in this way provokes Dasein to react. Kittler's essays perform a genealogy of our present by making possible, now and for us, moments/aspects of a "clearing" [*Momente einer "Lichtung"*]. On this path of thinking, which Heidegger opened, no one seems to have traveled farther than Kittler. Kittler arrived at the vantage point from which the technology of the present (and its past) could be discerned—a perspective that necessarily remained closed to the philosopher of the History of Being (and especially in relation to electronics). This vantage point "calls for a thinking that takes stock of the pathways of Technology [*Technik*] in its entirety: from the very beginning, with the Greek notion of *technē*, up to its completion in modern computing systems" (p. 299). Kittler's truth—preserving the clearing he made and "watching out for it"—is now the task of those who have survived him, and of the generations that will follow us. From the perspective of the History of Being, we cannot afford to forget him. This was the intuition that was "recorded" [*sich "anschrieb"*] in the intensive reactions to Friedrich Kittler's death in October 2011. Now—as an insight—may it keep the thinking to which his works give rise alive for the future.

Notes

Chapter 1

1. In lieu of constant indications of secondary literature, I refer the reader to the relevant theoretical and historical considerations presented in Philippe Ariès, "Le XIXe siècle et la révolution des mœurs familiales," *Renouveau des idées sur la famille*, ed. Robert Prigent (Paris: PUF, 1954), 111–18; Michel Foucault, *The History of Sexuality*, vol. 1: *An Introduction*, trans. Richard Hurley (New York: Vintage, 1990); Michel Foucault, *The Order of Things: An Archeology of the Human Sciences* (New York: Pantheon, 1971); Jacques Lacan, "La famille," *Encyclopédie française*, ed. A. de Monzie, vol. VIII (Paris: Société de Gestion de l'Encyclopédie française, 1938), 40.3–42.8.

2. Wolfram von Eschenbach, *Parzival and Titurel*, trans. Cyril Edwards (Oxford: Oxford University Press, 2009), II, 112, 21–28.

3. von Eschenbach, *Parzival* III, 126, 21–29.

4. von Eschenbach, *Parzival* IX, 476, 27–30.

5. Novalis, *Schriften*, ed. Paul Kluckhohn and Richard Samuel (Stuttgart: Kohlhammer, 1960), I, 345; hereafter, references to the text occur parenthetically. This section summarizes and revises (omitting lengthy demonstrations) my essay, "Die Irrwege des Eros und die absolute Familie. Psychoanalytischer und diskursanalytischer Kommentar zu Klingsohrs Märchen in Novalis' Heinrich von Ofterdingen," *Psychoanalytische und psychopathologische Literaturwissenschaft*, ed. Winfried Kudszus and Bernd Urban (Darmstadt: Wissenschaftliche Buchgesellschaft, 1981), 421–70. On the maternal imago, see also Gerhard Schulz, *Novalis* (Munich: Rowohlt, 1969).

6. Cf. Ingeborg Weber-Kellermann, *Die deutsche Familie: Versuch einer Sozialgeschichte* (Frankfurt a.M.: Suhrkamp, 1974), 110–12.

7. Novalis, *Schriften*, III, 296.

8. Cf. W. J. Fries, "Eros und Ginnistan: Ein Beitrag zur Symbolik in 'Heinrich von Ofterdingen,'" *Neophilologus* 38 (1963): 23–36.

9. Cf. Rolf Nägele, *Die Muttersymbolik bei Clemens Brentano* (Winterthur: Keller, 1959), 22.

Notes to Chapter 1

10. Edward Shorter, "Der Wandel der Mutter-Kind-Beziehung zu Beginn der Moderne," *Geschichte und Gesellschaft* 1 (1975): 256.
11. J.-B.-D. Bucquet (1804), quoted by Shorter, "Der Wandel," 261.
12. Clemens Brentano, *Werke*, ed. Friedhelm Kemp (Munich: Hanser, 1963), II, 138.
13. Heinrich Bosse, "The Marvellous and Romantic Semiotics," *Studies in Romanticism* 14 (1975): 228.
14. Novalis, *Schriften*, II, 672.
15. Johann Gottfried Herder, *Sämtliche Werke*, ed. Bernhard Suphan (Berlin: Wiedmann, 1877–1913), XXIX, 132, and VIII, 198.
16. Rousseau, *Confessions*, I.
17. Brentano, *Werke*, II, 613.
18. Cf. Gerhard Schaub, *Le Génie Enfant: Die Kategorie des Kindlichen bei Clemens Brentano* (Berlin/New York: de Gruyter, 1973).
19. Novalis, *Schriften*, 197ff.; E.T.A. Hoffmann, "The Sandman," *The Golden Pot and Other Tales*, trans. Ritchie Robertson (Oxford: Oxford University Press, 2009), 85–118.
20. Friedrich Schlegel, "Goethes Werke nach der Cottaschen Ausgabe von 1806," *Kritische Ausgabe*, ed. Ernst Behler (Paderborn: Schöningh, 1958), III, 113.
21. Brentano, *Werke*, II, 585–96.
22. Julia Kristeva, *Le texte du roman: Approche sémiologique d'une structure discursive transformationelle* (Den Haag: Mouton, 1970), 60.
23. Jacques Lacan, *On Feminine Sexuality, the Limits of Love and Knowledge (Encore: The Seminar of Jacques Lacan, Book XX)*, trans. Bruce Fink (New York: Norton, 1999), 99.
24. Schlegel, *Kritische Ausgabe*, V, 15.
25. E.T.A. Hoffmann, *Werke*, ed. Walter Müller-Seidel (Darmstadt: Wissenschaftliche Buchgesellschaft, 1961–65), 288.
26. E.T.A. Hoffmann, *Späte Werke* (Darmstadt: Wissenschaftliche Buchgesellschaft, 1965), 765.
27. Indeed, according to Jan Hendrik van den Berg, the renewal and "literarization" of the fairy tale is what effects this separation. Cf. *Metabletica: Über die Wandlung des Menschen* (Göttingen: Vandenhoeck & Ruprecht, 1960), 81f.
28. Ludwig Tieck, *Werke*, ed. Marianne Thalmann (Munich: Winkler, 1964), II, 179.
29. Tieck, *Werke*, 16f.
30. Tieck, *Werke*, 14.
31. Achim von Arnim, *Sämtliche Romane und Erzählungen*, ed. Walter Migge (Darmstadt: Wissenschaftliche Buchgesellschaft, 1963), II, 523.
32. Tieck, *Werke*, 20.

33. Tieck, *Werke*, 9.
34. Tieck, *Werke*, 21.
35. Jacques Lacan, *Écrits: The First Complete Edition in English*, trans. Bruce Fink (New York: Norton, 2007), 30.
36. Tieck, *Werke*, 26.
37. Gerhard Kaiser's lectures on the novella (Freiburg im Breisgau, 1969–70) focused on this matter.
38. Lacan (orally).
39. E.T.A. Hoffmann, "Die Marquise de la Pivardiere," *Späte Werke* (Darmstadt: Wissenschaftliche Buchgesellschaft, 1965), 354f.
40. This point is demonstrated, apropos of Hoffmann's "Sandman," in my essay, "'Das Phantom unseres Ichs' und die Literaturpsychologie," *Urszenen: Literaturwissenschaft als Diskursanalyse und Diskurskritik*, ed. Friedrich A. Kittler and Horst Turk (Frankfurt a.M.: Suhrkamp, 1977), 160f.
41. Cf. Klaus D. Post, "Kriminalgeschichte als Heilsgeschichte: Zu E. T. A. Hoffmanns Erzählung 'Das Fräulein von Scuderi,'" *Zeitschrift für deutsche Philologie* 95 (1976): *Sonderheft E. T. A. Hoffmann*, 143.
42. E.T.A. Hoffmann, *The Life and Opinions of the Tomcat Murr*, trans. Althea Bell (New York: Penguin, 1999), 175.
43. Hoffmann, *Werke*, 274, 226.
44. Hoffmann, *Werke*, 271.
45. Cf. Richard Alewyn, "Ursprung des Detektivromans," *Probleme und Gestalten* (Frankfurt a.M.: Insel, 1974), 353.
46. Sigmund Freud, *An Outline of Psycho-Analysis*, trans. James Strachey (New York: Norton, 1989), 68.

Chapter 2

1. Cf. Philippe Lacoue-Labarthe, "Le détour (Nietzsche et la rhétorique)," *Poétique* 2 (1971): 53–76; here, 64. When not otherwise noted, quotations follow Friedrich Nietzsche, *Werke in drei Bänden*, ed. Karl Schlechta (Munich: Hanser, 1954–56). Inset references are to the following texts: Beyond = *Beyond Good and Evil* (1886); Birth = *The Birth of Tragedy from the Spirit of Music* (1872); Case = *The Case of Wagner* (1888); Collected Works = *Gesammelte Werke* (Munich: Musarion, 1922–28); Daybreak = *Daybreak* (1878–80); Future = "On the Future of Our Educational Institutions" (1872); Gay Science = *The Gay Science* (1882–87); Genealogy = *On the Genealogy of Morals* (1887); Greek Literature = *History of Greek Literature* (1874–76 lecture); Greek State = *The Greek State* (1873); Human = *Human, All Too Human* (1878–80); Nachlass = *Aus dem Nachlaß der Achtzigerjahre* (1880–89); Nietzsche contra Wagner = *Nietzsche contra Wagner* (1889); Rhetoric = *Rhetoric* (1874 lecture); Untimely = *Untimely Meditations* (1873–76); Truth = "On Truth and Lie in an Extra-

Moral Sense" (1873); *Twilight* = *Twilight of the Idols* (1889); *Works and Letters* = *Historisch-kritische Gesamtausgabe der Werke und Briefe* (Munich: Beck, 1933–42 [uncompleted]).

2. Cf. Heinrich Bosse, "Herder (1744–1803)," *Klassiker der Literaturtheorie. Von Boileau bis Barthes*, ed. Horst Turk (Munich: Beck, 1979), 78–91.

3. For an exemplary misunderstanding of this pathway, cf. Jürgen Habermas, *Knowledge and Human Interests*, trans. Jeremy J. Shapiro (Boston: Beacon, 1972), 290–300.

4. Cf. Michel Foucault, "Nietzsche, Marx, Freud," *Friedrich Nietzsche, Cahiers de Royaumont, Philosophie*, no. 6 (1964): 189.

5. Martin Heidegger, *Nietzsche* (Pfullingen: Neske, 1961), I, 91–109.

6. Arthur Schopenhauer, *Die Welt als Wille und Vorstellung*, III § 52, *Sämtliche Werke*, ed. Wolfgang Freiherr von Löhneysen (Darmstadt: Wissenschaftliche Buchgesellschaft, 1974–76), I, 358.

7. Contra Eugen Fink, *Nietzsches Philosophie* (Stuttgart: Kohlhammer, 1960), 20–28.

8. Georg Wilhelm Friedrich Hegel, *Vorlesungen über Ästhetik* (Frankfurt a.M.: Suhrkamp, 1970), I, 26.

9. Cf. Odo Marquard, "Über einige Beziehungen zwischen Ästhetik und Therapeutik in der Philosophie des 19. Jahrhunderts," *Schwierigkeiten mit der Geschichtsphilosophie* (Frankfurt a.M.: Suhrkamp, 1973).

10. Cf. Pierre Klossowski, *Nietzsche et la cercle vicieux* (Paris: Mercure de France, 1969).

11. This is why Foucault has renewed the reading of Nietzsche. Cf. Michel Foucault, "Nietzsche, Genealogy, History," *Language, Counter-Memory, Practice: Selected Essays and Interviews*, ed. Donald F. Bouchard (Ithaca: Cornell University Press, 1980), 136–64.

12. Cf. Horst Turk, "Hegel (1770–1831)," *Klassiker der Literaturtheorie: Von Boileau bis Barthes*, ed. Horst Turk (Munich: Beck, 1979), 122–32.

13. Cf. Gerhard Rupp, *Rhetorische Strukturen und kommunikative Determinanz. Studien zur Textkonstitution des philosophischen Diskurses im Werk F. Nietzsches* (Bern: Peter Lang, 1976).

14. Cf. Martin Heidegger, "The Origin of the Work of Art," *Poetry, Language, Thought*, trans. Albert Hofstadter (New York: Perennial, 2001), 15–86.

15. Cf. H. Koller, *Die Mimesis in der Antike: Nachahmung, Darstellung, Ausdruck*. Diss. phil., Bern, 1954.

16. Cf. Gilles Deleuze, *Nietzsche and Philosophy*, trans. Janis Tomlinson (New York: Columbia University Press, 1983).

17. Cf. Bettina Rommel, "Transformationen des Ästhetizismus," Friedrich Kit-

tler/Horst Turk, *Urszenen. Literaturwissenschaft als Diskursanalyse und Diskurskritik* (Frankfurt a.M.: Suhrkamp, 1977), 323–54.

18. Cf. Elrud Kunne-Ibsch, *Die Stellung Nietzsches in der Entwicklung der modernen Literaturwissenschaft* (Tübingen: Niemeyer, 1972), 35–50.

19. Cf. Eric Blondel, "Les guillemets de Nietzsche," *Nietzsche aujourd'hui?* (Paris, 1973), II, 153–82.

20. For a brilliant account of how Dilthey's reading of Nietzsche concealed all of this in the name of "understanding" [*Einfühlung*], see Jan Kamerbeek Jr., "Dilthey versus Nietzsche," *Studia philosophica* 10 (1950): 52–84.

Chapter 3

1. Johann Wolfgang von Goethe, *Gespräche, Gesamtausgabe*, ed. Flodoard Freiherr von Biedermann (Leipzig: Insel, 1909–11), IV, 390.

2. Cf. Friedrich Kittler, "Über die Sozialisation Wilhelm Meisters," Gerhard Kaiser and Friedrich A. Kittler, *Dichtung als Sozialisationsspiel* (Göttingen: Vandenhoeck & Ruprecht, 1978), 103–6.

3. Johann Wolfgang von Goethe, *Wilhelm Meister's Apprenticeship*, trans. Eric Blackall (Princeton: Princeton University Press, 1995), 44.

4. Johann Wolfgang von Goethe, *Briefe und Tagebücher*, ed. Hans Gerhard Gräf (Leipzig: Insel, [no year]), II, 712 (entry of 27 August 1831).

5. Cf. Bernhard Siegert, *Relays: Literature as an Epoch of the Postal System*, trans. Kevin Repp (Stanford: Stanford University Press, 1999).

6. Johann Wolfgang von Goethe, *Wilhelm Meisters theatralische Sendung* (1777–85), ed. Wilhelm Haupt (Leipzig: Insel, 1959), 73.

7. Walter Benjamin, *Deutsche Menschen. Eine Folge von Briefen* (1936), *Gesammelte Schriften*, ed. Rolf Tiedemann und Hermann Schweppenhäuser (Frankfurt a.M.: Suhrkamp, 1972–89), IV/1, 211.

8. Jacques Lacan, *The Seminar: Book 2: The Ego in Freud's Theory and in the Technique of Psychoanalysis* (Cambridge: Cambridge University Press, 1988), 115.

9. Emil Staiger, *Grundbegriffe der Poetik* (Zurich, Freiburg/Br.: Atlantis, 1963), 13.

10. Martin Heidegger, *Sein und Zeit. Erste Hälfte* (1927) (Halle: Niemeyer, 1931), 165. On "Wanderer's Night Song," cf. Hermann A. Müller-Solger, "Kritisches Lesen. Ein Versuch zu 'Wandrers Nachtlied II,'" *Seminar. A Journal of Germanic Studies* 10 (1974): 257.

11. Staiger, *Grundbegriffe*, 16.

12. Sigmund Freud, "Formulations on the Two Principles of Mental Functioning," *Papers on Metapsychology; Papers on Applied Psycho-Analysis* (London: Hogarth Press, 1953), 20.

Notes to Chapter 3

13. Cf. Jacques Lacan, "The Subversion of the Subject and the Dialectic of Desire in the Freudian Unconscious," *Écrits: The First Complete Edition in English*, trans. Bruce Fink (New York: Norton, 2007), 688: "No authoritative statement has any guarantee here other than its very enunciation."

14. Ibid., 683.

15. Johann Wolfgang von Goethe, *The Sorrows of Young Werther*, trans. David Constantine (Oxford: Oxford University Press, 2012), 78.

16. Jacques Lacan, "On a Question Prior to Any Possible Treatment of Psychosis," *Écrits*, 463.

17. Goethe, *Sorrows of Young Werther*, 105.

18. Goethe, *Sorrows of Young Werther*, 7 (letter of 13 May).

19. Cf. Edward Shorter, "Der Wandel der Mutter-Kind-Beziehung zu Beginn der Moderne," *Geschichte und Gesellschaft. Zeitschrift für Historische Sozialwissenschaft* 1 (1975): 27.

20. C. Pfeufer, "Über das Verhalten der Schwangeren, Gebährenden und Wöchnerinnen auf dem Lande, u. ihre Behandlungsart der Neugeborenen und Kinder in den ersten Lebensjahren," *Jahrbuch der Staatsarzneikunde* 3 (1810): 63; quoted in Shorter, "Der Wandel der Mutter-Kind-Beziehung," 259.

21. Cf., for example, Jean-Jacques Rousseau, *Émile: Or, On Education*, trans. Allan Bloom (New York: Basic, 1979), 37: "The first education is the most important, and this first education belongs incontestably to women."

22. Ibid., 9: "There is no substitute for maternal solicitude." In an altogether similar vein, see Johann Heinrich Pestalozzi, "Weltweib und Mutter," *Sämtliche Werke*, ed. Artur Buchenau, Eduard Spranger, and Hans Stettbacher (Berlin/Leipzig: de Gruyter, 1927–76), XVI, 352. On the phantasm of such irreplaceability in Rousseau, cf. Jacques Derrida, *Of Grammatology*, trans. Gayatri Chakravorty Spivak (Baltimore: Johns Hopkins University Press, 1998), 265ff. All the same, Derrida takes philosophically—that is, for better and for worse—the irreplaceability of the mother to be merely an example of the category of "irreplaceability" itself, instead of analyzing the category on the basis of its discursive instances.

23. Johann Wolfgang von Goethe, *Dichtung und Wahrheit*, *Goethes Werke. Hamburger Ausgabe*, ed. Erich Trunz (Hamburg: Wegner, 1948–60), X, 74 (§ III.15).

24. Wolfgang Scheibe, *Die Strafe als Problem der Erziehung. Eine historische und systematische pädagogische Untersuchung* (Weinheim/Berlin: Julius Beltz, 1967), 44.

25. Johann Heinrich Pestalozzi, "Über den Sinn des Gehörs, in Hinsicht auf Menschenbildung durch Ton und Sprache," *Sämtliche Werke*, XVI, 266. Hereafter cited parenthetically.

26. Johann Heinrich Pestalozzi, "Vorrede," *Sämtliche Werke*, XV, 347.

27. Cf. (long before Lacan) Bruno Snell, *Die Entdeckung des Geistes bei den*

Griechen. Studien zur Entstehung des europäischen Denkens bei den Griechen (Hamburg: Claassen, 1948), 21.

28. Cf. Jacques Lacan, *De la psychose paranoïaque dans ses rapports avec la personnalité* (Paris: Seuil, 1975), 326: "The question arises, whether all knowledge [*connaissance*] is not, first of all, knowledge concerning a person before being knowledge about an object, and whether the very notion of object is not a secondary acquisition for human beings."

29. Cf. also Gerhard Kaiser, *Geschichte der deutschen Lyrik von Goethe bis Heine. Ein Grundriß in Interpretationen* (Frankfurt a.M.: Suhrkamp, 1988), I, 249, 254.

30. Cf., for example, Rousseau, *Émile*, 65: "All our languages are works of art. Whether there was a language natural and common to all men has long been a subject of research. Doubtless there is such a language, and it is the one children speak before knowing how to speak. This language is not articulate, but it is accented, sonorous, intelligible. [...] Nurses are our masters in this language. They understand everything their nurslings say; they respond to them; they have quite consistent dialogues with them; and, although they pronounce words, these words are perfectly useless; it is not the sense of the word that children understand but the accent which accompanies it."

31. Johann Gottfried Herder, "Das Ich" (1797), *Sämmtliche Werke*, ed. Bernhard Suphan (Berlin: Weidmannsche Buchhandlung, 1877–1913), XXIX, 132.

32. Lacan, "Subversion of the Subject," 692.

33. Johann Heinrich Pestalozzi, *Ältere Fassung* (1803), *Sämtliche Werke*, XVI, 1.

34. Johann Wolfgang von Goethe, "An den Mond," *Goethes Werke. Hamburger Ausgabe*, I, 130.

35. Joseph von Eichendorff, "Nachts," *Neue Gesamtausgabe der Werke und Schriften*, ed. Gerhard Baumann und Siegfried Grosse (Stuttgart: Cotta, 1957), I, 12.

36. Clemens Brentano, "Lureley," *Werke*, ed. Friedhelm Kemp. (Munich: Hanser, 1963–68), I, 258.

37. Theodor W. Adorno, "In Memory of Eichendorff," *Notes to Literature*, trans. Shierry Weber Nicholson (New York: Columbia University Press, 1991), I, 69.

38. Quoted in Wulf Segebrecht, *J. W. Goethe, "Über allen Gipfeln ist Ruh." Texte, Materialien, Kommentar* (Munich: Hanser, 1978), 64. Here, too, otherwise unjustified doubts are voiced.

39. Bruno Jöckel, "Der Erlebnisgehalt des Wiegenliedes," *Berliner Hefte für geistiges Leben* 3 (1948): 414.

40. Jacob Grimm/Wilhelm Grimm, *Deutsches Wörterbuch* (Leipzig: Hirzel, 1954).

41. Jöckel, "Der Erlebnisgehalt," 412.

42. Müller-Solger, "Kritisches Lesen," 262.

43. Lacan, *Seminar*, 47.

44. Gotthilf Heinrich Schubert, *Symbolik des Traumes* (Heidelberg: Schneider, 1968), 16n.

45. Richard Alewyn, "Clemens Brentano, 'Der Spinnerin Lied,'" *Probleme und Gestalten. Essays* (Frankfurt a.M.: Insel, 1974), 198.

46. Cf. Clemens Brentano, "Chronika des fahrenden Schülers Johannes Laurenburger zu Polsnich an der Lahn," *Werke*, ed. Friedhelm Kemp (Munich, 1963–68), II, 612–15.

47. Bettina von Arnim, *Günderode* (Boston: Burnham, 1861), 284.

48. Richard Wagner, *My Life* (Middlesex: Echo, 2007), 146.

49. Jacques Lacan, "Overture to This Collection," *Écrits*, 3.

50. Hans Dieter Zimmermann, *Vom Nutzen der Literatur. Vorbereitende Bemerkungen zu einer Theorie der literarischen Kommunikation* (Frankfurt a.M.: Suhrkamp, 1977), 112.

51. Cf. Kaiser, *Geschichte der deutschen Literatur*, I, 267: "Thus, ['Wanderer's Night Song'], while speaking of language ending, also speaks of it beginning."

52. Johann Gottfried Herder, "Vom Erkennen und Empfinden der menschlichen Seele" (1778), *Sämmtliche Werke*, VIII, 198.

53. On the material history of the text, the wall, and the hut (which burned down in 1870), cf. Segebrecht, *J. W. Goethe, "Über allen Gipfeln ist Ruh,"* 15–21.

Chapter 4

1. Pierre Klossowski, *Diana at Her Bath: The Women of Rome*, trans. Stephen Sartarelli and Sophie Hawkes (Boston: Eridanos, 1990), 30.

2. Walter J. Ong, *Orality and Literacy. The Technologizing of the Word* (New York: Methuen, 1982), 72. Here it stands in full:

> Vision comes to a human being from one direction at a time: to look at a room or a landscape, I must move my eyes around from one part to another. When I hear, however, I gather sound simultaneously from every direction at once: I am at the center of my auditory world, which envelops me, establishing me at a kind of core of sensation and existence. This centering effect of sound is what high-fidelity sound reproduction exploits with intense sophistication. You can immerse yourself in hearing, in sound. There is no way to immerse yourself similarly in sight.

3. Cf. *Der Spiegel* 51/1979, 176.

4. David Gilmour, quoted in Jean-Marie Leduc, *Pink Floyd* (Paris, 1973), 54.

5. Cf. Gilles Deleuze and Félix Guattari, *Anti-Oedipus: Capitalism and Schizophrenia*, trans. Robert Hurley, Mark Seem, and Helen R. Lane (Minneapolis: University of Minnesota Press, 1983), 240ff.

6. E.g., Gottfried Benn, *Roman des Phänotyp* (Wiesbaden: Limes, 1959–61), 174.

7. On architects, cf. Wolfgang Scherer, *BABELLOGIK. Soundproduktion bei Patti Smith* (Frankfurt a.M.: [no pub.], 1982).

8. Jacques Lacan, *Le séminaire, livre XI: Les quatre concepts fondamentaux de la psychanalyse* (Paris, 1973), 178. Cf., already, August Ferdinand Bernhardi, *Sprachlehre* (Berlin: Frölich, 1801–3), I, 24.

9. The details are provided in Walter Bruch, "Von der Tonwalze zur Bildplatte. 100 Jahre Ton- und Bildspeicherung," *Funkschau*, Sonderheft, 1979, [unpaginated].

10. Cf. Philippe Auguste Mathias, Comte de Villiers de l'Isle-Adam, *L'Ève future* (Paris: Corti, 1977), 29.

11. Wildenbruch's words—which (sensibly enough) are not included in the poet's collected works—are transcribed in Bruch, "Von der Tonwalze zur Bildplatte."

12. Cf. Roland Gelatt, *The Fabulous Phonograph. From Edison to Stereo* (New York: Appleton-Century, 1965), 234.

13. Cf. ibid., 282f.

14. Cf. Steve Chapple and Reebee Garofalo, *Rock 'n' Roll Is Here to Pay* (Chicago: Nelson-Hall, 1977), 53.

15.

Thus in Scene Three, Alberich puts on the magic cap, disappears, and then thrashes the unfortunate Mime. Most stage productions make Alberich sing through a megaphone at this point, the effect of which is often less dominating than that of Alberich in reality. Instead of this, we have tried to convey, for thirty-two bars, the terrifying, inescapable presence of Alberich: left, right, or center there is no escape for Mime. (John Culshaw, quoted in Gelatt, *Fabulous Phonograph*, 316)

16. Cf. David Gilmour, "Interview with Gary Cooper," *Wish You Were Here. Songbook* (London: Music Sales, 1975), 77:

When the track disappears into a thin, reedy transistor radio sound which is then joined by a plainly recorded acoustic guitar, there has obviously been a lot of thought behind the end product. How did they tackle that one? —"When it sounds like it's coming out of a radio, it was done by equalization. We just made a copy of the mix and ran it through eq. to make it very middly, knocking out all the bass and most of the high top so that it sounds radio-like."

17. Cf. Gilles Deleuze/Félix Guattari, *Mille plateaux. Capitalisme et Schizophrénie* (Paris: Minuit, 1980), 121, 424.

18. Christian Müller (ed.), *Lexikon der Psychiatrie* (Berlin, Heidelberg, New York: Springer, 1973), entry: "Halluzination."

Notes to Chapters 4 and 5

19. E.g., Eugen Bleuler, *Lehrbuch der Psychiatrie*, ed. Manfred Bleuler (Berlin, Heidelberg, New York: Springer, 1969), 32.

20. Cf. Deleuze/Guattari, *Mille plateaux*, 103.

21. Cf. Alain Dister, Udo Woehrle, and Jacques Leblanc, *Pink Floyd* (Bergisch-Gladbach: Böhler, 1978) [unpaginated].

22. Quoted in Nicholas Schaffner, *Saucerful of Secrets: The Pink Floyd Odyssey* (New York: Random House, 1991), 55.

23. On the two kinds of music, cf. also *The Wall*, where the maximization of wattage is followed, at the end, by a little piece for accordion, clarinet, and toy drums—Merry Old England, once more.

24. Benn, *Roman des Phänotyp*, 182.

25. Michel Foucault, *Les mots et les choses. Une archéologie des sciences humaines* (Paris: Gallimard, 1966), 396.

26. Friedrich Nietzsche, *Beyond Good and Evil*, trans. Helen Zimmern (Rockville: Serenity, 2008), 137.

27. On the preceding, cf. Deleuze and Guattari, *Mille plateaux*, 416–28.

28. Gottfried Benn, "Probleme der Lyrik," *Gesammelte Schriften*, ed. Dieter Wellershoff (Wiesbaden: Limes, 1959–61), I, 518.

29. Friedrich Nietzsche, *The Gay Science*, ed. Bernard Williams (Cambridge: Cambridge University Press, 2001), 83.

30. See Chapter 2.

31. For an illustration on the example of Schubert's settings of Goethe to music, see Thrasybulos Georgiades, "Sprache als Rhythmus," *Kleine Schriften* (Tutzing: Hans Schneider, 1977), 81–96.

32. Cf. Jean Lescure, "Radio et littérature," *Encyclopédie de la Pléiade. Histoire des littératures*, vol. III, ed. Raymond Queneau (Paris: Gallimard, 1958), 1705–8. Nothing illustrates the technical coupling of words and music more nicely (and therefore more philologically) than two quotations hidden on *Dark Side of the Moon*. "Look around, choose your own ground" refers, of course, to Don Juan's first instruction to his pupil Castaneda; but the mysterious command, "Run, rabbit, run!" is a direct quote (cf. Carlos Castaneda, *Journey to Ixtlan. The Lessons of Don Juan* [Harmondsworth: Penguin, 1973], 154). In this way (as in the Beatles' "Revolution 9"), the record became the medium of secret messages.

33. Deleuze and Guattari, *Anti-Oedipus*, 258.

34. Roger Waters, "Interview with Nick Sedgewick," *Wish You Were Here. Songbook*, 13.

Chapter 5

1. Cf. Michel Foucault, *History of Madness*, trans. Jonathan Murphy (London: Routledge, 2006), xxxi–xxxii.

2. Jacques Lacan, *Écrits: The First Complete Edition in English*, trans. Bruce Fink (New York: Norton, 2007), 232.

3. Daniel Paul Schreber, *Memoirs of My Nervous Illness*, trans. Ida Macalpine and Richard A. Hunter (New York: NYRB Classics, 2000), 307.

4. Sigmund Freud, *Dora: An Analysis of a Case of Hysteria* (New York: Touchstone, 1997), 8.

5. Cf. Jacques Lacan, *The Seminar: The Psychoses (Book III)*, trans. Russell Grigg (New York: Norton, 1997), 109.

6. Sigmund Freud, *Three Case Histories* (New York: Touchstone, 1996), 154.

7. Cf. Jens Schreiber, "Strahlenverkehr," *ZETA 02. Mit Lacan*, ed. Dieter Hombach (Berlin, 1982), 155.

8. Sigmund Freud, *Introductory Lectures on Psychoanalysis*, trans. James Strachey (New York: Norton, 1989), 19.

9. Sigmund Freud, *An Outline of Psycho-Analysis*, trans. James Strachey (New York: Norton, 1989), 82.

10. Georg Hirth, *Die Lokalisationstheorie angewandt auf psychologische Probleme. Beispiel: Warum sind wir "zerstreut"?* (Munich: Hirth, 1895), 33.

11. Ibid., xii.

12. Cf. Carl Gehrmann, *Körper, Gehirn, Seele, Gott. Vier Theile in drei Bänden* (Berlin: Dames, 1893). (The title alone prompts one to anticipate a few thousand pages.)

13. Cf. Jacques Lacan, "L'étourdit," *Scilicet* 4 (1973): 16.

14. Freud, *Three Case Histories*, 154.

15. Sigmund Freud, "Der Wahn und die Träume in W. Jensens 'Gradiva,'" *Gesammelte Werke, chronologisch geordnet* (Frankfurt a.M.: Fischer, 1946–68), VII, 120. Thus does Freud formulate the relationship between psychoanalysis and writers—who, according to his theory, discern the laws of the Unconscious through endopsychic perception (like Schreber).

16. June 1895, quoted in Franz Baumeyer, "Der Fall Schreber," *Psyche* 9 (1955/56): 517.

17. Sigmund Freud, "Charcot," *Gesammelte Werke*, I, 25.

18. Paul Flechsig, *Die körperlichen Grundlagen der Geistesstörungen. Vortrag gehalten beim Antritt des Lehramts an der Universität Leipzig am 4. März 1882* (Leipzig, 1882), 3.

19. Ibid., 21.

20. Schreber, *Memoirs*, 350. Cf. also 15, 221.

21. Cf. Charles E. McClelland, *State, Society, and University in Germany 1700–1914* (Cambridge: Cambridge University Press, 1980), 212–23.

22. Paul Flechsig, *Meine myelogenetische Hirnlehre mit biographischer Einleitung* (Berlin: Springer, 1927), 26f.

23. Ibid., 41.
24. Ibid.
25. Cf. Georg Hirth, *Aufgaben der Kunstphysiologie* (Munich: Hirth, 1897), 264ff.
26. Flechsig, *Hirnlehre*, 20ff.
27. Paul Flechsig, "Über die Associationscentren des menschlichen Gehirns. Mit anatomischen Demonstrationen," *Dritter Internationaler Congress für Psychologie in München vom 4. bis 7. August 1896* (Munich: Lehmann, 1897), 57. Cf. Lacan's earliest formulation of the "mirror stage" and the *corps morcelé* (*Écrits*, 73–74).
28. Paul Flechsig, *Die Grenzen geistiger Gesundheit und Krankheit. Rede, gehalten zur Feier des Geburtstages Sr. Majestät des Königs Albert von Sachsen am 23. April 1896* (Leipzig,1896), 18.
29. Flechsig, *Die körperlichen Grundlagen*, 9.
30. Ibid., 11.
31. Schreber, *Memoirs*, 158.
32. Flechsig (?), quoted in Baumeyer, "Der Fall Schreber," 514.
33. Quoted in ibid., 521.
34. Quoted in ibid., 516.
35. Schreber, *Memoirs*, 9.
36. Ibid., 9 and passim. Ellen Key's *The Century of the Child*, incidentally, uses the same term to refer to the effect of school on children.
37. Schreber, *Memoirs*, 37. Cf. Roberto Calasso, *Die geheime Geschichte des Senatspräsidenten Dr. Daniel Paul Schreber* (Frankfurt a.M.: Suhrkamp, 1980), 20.
38. Schreber, *Memoirs*, 307.
39. Cf. Baumeyer, "Der Fall Schreber," 522, on this "very thorough autopsy report" (which, incidentally, found none of the transformations of nervous tissue that Schreber feared—or hoped for).
40. Schreber, *Memoirs*, 48.
41. Ibid., 66.
42. Lacan (*Seminar*) makes this point repeatedly in his seminar on psychosis. On "nerve-language": "We have here a highly elaborate theory whose position it would not be difficult to encounter, even if it were only as a stage of the discussion, in standard scientific works" (66). On hallucinations: "Not only is he on the whole a good witness, but he commits no theological mistakes. Moreover, he's well informed, I would even say he's a good classical psychiatrist" (125). It is only necessary, then, to name the "good classical psychiatrist" who repeats Schreber's every word. All the same, Lacan—despite his insights that Flechsig stands at the center of the delirium (26) and that scientific language represents the modern form of oppression (Lacan, *Écrits*, 80)—did not do so.
43. Schreber, *Memoirs*, 31.
44. Schreber, *Memoirs*, 86. This passage, along with many others, makes it ab-

solutely clear that all the features of Schreber's God at the beginning of his confinement were attributes of the patient's first psychiatrist.

45. Flechsig, *Die körperlichen Grundlagen*, 11.

46. The fact that Wilhelm Griesinger had posited this same dependency twenty years before Flechsig does not mean, of course, that it could be demonstrated anatomically and physiologically at the time.

47. Schreber, *Memoirs*, 54ff.

48. Schreber, *Memoirs*, 25n. Yet another—and for all that, overlooked—reference to Flechsig's exchanges with Schreber as the source of the latter's newfound knowledge.

49. Schreber, *Memoirs*, 54–55.

50. Saussure's "image acoustique" corresponds exactly to the contents of the acoustic center that, since the research of Carl Wernicke, had been located neurologically.

51. Schreber, *Memoirs*, 192. Cf. Paul Flechsig, *Gehirn und Seele. Rede, gehalten am 31. Oktober 1894 in der Universitätskirche zu Leipzig* (Leipzig, 1896), 43ff. (on receptive aphasia); as well as Lacan, *Seminar*, 219 (on Schreber and Wernicke's aphasia).

52. Schreber, *Memoirs*, 24. The fact that Schreber's "rays" [*Strahlen*] and Flechsig's "radiation" [*Strahlungen*] are both nerves disproves the argument that the patient's words have the same, conventionally religious sense that they possess in his father's writings (cf. Morton Schatzman, *Soul Murder: Persecution in the Family* [New York: Signet, 1974]).

53. Schreber, *Memoirs*, 273.

54. Ibid., 132.

55. Schreber, *Memoirs*, 277.

56. As Octave Mannoni recognized; see "Schreber als Schreiber," *Clefs pour l'Imaginaire ou l'Autre Scène* (Paris: Seuil, 1969), 91.

57. Flechsig, *Gehirn und Seele*, 45–47.

58. On this connivance between Flechsig and Freud, cf. Calasso, *Die geheime Geschichte*, 22f.

59. Cf. Lacan, *Seminar*, 284.

60. Schreber, *Memoirs*, 62; cf. Samuel M. Weber, "Schreber," *Denkwürdigkeiten eines Nervenkranken* (Frankfurt a.M.: Ullstein, 1973), 490.

61. As is well known, Foucault (with a sideways glance at National Socialist biotechnologies) held this "to the political credit of psychoanalysis" (Michel Foucault, *The History of Sexuality, Vol. 1: An Introduction*, trans. Robert Hurley [New York: Vintage, 1990], 150).

62. Schreber, *Memoirs*, 147.

63. Ibid., 367. For Schreber, it necessarily follows that "every judgment" he has

reached as a writer and "judge" permits one to discern a "human being who is perfectly lucid intellectually." As far as his view of Flechsig is concerned, this statement is certainly true.

Chapter 6

1. Adalbert von Chamisso, "Erscheinung," *Gesammelte Werke in vier Bänden* (Stuttgart [no year]), II, 13–15.

2. Cf. Sigmund Freud, "The 'Uncanny,'" *The Standard Edition of the Complete Psychological Works of Sigmund Freud*, vol. XVII, ed. and trans. James Strachey (London: Hogarth, 1955), 219–26. In taking language literally, Freud follows in the steps of his forebear. Ernst Jentsch, whom the countless interpreters of Freud today no longer read (of course), considers "the spirit of language" not to be a "particularly strong psychologist" in general, but he concedes that in the case of the "word *unheimlich*," German has arrived at a "reasonably felicitous formulation" (Ernst Jentsch, "Zur Psychologie des Unheimlichen," *Psychiatrisch-neurologische Wochenschrift* 22 [1906]: 195).

3. Otto Rank, *The Double: A Psychoanalytic Study*, trans. and ed. Harry Tucker Jr. (Chapel Hill: University of North Carolina Press, 1971), 35–37.

4. Ibid., 76–77.

5. Ibid., 69. Cf. Friedrich Kittler, "'Das Phantom unseres Ichs' und die Literaturpsychologie," *Urszenen. Literaturwissenschaft als Diskursanalyse und Diskurskritik*, ed. F. A. Kittler and Horst Turk (Frankfurt a.M.: Suhrkamp, 1977), 139–66.

6. The exception—Goethe's famous encounter with himself when leaving Sesenheim and Friederike Brion—did not escape Freud's acumen: the "pike-gray" frock, which the doppelgänger wore in 1771 and Goethe himself would don only in 1779, when paying another visit, represents the "official uniform" [*Staatskleid*] of a successful bureaucrat who draws up documents first and poems second. Cf. Rank, *Double*, 39–40n. On the historical conditions framing Goethe's doppelgänger (the modern, nuclear family and narcissism), see also Jacques Lacan, "Der Individualmythos des Neurotikers," *Der Wunderblock* 5/6 (1980): 61–68.

7. Rank, *Double*, 39. The primary source for this information is Paul Sollier, *Les phénomènes d'autoscopie* (Paris: Félix Alcan, 1903).

8. Johann Wolfgang von Goethe, *Wilhelm Meister's Apprenticeship*, trans. Eric Blackall (Princeton: Princeton University Press, 1995), 103.

9. Friedrich Wilhelm Riemer, *Mitteilungen über Goethe*, ed. Arthur Pollmer (Leipzig, 1921), 261.

10. Goethe, *Wilhelm Meister's Apprenticeship*, 112.

11. Daniel Jenisch, *Über die hervorstechendsten Eigenthümlichkeiten von Meisters Lehrjahren; oder über das, wodurch dieser Roman ein Werk von Göthen's Hand ist. Ein ästhetisch-moralischer Versuch* (Berlin: Langhoff, 1797), 14. According to

Friedrich Schlegel, too, "the characters in this novel—because of the way they are portrayed—are like a portrait [!], but are essentially more or less general and allegorical" ("Über Goethe's Meister," *Kritische Friedrich-Schlegel-Ausgabe*, ed. Ernst Behler [Paderborn: Schöningh, 1958–], II, 143).

12. Jenisch, *Eigenthümlichkeiten*, 14f. On writing and reading techniques of identification in general, cf. Friedrich Kittler, "Über die Sozialisation Wilhelm Meisters," *Dichtung als Sozialisationsspiel*, ed. Gerhard Kaiser and Friedrich A. Kittler (Göttingen: Vandenhoeck & Ruprecht, 1978), 99–114.

13. Cf. Schlegel, "Über Goethe's Meister," 136, 141f.

14. "Fragment von 1798," in Novalis, *Schriften. Die Werke Friedrich von Hardenbergs*, ed. Paul Kluckhohn and Richard Samuel (Stuttgart, 1960–), III, 377.

15. Alfred de Musset, "La nuit de décembre," *Œuvres complètes*, ed. Philippe Van Tieghem (Paris, 1963), 153; "The December Night," *The Complete Writings of Alfred de Musset*, trans. George Santayana, Emily Shaw Forman, and Marie Agathe Clarke (New York: Edwin C. Hill, 1907), II, 336.

16. Freud, "Uncanny," 248n.

17. Stéphane Mallarmé, "Sur le beau et l'utile," *Œuvres complètes*, ed. Henri Mondor and G. Jean-Aubry (Paris: Gallimard, 1961), 880. Mallarmé's prose poem "Le nénuphar blanc" (*Œuvres complètes*, 283–86), offers a pretechnological version of the "travel shot." On cinema and automotive travel in general, see Paul Virilio, *L'insécurité du territoire* (Paris: Galilée, 1976), 251–57.

18. Stéphane Mallarmé, "Sur le livre illustré," *Œuvres complètes*. 878.

19. Georg Büchner, *Leonce und Lena, Werke und Briefe, Gesamtausgabe*, ed. Fritz Bergemann (Wiesbaden, 1958), 447. Police warrants—which Büchner parodied on the basis of personal experience—seem to go back to the age of high Absolutism.

20. Cf. the details in Carlo Ginzburg, "Clues: Morelli, Freud, and Sherlock Holmes," *The Sign of Three: Dupin, Holmes, Peirce*, ed. Umberto Eco and Thomas Sebeok (Bloomington: Indiana University Press, 1984), 81–118.

21. Jentsch, "Zur Psychologie des Unheimlichen," 205.

22. Cf. Walter Görlitz, *Kleine Geschichte des deutschen Generalstabes* (Berlin: Hause & Spener, 1967), 194f. Ludendorff's exact words are quoted in Ludwig Greve, Margot Pehle, Heidi Westhoff (eds.), *Hätte ich das Kino! Die Schriftsteller und der Stummfilm* (*Sonderausstellungen des Schiller-Nationalmuseums*) (Marbach, 1976), 75.

23. Cf. Michael Herr, *Dispatches* (New York: Knopf, 1977).

24. Cf. Henri Bergson, *L'Évolution créatrice* (Paris: Presses universitaires de France, 1923), 330ff.

25. Cf. Michel Foucault, *The History of Sexuality. Volume 1: An Introduction*, trans. Robert Hurley (New York: Vintage, 1990), 56n.

26. Cf. Jean Martin Charcot, *Œuvres completes*, vol. 1 (Paris, 1886).

Notes to Chapter 6

27. Information about Albert Londe (1858–1917) follows Hrayr Terzian, "La fotografia psichiatrica," *Nascita della fotografia psichiatrica*, ed. Franco Cagnetta (Venice, 1981), 39. Facts concerning his "films" of hysteria follow Joël Farges, "L'image d'un corps," *Communications* 23 (1975): 89.

28. Rank, *Double*, 3ff; translation modified.

29. Rank, *Double*, 65n.

30. Tzvetan Todorov, *The Fantastic: A Structural Approach to a Literary Genre*, trans. Richard Howard (Ithaca: Cornell University Press, 1975), 161.

31. Hugo Münsterberg, *The Photoplay: A Psychological Study*, reprinted with an introduction by Richard Griffith as *The Film: A Psychological Study. The Silent Photoplay in 1916* (New York, 1970), 15: "Rich artistic effects have been secured, and while on the stage every fairy play is clumsy and hardly able to create an illusion, in the film we really see the man transformed into a beast and the flower into a girl. There is no limit to the trick pictures which the skill of the experts invent. [...] Every dream becomes real." Münsterberg's thesis is unambiguously confirmed by the kind of literature that the feature film replaced after 1895. As is well known, in the most Romantic of all Romantic novels—Hardenberg's *Heinrich von Ofterdingen*—the hero dreams of a blue flower. "Finally, he sought to approach it, when, all of a sudden, it began to move and change: the leaves became more luminous and nestled on the growing stalk, the Flower inclined toward him, and the petals unfolded a blue collar in which a delicate face was floating" (Novalis, *Schriften. Die Werke Friedrich von Hardenbergs*, ed. Paul Kluckhohn and Richard Samuel [Darmstadt: Wissenschaftliche Buchgesellschaft, 1960–68], I, 197).

32. Rank, *Double*, 7.

33. Quoted in Greve, Pehle, and Westhoff, *Hätte ich das Kino!* 110. On *The Student of Prague* as a filming of film itself, cf. also Jean Baudrillard, *Symbolic Exchange and Death* (London: Sage, 2006), 84; and a forthcoming study of film by Michael Zeh (Freiburg).

34. This discussion of the Gerhard Hauptmann film *Phantom* (1922) is quoted in Greve, Pehle, and Westhoff, *Hätte ich das Kino!* 172.

35. Paul Lindau's *Schauspiel in vier Akten*, upon which the film is based—and from which necessity compels me to cite—employs photographs as metaphors for film. Cf. *Der Andere*, (Leipzig, ca. 1906), 22, 81. Lindau, incidentally, who numbered among the first German writers to use a typewriter, was an author Freud read in his youth. Cf. Ernest Jones, *Sigmund Freud—Leben und Werk*, ed. Lionel Trilling and Steven Marcus (Frankfurt a.M.: Fischer, 1969), 182.

36. Georg Seeßlen and Claudius Weil, *Kino des Phantastischen. Geschichte und Mythologie des Horror-Films* (Reinbek: Rowohlt, 1980), 48.

37. Cf. Chapter 9.

38. Edgar Morin, *The Cinema, or The Imaginary Man*, trans. Lorraine Mortimer (Minneapolis: University of Minnesota Press, 2005), 134–35.

39. Gustav Meyrink, *The Golem*, trans. Mike Mitchell (Sawtry: Dedalus, 2008), 23; translation slightly modified.

40. Cf. Bergson, *L'Évolution créatrice*, 330ff.; and Gilles Deleuze, *Cinema 1: The Movement-Image*, trans. Hugh Tomlinson (Minneapolis: University of Minnesota Press, 1986).

41. Meyrink, *Der Golem*, 25 (thanks to Michael Müller).

42. Cf. Daniel Paul Schreber, *Memoirs of My Nervous Illness*, trans. Ida Macalpine and Richard A. Hunter (New York: NYRB Classics, 2000), 100 and passim. The context proves clearly enough that hordes of doubles that lack identity and appear in series are also commuters for Schreber.

43. Cf. Hugo Münsterberg, *Grundzüge der Psychotechnik* (Leipzig: Barth, 1914): 767 pages that are just as great as they are now forgotten.

44. Münsterberg, *The Photoplay*, 31–48 ("The Psychology of the Photoplay").

45. Cf. Jones, *Sigmund Freud*, 350.

46. Biographical information follows Richard Griffith's introduction to the reprint edition.

Chapter 7

1. *Frankfurter Allgemeine Zeitung*, 3 November 1983, 12.

2. On the strategy of roadway design from the First World War on, cf. Friedrich Kittler, "Autobahnen," *Kulturrevolution. Zeitschrift für angewandte Diskurstheorie* 5 (1984): 42–44.

3. Cf. Erik Bergaust, *Wernher von Braun. Ein unglaubliches Leben* (Düsseldorf/Vienna: Econ, 1976), 111.

4. Cf. Pink Floyd, *The Final Cut: A Requiem for the Post War Dream* (London, 1983), side 1.

5. Thomas Pynchon, *Gravity's Rainbow* (New York: Viking, 1983), 39. Further references occur parenthetically in the text.

6. Cf. Bernd Ruland, *Wernher von Braun. Mein Leben für die Raumfahrt* (Offenburg: Burda, 1969), 141; needless to say, the account differs in Walter Dornberger, *V2—Der Schuß ins Weltall. Geschichte einer großen Erfindung* (Esslingen: Bechtle, 1953), 120f.

7. Ruland, *Wernher von Braun*, 268.

8. On Nordhausen, the largest known subterranean factory, cf. Manfred Bornemann, *Geheimprojekt Mittelbau. Die Geschichte der deutschen V-Waffen-Werke* (Munich: Lehmanns, 1971).

9. On absolute enmity, total mobilization, and Kleist's "partisan" literature, cf.

Notes to Chapter 7

Carl Schmitt, *Theorie des Partisanen. Zwischenbemerkung zum Begriff des Politischen* (Berlin: Duncker & Humblot, 1963).

10. Here is the wording of the "tame" *Xenion* that treats the introduction of compulsory army service neither in military nor in ideological fashion, but in terms of discourse analysis:

Hatte sonst einer ein Unglück getragen,	Once, if misfortune befell you,
So durft' er es wohl dem andern klagen;	You could complain to others;
Mußte sich einer im Felde quälen,	If you toiled on the field,
Hatt' er im Alter was zu erzählen.	You had something to tell in old age.
Jetzt sind sie allgemein, die Plagen,	Now, the affliction is general,
Der einzelne darf sich nicht beklagen;	And no one may complain.
Im Felde darf nun niemand fehlen—	No one may be absent from the field—
Wer soll denn hören, wenn sie erzählen?	Who will be left when they tell their tales?

(Johann Wolfgang Goethe, *Sämtliche Werke, Jubiläums-Ausgabe*, ed. Eduard von der Hellen [Stuttgart: Cotta, 1904–5], IV, 131)

11. Cf. Paul Fussell, *The Great War and European Memory* (Oxford: Oxford University Press, 1975).

12. Cf. Ernst Jünger, *Der Kampf als inneres Erlebnis* (Berlin: Mittler & Sohn, 1922), 98 and 92 (the First World War as "strangler of our literature").

13. Cf. ibid., 12, 28, 50, 107ff.

14. For the V-2, the delay lasted sixteen seconds; cf. Ruland, *Wernher von Braun*, 221.

15. Dietmar Kamper, "Atlantis—vorgeschichtliche Katastrophe, nachgeschichtliche Dekonstruktion" (lecture manuscript [Paris, 1984]).

16. Cf. William Stevenson, *A Man Called Intrepid: The Secret War* (New York: Simon & Schuster, 1977), as well as the inadequate study by Joseph Borkin, *The Crime and Punishment of I. G. Farben* (New York: Pocket Books, 1978).

17. On Flaubert's *La Tentation de Saint-Antoine*, cf. Michel Foucault, "La bibliothèque fantastique," *Travail de Flaubert*, ed. Raymond Debray-Genette et al. (Paris: Seuil, 1983), 103–22.

18. Cf. Sigmund Freud, *Gesammelte Werke, chronologisch geordnet*, ed. Anna Freud et al. (Frankfurt a.M.: Fischer, 1946–68), X, 302.

19. Cf. Charles William Morris, *Foundation of the Theory of Signs* (Chicago: University of Chicago Press, 1972), 67ff.

20. On Staver, cf. Ruland, *Wernher von Braun*, 249. Further name games between fact and fiction: "Höhler," the architect of the central works at Nordhausen (Bornemann, *Geheimprojekt Mittelbau*, 23), becomes "Ötsch" (*Gravity's Rainbow*, 298–302); "Enzian," the code name for a rocket project at Peenemünde (Ruland,

Wernher von Braun, 261), becomes the name of the leader of the fictitious Hereros in the Waffen-SS; finally, "Max" and "Moritz," the two engineers at the launch of the manned V-2 (*Gravity's Rainbow*, 757f.), refer to Braun's A2 launch of November 1934 (Ruland, *Wernher von Braun*, 89ff.). Readers are encouraged to keep looking.

21. Cf. Claude E. Shannon and Warren Weaver, *The Mathematical Theory of Communication* (Urbana: University of Illinois Press, 1964), 12.

22. On Speer's theory that all architecture must reckon with its future "value as ruins," see Paul Virilio, *War and Cinema*, trans. Patrick Camiller (New York: Verso, 1989), 55.

23. Cf. Ellic Howe, *The Black Game: British Subversive Operations against the Germans During the Second World War* (London: Queen Anne, 1988).

24. After witnessing this film about rocket travel to the moon, a character muses that "real flight and dreams of flight go together" (159). On UFA and Professor Oberth's first project for liquid-fuel rockets, cf. Ruland, *Wernher von Braun*, 57–67. On Lang's movie and the power of the medium, cf. Virilio, *War and Cinema*, 58f.:

> The film came out on 30 September 1929, but without the intended publicity of a real rocket launch from the beach of Horst in Pomerania to an altitude of forty kilometers. By 1932 jet technology, . . . developed at Dornberger's newly opened Kummersdorf West Research Center, was set to become one of the main military secrets of the Third Reich, and the German authorities of the time seized Lang's film on the grounds that it was *too close to reality*. A decade later, on 7 July 1943, von Braun and Dornberger presented Hitler with the film of the real launch of the A4 [V-2] rocket. The Führer was in a bitter mood: "Why was it I could not believe in the success of your work? If we'd had these rockets in 1939 we'd never have had this war."

There is no clearer proof of the power of film: Hitler, the cineaste who was bored by all real-life demonstrations of the V-2 (cf. Dornberger, *V2—Der Schuß ins Weltall*, 73–77, 99–101), was convinced by a movie.

25. Cf. Virilio, *War and Cinema*, 11, for Marey's chrono-photographic firearm.

26. Cf. the admissions made by the head of this division in Reginald V. Jones, *Most Secret War* (London: Penguin, 2009).

27. For the function—and a picture—of this device, see Sigfried Giedion, *Mechanization Takes Command: A Contribution to Anonymous History* (Oxford: Oxford University Press, 1948), 21f.

28. See Chapter 6.

29. Cf. Virilio, *War and Cinema*, 81.

30. Cf. Dornberger, *V2—Der Schuß ins Weltall*, 259.

31. On Kammler's construction activities (also in Nordhausen) as part of the Chief Economic Administration Office [*Wirtschafts-Verwaltungshauptamt*] of the

SS, cf. Enno Georg, *Die wirtschaftlichen Unternehmungen der SS* (*Schriftenreihe der Vierteljahreshefte für Zeitgeschichte Nr. 7*) (Stuttgart: Deutsche Verlags-Anstalt, 1963), 37f.; as well as Bornemann, *Geheimprojekt Mittelbau*, 43, 82f., 125. For events in his life until 1932 (*Grenzschutz Ost, Sturmabteilung Roßbach, Siedlungsamt Danzig, Reichsarbeitsministerium*, etc.), see Hans Kammler, *Zur Bewertung von Geländeerschließungen für die großstädtische Besiedlung*, Diss. Ing., TH Hannover, 1932. Should readers have any further information...

32. Cf. Ruland, *Wernher von Braun*, 170, where nothing further is said about Kammler's motivations. As far as Pynchon is concerned, though, readers will recall the question he asks in *Gravity's Rainbow*: "Is that who you are, that vaguely criminal face on your ID card, its soul snatched by the government camera as the guillotine shutter fell?" (134).

33. As the model for the dialogue between Weissmann and Pökler, read the lengthy exchange Dornberger records between himself and Dr. Ernst Steinhoff, the electronics specialist at Peenemünde (Dornberger, *V2—Der Schuß ins Weltall*, 147–49).

34. Cf., for example, ibid., 286:

Kammler refused to believe that collapse was imminent. He rushed from the front in Holland and the Rhineland to Thuringia and Berlin. He traveled day and night. Back and forth, again and again. Conferences were called for one o'clock in the morning, somewhere in the Harz Mountains, or else we met at midnight, somewhere on the autobahn, to exchange information and opinions quickly before driving back to work. Inhuman nervous tension held us in suspense. We were irritated, nervous, overworked. We didn't take everything that was said at face value. When things weren't going fast enough for him, Kammler woke up the accompanying officers with a blast from his automatic before driving on.

35. Cf. Ruland, *Wernher von Braun*, 282f.:

Since the SS-General Kammler, Hitler's special deputy [*Sonderbevollmächtigter*] for V-weapons, could not be found [after the war], London wanted to put Dornberger on trial in his stead. No one knew at the time, what had become of Kammler. A few years later only one thing was certain: on 4 May 1945, Kammler showed up in Prague with an airplane. On May 9, he defended a bunker with twenty-one SS men against 600 Czech partisans. Exultant, Kammler abandoned the bunker and fired at the onrushing Czechs with his machine gun. Kammler's adjutant, *Sturmbannführer* Starck, had been ordered, months earlier, not to let his commander fall into the hands of the enemy. He had been ordered always to follow him at ten paces—"shooting distance." Now, in this hopeless situation, Starck fired a cluster of bullets from his machine gun into the back of the SS-General's head.

Chapter 8

1. Quotations follow Novalis, *Schriften. Die Werke Friedrich von Hardenbergs*, ed. Paul Kluckhohn und Richard Samuel (Darmstadt: Wissenschaftliche Buchgesellschaft, 1960–68). Simple page numbers refer to the second edition of the first volume; Roman numerals and page numbers refer to the other volumes.

2. On Klingsohr's tale, see Friedrich Kittler, "Die Irrwege des Eros und die 'absolute Familie.' Psychoanalytischer und diskursanalytischer Kommentar zu Klingsohrs Märchen in Novalis' 'Heinrich von Ofterdingen,'" *Psychoanalytische und psychopathologische Literaturinterpretation*, ed. Bernd Urban and Winfried Kudszus (Darmstadt: Wissenschaftliche Buchgesellschaft, 1981), 421–70. Whereas the latter study discusses what is necessarily unconscious in the text—its psychohistorical preconditions—the one at hand treats its surface: word processing [*Textverarbeitung*] as such.

3. Michel Foucault, *History of Madness*, trans. Jonathan Murphy (London: Routledge, 2006), xxxi–xxxii.

4. Cf. Claude E. Shannon and Warren Weaver, *The Mathematical Theory of Communication* (Urbana: University of Illinois Press, 1964), 33–35.

5. On the development of the notion of place value in Schleiermacher's hermeneutics, cf. Manfred Frank, "Einleitung," F.D.E. Schleiermacher, *Hermeneutik und Kritik* (Frankfurt a.M.: Suhrkamp, 1977), 34–37.

6. Johann Wolfgang Goethe, "Über Philostrats Gemählde," *Werke, Weimarer Ausgabe* (Weimar: Böhlau, 1887–1919), XLIX/1, 142.

7. On the "symbol of intuition," cf. Horst Turk, "Goethes Wahlverwandtschaften: 'der doppelte Ehebruch der Phantasie,' *Urszenen. Literaturwissenschaft als Diskursanalyse und Diskurskritik*, ed. Friedrich A. Kittler and Horst Turk (Frankfurt a.M.: Suhrkamp, 1977), 204–7.

8. For documentation and media-technical conclusions, cf. Friedrich Kittler, *Discourse Networks, 1800/1900*, trans. Michael Metteer with Chris Cullens (Stanford: Stanford University Press, 1992), 113–19.

9. G.W.F. Hegel, *Phänomenologie des Geistes* (Hamburg: Meiner, 1980), 433.

10. Cf. Michel Foucault, "La bibliothèque fantastique," *Travail de Flaubert*, ed. Raymond Debray-Genette et al. (Paris: Seuil, 1983), 103–22.

11. On literacy programs and "automatic" reading, cf. Joachim Gessinger, "Schriftspracherwerb im 18. Jahrhundert. Kulturelle Verelendung und politische Herrschaft," *Osnabrücker Beiträge zur Sprachtheorie* 11 (1979): 39.

12. G.W.F. Hegel, *System der Philosophie (= Encyclopädie)*, ed. Hermann Glockner (Stuttgart: Frommann-Holzboog, 1927–40), 351.

13. Cf. Alain Montandon, "Écriture et folie chez E. T. A. Hoffmann," *Romantisme* 24 (1979): 12.

14. Cf. Jacques Lacan, *On Feminine Sexuality, the Limits of Love and Knowledge (Encore)*, trans. Bruce Fink (New York: Norton, 1999), 35:

> Everything that is written stems from the fact that it will forever be impossible to write, as such, the sexual relationship. It is on that basis that there is a certain effect of discourse, which is called writing. One could, at a pinch, write $x\,R\,y$, and say x is man, y is woman, and R is the sexual relationship. Why not? The only problem is that it's stupid, because what is based on the signifier function (*la fonction de signifiant*) of "man" and "woman" are mere signifiers.

15. Hugo Münsterberg, *The Photoplay: A Psychological Study*, reprinted as *The Film: A Psychological Study. The Silent Photoplay in 1916*, ed. Richard Griffith (New York: Dover, 1970), 15.

16. On this relationship between seeing and being seen, cf. Jacques Lacan, *Le séminaire, livre XI: Les quatre concepts fondamentaux de la psychanalyse* (Paris: Seuil, 1973), 88ff.

17. Cf. Anke Bennholdt-Thomsen, *Stern und Blume. Untersuchungen zur Sprachauffassung Hölderlins* (Bonn: Bouvier, 1967).

18. Cf. Friedrich Schlegel, "Über die Philosophie. An Dorothea," *Kritische Friedrich-Schlegel-Ausgabe*, ed. Ernst Behler (Paderborn: Schöningh, 1958–), VIII, 42.

19. Johannes Mahr, *Übergang zum Endlichen. Der Weg des Dichters in Novalis' "Heinrich von Ofterdingen"* (Munich: Fink, 1970), 172.

20. A novel written in 1808, which continued and completed the fragmentary *Ofterdingen*, turned the matter into plain speech and pragmatic plot points: "The Princess often seemed to have forgotten that she was mute, for none of her words escaped her father" (Ferdinand August Otto Heinrich Graf von Loeben [= Isidorus Orientalis], *Guido* [Mannheim: Schwan und Götz, 1808], 13).

21. Cf., once more, the parallel in Schlegel, "Über die Philosophie," VIII, 42.

22. On the function of these two deletions, cf. Kittler, *Discourse Networks*, 121–22.

23. Cf. Kittler, "Die Irrwege des Eros," 449–63.

24. Johann Wolfgang Goethe, *Aus meinem Leben. Dichtung und Wahrheit, Sämtliche Werke. Jubiläums-Ausgabe*, ed. Eduard von der Hellen (Stuttgart: Cotta, 1904–5), XXIV, 161.

25. Johann Christian Reil, *Rhapsodieen über die Anwendung der psychischen Curmethode auf Geisteszerrüttungen* (Halle: Curtsche Buchhandlung, 1803), 55.

26. Johann Gottlieb Fichte, "Deducirter Plan einer zu Berlin zu errichtenden höheren Lehranstalt," *Sämmtliche Werke*, ed. Immanuel Hermann Fichte (Berlin: Veit, 1845–46), VIII, 98.

27. Johann Gottlieb Fichte, *Die Grundzüge des gegenwärtigen Zeitalters, Sämmtliche Werke*, VII, 109.

28. Ibid.

29. Fichte, "Deducirter Plan," 100ff.

30. Cf., for example, the two encyclopedic reference works that Johann Christoph Männling compiled, around 1700, for Lohenstein's novel.

31. Cf. E.T.A. Hoffmann, "Der goldne Topf," *Fantasie- und Nachtstücke*, ed. Walter Müller-Seidel (Munich: Winkler, 1960), 215.

32. Heinrich Bosse, "'Die Schüler müßen selbst schreiben lernen' oder Die Einrichtung der Schiefertafel," *Schreiben—Schreiben lernen, Rolf Sanner zum 65. Geburtstag*, ed. Dietrich Boueke and Norbert Hopster (Tübingen: Narr, 1985), 194.

33. Ibid., 195.

34. Hegel, *Phänomenologie des Geistes*, 25.

35. Compare formulations concerning the inversely proportional relation between the "Astrologist" and the "Miner" (260) with the innumerable passages in the *Brouillon*, where two or more sciences are chalked up as analogical proportions.

36. Friedrich Paulsen, *Geschichte des gelehrten Unterrichts auf den deutschen Schulen und Universitäten vom Ausgange des Mittelalters bis zur Gegenwart. Mit besonderer Rücksicht auf den klassischen Unterricht* (Berlin, Leipzig: Veit, 1919–21), II, 166.

37. Cf. Ferdinand Bünger, *Entwicklungsgeschichte des Volksschullesebuches* (Leipzig: Dürr, 1898), esp. 231.

38. Schleiermacher, "Gutachten vom 14. Dezember 1810," quoted in Paul Schwartz, "Die Gründung der Universität Berlin und der Anfang der Reform der höheren Schulen im Jahre 1810," *Mitteilungen der Gesellschaft für deutsche Erziehungs- und Schulgeschichte* 20 (1910): 173.

39. Ibid., 196.

40. Cf. the keywords on the planned continuation: "Conversation with the Emperor on government, empire, etc." (340); "Conversation with the Emperor on government, etc. Mystical Emperor. Book *de Tribus Impostoribus*" (341).

41. Jens Schreiber, *Das Symptom des Schreibens. Roman und absolutes Buch in der Frühromantik (Novalis/Schlegel)* (Frankfurt/M., Bern, New York: Peter Lang, 1983), 212.

42. Cf. Gerhard Schulz, *Novalis* (Reinbek: Rowohlt, 1969), 141ff.

43. Hegel, quoted in Karl Löwith, *Von Hegel zu Nietzsche. Der revolutionäre Bruch im Denken des 19. Jahrhunderts* (Stuttgart: Kohlhammer, 1950), 34.

44. K. Fricke, "Die geschichtliche Entwicklung des Lehramts an den höheren Schulen," K. Fricke and K. Eulenburg, *Beiträge zur Oberlehrerfrage* (Leipzig: Teubner, 1903), 16.

45. Cf. Schulz, *Novalis*, 139–41.

46. Georg Lukács, *The Theory of the Novel: A Historico-Philosophical Essay on the Forms of Great Epic Literature*, trans. Anna Bostock (Cambridge: MIT Press, 1971), 140.

47. Ibid.

48. Cf. Kittler, *Discourse Networks*, 154ff.

49. Johann Gottlieb Fichte, "Beweis der Unrechtmässigkeit des Büchernachdrucks. Ein Räsonnement und eine Fabel," *Sämmtliche Werke*, VIII, 227. On the preparatory stages for such authorship—namely, when Ofterdingen learns to identify as the protagonist of the poetic novels of education [*Dichterbildungsromane*] that he reads—cf. Kittler, *Discourse Networks*, 119ff.

Chapter 9

1. For a systematic development of this argument, see Rudolf Arnheim, *Kritiken und Aufsätze zum Film*, ed. H. H. Dieterichs (Munich: Hanser, 1977), 27.

2. "The Artwork of the Future," *Richard Wagner's Prose Works*, trans. W. A. Ellis, 8 vols. (London: Routledge & Kegan Paul, 1899; New York: Broude Brothers, 1966), I, 67–213.

3. On the categories of classical aesthetics as rewritings of new and complete literacy, see Friedrich Kittler, *Discourse Networks, 1800/1900*, trans. Michael Metteer with Chris Cullens (Stanford: Stanford University Press, 1990), 73–87. On the obsolescence of inherited art forms, see Nietzsche's *Unmodern Observations*, trans. Gary Brown (New Haven: Yale University Press, 1990), 227–304: "For if anything at all distinguishes [Wagner's] art from all arts of the modern age, it is this: it no longer speaks the language of caste culture, and in general no longer recognizes the difference between cultivated and uncultivated. It thereby stands in opposition to all Renaissance culture which has until now enveloped us moderns in its light and shadow" (299). A century after Nietzsche, one may replace some terms of his exceedingly precise definition of modern media: "Gutenberg" for "Renaissance," and "paper and printer's ink" for "light and shadow."

4. "In Bayreuth the darkened room was the aim. At the time, this was a surprising stylistic device, as well. 'The deepest night was created in the auditorium, and so one could not even recognize one's neighbor,' wrote Wagner's nephew, Clemens Brockhaus, on the occasion of the Emperor's visit to Bayreuth in 1876. 'And from the depths, the wonderful orchestra began!'" (Georg Gustav Wiessner, *Richard Wagner der Theaterreformer: Vom Werden des deutschen Nationaltheaters im Geiste des Jahres 1848* [Emstetten: Lechte, 1951], 115). And so, with even emperors and kings now in darkness—parties whom the very architecture of old-European theater had guaranteed representative visibility—the new era was announced to the public. Hardly four decades later, in 1913, a cinema in Mannheim would advertise with the slogan: "Come right on in! We have the darkest cinema in town!" (quoted in Silvio Vietta, "Expressionistische Literatur und Film: Einige Thesen zum wechselseitigen Einfluss ihrer Darstellung und Wirkung," *Mannheimer Berichte* 10 (1975): 295.

5. According to André Glucksmann (*The Master Thinkers*, trans. Brian Pearce

[New York: Harper & Row, 1980], 261): "Behind the theft of the Ring is Wotan's enterprise. Behind the phantasm of capital is the question of power. The gods with long teeth have need of final battles. If Valhalla, power, the Forbidden City, the palace of the Central Committee burn, then everything burns."

6. See Chapter 4 for a discussion of the stereophonic effects employed by John Culshaw and Georg Solti to represent Mime wrestling with his masked brother in their recording of *The Rhine Gold*.

7. An exercise in statistical analysis concluded that in the *Ring* alone the stage directions offer 220 acoustic and 190 optical instructions. These findings relativize the conclusion, made in the same study, that Wagner preferred optical to acoustic data "because the musical treatment does justice to auditory needs on its own" (Karl Gross, "Die Sinnesdaten im *Ring des Nibelungen*: Optisches und Akustisches Material," *Archiv für die gesamte Psychologie* 22 [1912]: 401–22).

8. Richard Wagner, *Die Musikdramen* (Munich: Deutscher Taschenbuch Verlag, 1978), 727–29.

9. Cf. Wolfgang Scherer, "Klaviaturen. Visible Speech und Phonographie," *Diskursanalysen I: Medien*, ed. Friedrich A. Kittler, Manfred Schneider, and Samuel Weber (Opladen: Westdeutscher Verlag, 1986), 37–54.

10. Wagner, *Die Musikdramen*, 731ff.

11. On "respiratory erogeneity," see Jacques Lacan, "The Subversion of the Subject and the Dialectic of Desire in the Freudian Unconscious," *Écrits: The First Complete Edition in English*, trans. Bruce Fink (New York: Norton, 2007), 692. The concept of a partial object voice outlined here derives solely from Lacan.

12. The symmetry between the two lovers in *The Valkyrie* already suggests that music drama as a whole (and as illustrated here otherwise only in reference to *Tristan and Isolde*) may be analyzed as the arc of a single, great breath. The orchestral Prelude begins with the superhuman breath of the same storm that leads to Siegmund's unconsciousness and breathlessness at the outset of the dramatic action proper. After Sieglinde has restored, by song, life and breath to the fugitive, their love for each other grows into the natural "breath" of a "springtide" [*Lenz*] that practically commands them to engage in sibling incest. (Cf. 598ff. and also the "springtide," its "echo," crescendo, and feedback through love in *Die Meistersinger* [Wagner, *Musikdramen*, 422ff.].) Conversely, when Wotan's intervention destroys the possibility of this love, unconsciousness and breathlessness befall Sieglinde. That is why, in Act III, the superhuman storm breath of the beginning triumphs once more—a matter evident both in the Valkyries' "Hojotoho!" and in Wotan's "thunderstorm" (630–34).

13. Wagner, *Die Musikdramen*, 366, 380.

14. Ibid., 686.

15. Ibid., 702ff.

Notes to Chapter 9

16. Ibid., 778ff.
17. Cf. ibid., 190 and 196ff.
18. Ibid., 270ff.
19. Ibid., 840. All that Kundry says might be examined as a hysterical speech disorder in the technical, psychoanalytic sense—that is, as the separation of head and body as intersected by the throat (see Lucien Israël, *Die unerhörte Botschaft der Hysterie* [Munich/Basel: Renst Reinhardt, 1983]). This is also why Gurnemaz—a therapist avant la lettre—listens so attentively to Kundry's "dull groaning" (which, as ever in Wagner, signifies non-death [854ff.]).
20. See Chapter 4.
21. Compare, for example, Bettina Brentano's "Goethes Briefwechsel mit einem Kinde":

> The stars set in a sea of colors; flowers blossomed and grew forth into the heights; faraway golden shadows shielded them from a higher, white light. And so, in this inner world, one apparition drew up after the other; thereby, my ears perceived a fine, silvery tone; gradually it turned into a sound that grew louder and stronger the more closely I listened to it; I was happy, for it strengthened me; it strengthened my spirit to harbor this great tone in my ear. (*Werke*, ed. Gustav Konrad, 5 vols. [Frechen: Bartmann, 1959–61], II, 51–52)

For Schelling, compare the opening of the dialogue "Bruno."

22. On all of this, cf. Wagner, *My Life* (New York: Dodd, Mead, 1911), 361 (echo), 369f., 401f. (feedback), 436, 648f. (fading). The well-known legend of how the Prelude to *The Rhine Gold* occurred to Wagner in his sleep is an acoustic hallucination (603). Ultimately, an echo triggers incest between Siegmund and Sieglinde (cf. *Die Musikdramen*, 600ff.).
23. For a detailed analysis of literality and, as the case may be, *sound* in the two versions of *Tristan*, see Norbert W. Bolz, "*Tristan und Isolde*—Richard Wagner als Leser Gottfrieds," *Mittelalter Rezeption: Gesammelte Vorträge des Salzburger Symposions*, ed. J. Kühnel, H.-D. Mück, and U. Müller (Göppingen: Kummerle, 1979), 279–84.
24. Wagner, *Die Musikdramen*, 344.
25. Richard Wagner, *Tristan und Isolde* (London, Zürich, Mainz, New York: Eulenberg), 323, 328.
26. Richard Wagner, "Brief an Mathilde Wesendonck, 3. März 1860," *Richard Wagner und Mathilde und Otto Wesendonck, Tagebuchblätter und Briefe*, ed. Julius Kapp (Leipzig: Hesse & Becker, 1915), 293.
27. Wagner, *Die Musikdramen*, 321.
28. Ibid., 322.
29. Wagner, *Tristan und Isolde*, 37.

30. Ibid., 994.

31. Wagner, *Die Musikdramen*, 383f.

32. This is a form of acoustics that (in a sense requiring further specification) stands strictly opposed to the philosophical phantasm of hearing-oneself-speak as the substrate of all theories of consciousness. Cf. Jacques Derrida, *Of Grammatology*, trans. Gayatri Chakravorty Spivak (Baltimore: Johns Hopkins University Press, 1976), 240.

33. This (more than so-called intellectual history) would offer a point of departure for reading Wagner with Lacan. Cf. Jochen Hörisch, "Wagner mit Homer. Zur Dialektik von Wunsch und Wissen in Wagners Musikdrama," *Der Wunderblock, Zeitschrift für Psychoanalyse* 3 (1979): 20–32.

34. Wagner, *Die Musikdramen*, 807. On Siegfried's last words, see also Friedrich A. Kittler, "Forgetting," *Discourse* 3 (1981): 88–121; 113.

35. Jimi Hendrix, *Electric Ladyland*, Polydor LP 2335 204, Side A.

36. Wagner, *Die Musikdramen*, 384.

37. Wagner, letter to M. Wesendonck, April 1859, *Richard Wagner und Mathilde und Otto Wesendonck*, 185.

38. The exclusion of visuality at the end of *Tristan* also offers an instance of eradicated phallocentrism. Wagner's first draft contained five lines with an unambiguous meaning: "How he shines / how dearly [*minnig*], / always stronger [*mächt'ger*], / bathed in starlight / he raises himself up." In the process of composition, the two middle lines—with the key words *minnig* and *mächt'ger*—disappeared.

39. Wagner, *Die Musikdramen*, 321.

40. Ibid., 576ff.

41. Nietzsche, "Richard Wagner in Bayreuth," 289.

42. Cf. Werner Wahle, *Richard Wagners szenische Visionen und ihre Ausführung im Bühnenbild. Ein Beitrag zur Problematik des Wagnerstils* (Zeulenroda: Sporn, 1937), 93n.77. "Inventing new scenic lighting effects" was already one of the orders Wagner gave his architect Semper. Cf. Richard Wagner, "A Music School for Munich," *Prose Works*, IV, 171–224; 179.

43. Paul Pretsch (ed.), *Cosima Wagner und Houston Stewart Chamberlain im Briefwechsel 1888 bis 1908* (Leipzig: Reclam, 1914), 146. Adorno's commentary on this passage—that "Nietzsche, in his youthful enthusiasm, failed to recognize the artwork of the future in which we witness the birth of film out of the spirit of music"—only shows, all acumen notwithstanding, that Adorno failed to recognize Nietzsche's enthusiasm and likewise failed to understand his writings (cf. Theodor Adorno, *In Search of Wagner*, trans. Rodney Livingstone [London: Verso, 2005], 96). The account of Attic tragedy as "a bright image projected on a dark wall" (Friedrich Nietzsche, *The Birth of Tragedy*, trans. Walter Kaufmann [New York:

Random House, 1967], 67) is altogether cinematic—and appears in a book published in 1871.

44. Cf. Richard Wagner, "The Destiny of Opera," *Prose Works*, V, 127–56.

45. For music history, see Paul Bekker, *The Story of the Orchestra* (New York: Norton, 1936); for the military, cf., for example, Hansjürgen Usczeck, *Scharnhorst. Theoretiker, Reformer, Patriot* (Berlin: Militärverlag, 1974), 31–35. For all that, the historical parallels between music and strategy remain to be drawn. In the meanwhile, orders such as Isolde's command to the winds or Elsa's to Lohengrin provide the clearest evidence.

46. Wagner, *Die Musikdramen*, 718.

47. This argument is developed by Gilles Deleuze und Félix Guattari, *Mille plateaux. Capitalisme et Schizophrénie* (Paris: Minuit, 1980), 416–22.

48. Richard Wagner, *Opera and Drama*, *Prose Works*, II, 265.

49. Ernst Jünger, *Strahlungen* (Stuttgart: Klett, 1963), II, 159ff., 281.

50. Marcel Proust, *À la recherche du temps perdu*, ed. Pierre Clarac and André Ferré (Paris: Gallimard, 1954), III, 758. On Wagner, Proust, Jünger, and Coppola, cf. Norbert W. Bolz, "Vorschule der profanen Erleuchtung," *Walter Benjamin. Profane Erleuchtung und rettende Kritik*, ed. N. W. Bolz and Richard Faber (Würzburg: Königshausen & Neumann, 1985), 219ff. On Proust's description of the aerial battle, see—once more—Felix Philipp Ingold, *Literatur und Aviatik. Europäische Flugdichtung 1909–1927* (Frankfurt a.M.: Birkhauser, 1980), 259–61.

Chapter 10

1. Michel Foucault, *The History of Sexuality*, vol. 1, trans. Robert Hurley (New York: Vintage, 1978), 88f.

2. Gilles Deleuze and Félix Guattari, *Capitalisme et Schizophrénie: Mille plateaux* (Paris: Minuit, 1980), 539.

3. Martin Heidegger, *An Introduction to Metaphysics*, trans. Ralph Manheim (New York, 1961), 152.

4. Didier Gille, "Maceration and Purification," *Zone Magazine* 1, no. 2 (1986).

5. Christopher Alexander, "A City Is not a Tree," *Design Magazine* (1965).

6. Paul Valéry, "Présence de Paris," *Oeuvres*, ed. J. Hytier (Paris: Gallimard, 1957–60), II, 1015.

7. Johannes Lohmann, "Die Geburt der Tragödie aus dem Geiste der Musik," *Archiv für Musikwissenschaft* 37 (1980): 167–86.

8. Harold Adams Innis, *Empire and Communications* (Oxford: Oxford University Press, 1950).

9. Lewis Mumford, *The City in History: Its Origins, Its Transformations, and Its Prospects* (New York: Mariner, 1968), 569.

10. Alan Turing, "On Computable Numbers, with an Application to the Ents-

cheidungsproblem," *Proceedings of the London Mathematical Society* 2, no. 42 (1936): 23–165.

11. August Herrmann, *Geschichte der Stadt St. Pölten*, 2 vols. (St. Pölten: Sydy 1917–30).

12. Sven Hedin, *Ein Volk in Waffen. Den deutschen Soldaten gewidmet* (Leipzig: Brockhaus, 1915).

13. Walter Benjamin, *The Arcades Project*, trans. Howard Eiland (Cambridge: Harvard University Press, 1999), 602.

14. Richard Euringer, *Chronik einer deutschen Wandlung, 1925–1935* (Hamburg: Hanseatische Verlagsanstalt, 1936).

15. Hans Magnus Enzensberger, *Mausoleum: Thirty-Seven Ballads for the History of Progress*, trans. Joachim Neugroschel (New York: Urizen, 1976), 4.

16. See Jacques Saint-Germain, *La Reynie et la police du Grand siècle d'après de nombreux documents inédits* (Paris: Hachette, 1962).

17. James Joyce, *A Portrait of the Artist as a Young Man* (New York: B. W. Huebsch, 1922), 11f.

18. T. S. Eliot, "The Waste Land," *Selected Poems* (London: Faber & Faber, 1954), 179–81.

19. Wolfgang Schivelbusch, *Intellektuellendämmerung: Zur Lage der Frankfurter Intelligenz in den zwanziger Jahren* (Frankfurt a.M.: Suhrkamp, 1983).

20. Kurt Kaftan, *Der Kampf um die Autobahnen: Geschichte und Entwicklung des Autobahngedankens in Deutschland von 1907–1935 unter Berücksichtigung ähnlicher Pläne und Bestrebungen im übrigen Europa* (Berlin: Wigankow, 1955), 13.

21. Deleuze and Guattari, *Mille plateaux*, 494.

22. Otto Hintze, "Der österreichische und preußische Beamtenstaat im 17. und 18. Jahrhundert: Eine vergleichende Betrachtung," *Historische Zeitschrift* 86 (1901).

23. Rolf Oberliesen, *Information, Daten und Signale: Geschichte technischer Informationsverarbeitung* (Reinbek bei Hamburg: Deutsches Museum, 1982), 241.

24. Herrmann, *Geschichte der Stadt St. Pölten*.

25. Manfred Durth, *Deutsche Architekten: Biographische Verflechtungen 1925–1970* (Munich: DTV, 1987), 252–68.

26. Mumford, *City*, 569.

Chapter 11

1. Friedrich Nietzsche, *The Gay Science*, ed. Bernard Williams (Cambridge: Cambridge University Press, 2001), 83f.

2. For the details, see Friedrich A. Kittler, *Gramophone, Film, Typewriter*, trans. Geoffrey Winthrop-Young and Michael Wutz (Stanford: Stanford University Press, 1999), 201–18.

3. Cf. Thrasybulos Georgiades, "Sprache als Rhythmus," *Kleine Schriften* (Tutzing: Hans Schneider, 1977), 81–96.

4. Cf. Kittler, *Gramophone, Film, Typewriter*, 221–31.

5. Cf. Thomas Pynchon, *Gravity's Rainbow* (New York: Viking, 1983), 245ff.

6. Cf. Adolf Slaby, *Entdeckungsfahrten in den elektrischen Ozean. Gemeinverständliche Vorträge* (Berlin: Leonhard Simion, 1911), 340–44.

7. William R. Blair, "Army Radio in Peace and War," *Radio*, ed. Irwin Stewart (Annals of the American Academy of Political and Social Sciences. Supplement to vol. CXLII, Philadelphia), 87.

8. Quoted in Johannes Ulrich (ed.), *Deutsches Soldatentum* (Stuttgart: Kröner, 1941), 266f.

9. Cf. Ernst Volckheim, *Die deutschen Kampfwagen im Weltkriege* (Berlin: Mittler, 1923), 14 (2. Beiheft zum 107. Jahrgang des Militär-Wochenblattes).

10. These figures are all found in the brilliant study by Winfried B. Lerg, *Die Entstehung des Rundfunks in Deutschland. Herkunft und Entwicklung eines publizistischen Mittels* (Frankfurt a.M: Knecht, 1970).

11. Hasso von Wedel, *Die Propagandatruppen der deutschen Wehrmacht* (Neckargmünd: Scharnorst, 1962), 12 (*Wehrmacht im Kampf*, Vol. 34).

12. *Berliner Börsen-Courier*, 1923, quoted in Lerg, *Entstehung des Rundfunks*, 162.

13. "Verteidigungsschrift in der Strafsache gegen Staatssekretär a.D. Dr. Hans Bredow, Berlin, den 5. September 1934," quoted in Lerg, *Entstehung des Rundfunks*, 53.

14. Cf. ibid., 213.

15. *Reichspostminister* Dr. Höfle, 1923, quoted in ibid., 188.

16. Cf. ibid., 159; and Kittler, *Gramophone, Film, Typewriter*, 253–63.

17. Quoted in Orrin E. Dunlap Jr., *Marconi. The Man and His Wireless* (New York: Macmillan, 1941), 353.

18. The Rolling Stones, *Beggar's Banquet* (London: Mirage Music, 1969), 4.

19. Cf. Steve Chapple and Reebee Garofalo, *Rock 'n' Roll Is Here to Pay* (Chicago: Nelson-Hall, 1977), 110ff.

20. Cf. Dermot Bradley, *Generaloberst Guderian und die Entstehungsgeschichte des modernen Blitzkrieges* (Osnabrück, 1978), 157ff. (*Studien zur Militärgeschichte, Militärwissenschaft und Konfliktforschung*, Vol. 16).

21. Cf. Karl Heinz Wildhagen (ed.), *Erich Fellgiebel—Meister operativer Nachrichtenverbindungen. Ein Beitrag zur Geschichte der Nachrichtentruppe* (Wennigsen, Hannover, 1970), 31ff.

22. Cf. Reginald V. Jones, *Most Secret War* (London: Coronet, 1978), 60–78.

23. Cf. Roland Gelatt, *The Fabulous Phonograph. From Edison to Stereo* (New York, 1977), 282.

24. Brian Southall, *Abbey Road: The Story of the World's Most Famous Recording Studio* (Cambridge: Haynes Manuals, 1982), 37.
25. Von Wedel, *Die Propagandatruppen*, 116f.
26. For further details, see Kittler, *Gramophone, Film, Typewriter*, 108.
27. Cf. Andrew Hodges, *Alan Turing: The Enigma* (New York: Simon & Schuster, 1983), 96ff.
28. On the strategic significance of Turing's computers, cf. Jürgen Rohwer and Eberhard Jäckel (eds.), *Die Funkaufklärung und ihre Rolle im Zweiten Weltkrieg. Eine internationale Tagung in Bonn-Bad Godesberg und Stuttgart vom 15. bis 18. September 1978* (Stuttgart: Motorbuch, 1979).
29. On the vocoder, cf. Hodges, *Alan Turing*, 273–88.

Chapter 12

1. Claude E. Shannon and Warren Weaver, *The Mathematical Theory of Communication* (Urbana: University of Illinois Press, 1964), 31.
2. Cf. ibid., 62f.
3. Cf. K. Beyrer, *Die Postkutschenreise*. Tübingen (= Untersuchungen des Ludwig- Uhland-Instituts der Universität Tübingen im Auftrag der Tübinger Vereinigung für Volkskunde, ed. Hermann Bausinger u. a., Bd. LXVI) 1985, 54.
4. That *person, individual, subject*, and other titles of "man" [*der Mensch*] do not refer to "the unity of an object" but simply to an address might be gathered from the words' traditional definitions—since Deconstruction, at the latest. For a more elegant derivation, however, see Niklas Luhmann, "How Can the Mind Participate in Communication?" *Materialities of Communication*, ed. Hans Ulrich Gumbrecht and K. Ludwig Pfeiffer, trans. William Whobrey (Stanford: Stanford University Press, 1994), 371–88.
5. Cf. Sven Hedin, *Ein Volk in Waffen. Den deutschen Soldaten gewidmet* (Leipzig: Brockhaus, 1915), 75.
6. That said, the author of these words—Wehrmacht rail engineer Blum—posits an exception that illuminates the status of literature under high-tech conditions: "Cf., moreover, the novel by Mitchell, *Gone with the Wind* (long-winded, but exceptionally instructive)" (Prof. Dr.-Ing. e. h. Blum, "Das neuzeitliche Verkehrswesen im Dienste der Kriegführung," *Jahrbuch für Wehrpolitik und Wehrwissenschaften* [Hamburg, 1939], 73–92; 73n).
7. Cf. Jacques Derrida, *The Post Card: From Socrates to Freud and Beyond*, trans. Alan Bass (Chicago: University of Chicago Press, 1987).
8. Claude E. Shannon, "Communication Theory of Secrecy Systems," *Bell System Technical Journal* (1949): 656–715; 657.
9. Cf. Jacques Lacan, *Écrits: The First Complete Edition in English*, trans. Bruce Fink (New York: Norton, 2007), 712f.

Notes to Chapter 12

10. Ibid., 17.

11. Cf. David A. Bell, *Information Theory and Its Engineering Applications* (New York, Toronto, London: Pitman, 1955), 35.

12. Cf. Jacques Lacan, *The Seminar: Book 2: The Ego in Freud's Theory and in the Technique of Psychoanalysis* (Cambridge: Cambridge University Press, 1988), 300ff.

13. Cf. Claude E. Shannon, "A Symbolic Analysis of Relay and Switching Circuits," *Transactions of the American Institute of Electrical Engineers* 57 (1938): 713–22.

14. Bell, *Information Theory*, 35.

15. Ibid., 97.

16. Cf. Shannon, "Communication Theory," 685.

17. As late as 1940, the leading mathematician at Cambridge could still write: "The 'real' mathematics of the 'real' mathematicians, the mathematics of Fermat and Euler and Gauss and Abel and Riemann, is almost wholly 'useless' (and this is true of 'applied' as of 'pure' mathematics). [. . .] It is the dull and elementary parts of applied mathematics, as it is the dull and elementary parts of pure mathematics, that work for good or ill" (G. H. Hardy, "A Mathematician's Apology," quoted in A. Hodges, *Alan Turing: The Enigma* [New York: Simon & Schuster, 1983], 120). Hodges has no difficulty showing how the Second World War definitively refuted such statements.

18. Cf. Alan M. Turing, *The Essential Turing*, ed. B. Jack Copeland (Oxford: Oxford University Press, 2004), 421.

19. Paul Valéry, *Œuvres*, ed. J. Hytier (Paris: Gallimard, 1957–60), II, 300f.

20. Lacan, *Écrits*, 551.

21. Cf. Shannon and Weaver, *Mathematical Theory*, 56: "Two extremes of redundancy in English prose are represented by Basic English and by James Joyce's book *Finnegan's Wake*. The Basic English vocabulary is limited to 850 words and the redundancy is very high. [. . .] Joyce on the other hand enlarges the vocabulary and is alleged to achieve a compression of semantic content."

22. Ibid., 43f.

23. Claude E. Shannon, "Communication in the Presence of Noise," *Proceedings of the Institute of Radio Engineers* 37 (1949): 10–21.

24. Klaus Sickert, *Automatische Spracheingabe und Sprachausgabe. Analyse, Synthese und Erkennung menschlicher Sprache mit digitalen Systemen* (Haar: Markt & Technik, 1983), 44.

25. Cf. Theodor W. Adorno, *Noten zur Literatur*, ed. Gretel Adorno and Rolf Tiedemann (Frankfurt a.M.: Suhrkamp, 1974), 536.

26. J. Lohmann, "Die Geburt der Tragödie aus dem Geiste der Musik," *Archiv für Musikwissenschaft* 37 (1980): 167–86; 174.

27. Cf. Niklas Luhmann, "The Individuality of the Individual: Historical Meanings and Contemporary Problems," *Reconstructing Individualism: Autonomy, Indi-*

viduality and the Self in Western Thought, ed. T. C. Heller, M. Sosna, and D. E. Wellbery (Stanford: Stanford University Press, 1986), 313–25; 321.

28. Roman Jakobson, "Linguistics and Poetics," *Language and Literature*, ed. Krystyna Pomorska and Stephen Rudy (Cambridge: Harvard University Press, 1987), 62–94; 70.

29. Johann Wolfgang von Goethe, *Sämtliche Werke. Jubiläums-Ausgabe*, ed. Eduard von der Hellen (Stuttgart/Berlin: Cotta, 1904–5), V, 13.

30. Cf. Friedrich Bird, *Notizen aus dem Gebiete der psychischen Heilkunde* (Berlin: Hirschwald, 1835), 7–15; and for commentary, Friedrich Kittler, "Ein Subjekt der Dichtung," *Das Subjekt der Dichtung. Festschrift für Gerhard Kaiser zum 60. Geburtstag*, ed. H. Turk, G. Buhr, and F. Kittler (Würzburg: Königshausen & Neumann, 1990).

31. Cf. Wilhelm Stauder, *Einführung in die Akustik* (Wilhelmshaven: Noetzel, 1976), 142–58.

32. Niklas Luhmann, "Intersubjektivität oder Kommunikation. Unterschiedliche Ausgangspunkte soziologischer Theoriebildung," *Diskus* 112 (1987): 24–33; 28.

33. Cf. Thomas Pynchon, *The Crying of Lot 49* (New York: HarperCollins, 1999), 113f; and *Gravity's Rainbow* (New York: Viking, 1983), 703ff.

34. Cf. Bernhard Siegert, *Die Posten und die Sinne*. Master's thesis. Freiburg/Br. 1986 (typescript), 185.

35. Cf. Sickert, *Automatische Spracheingabe*, 261f.

36. Salomo Friedlaender, "Der antibabylonische Turm," *Geschichten vom Buch*, ed. K. Schöffling (Frankfurt a.M.: Insel, 1985), 135–70; 156–70.

37. Cf. F. Kittler, *Gramophone, Film, Typewriter*, trans. Geoffrey Winthrop-Young and Michael Wutz (Stanford: Stanford University Press, 1999).

38. Accordingly, the musical apocatastasis of all noise sources in and since Wagner has led poets to indulge in impossible Fourier analyses when offering descriptions. On Strauss and Hofmannsthal's *Elektra*, Anton Wildgans reported from the Kleines Theater in Berlin: "One feels as if one were in a temple. Then, from its concealment, the orchestra sounds and the tragedy begins; it rushes past one in a movement uninterrupted by acts, like a heightened dream-experience, without relenting in tension, yet comparable to the line of wave, up and down. The whole time, one's own soul vibrates along with it [*Aber immer Schwingung und Mitschwingen der eigenen Seele*]" (Anton Wildgans, *Ein Leben in Briefen*, ed. Lilly Wildgans [Vienna: Frick, 1947], 55; reference thanks to Martin Stingelin, Basel).

39. Theodor W. Adorno, *In Search of Wagner*, trans. Rodney Livingstone (London: Verso, 2005), 125.

40. Richard Wagner, *Götterdämmerung* (London, Zürich, Mainz, New York: Eulenberg, 1970), 1273–77.

41. Cf. Goethe *Sämtliche Werke. Jubiläums-Ausgabe*, XXX, 329.

42. Cf. Steve Joshua Heims, *John von Neumann and Norbert Wiener. From Mathematics to the Technologies of Life and Death* (Cambridge, Mass.: MIT Press, 1982), 63f.

43. Ibid., 437.

44. Ibid., 70.

45. Johann Christian Reil, *Rhapsodieen über die Anwendung der psychischen Curmethode auf Geisteszerrüttungen* (Halle: Curtsche Buchhandlung, 1803), 417.

46. Cf. Sickert, *Automatische Spracheingabe*, 137f.

47. Lacan, *Seminar*, 83.

48. Cf. Lacan, *Écrits*, 38ff.

49. Lacan, *Seminar*, 316ff.

50. Cf. Lacan, *Écrits*, 730.

51. Norbert Wiener, *Cybernetics or Control and Communication in the Animal and the Machine* (Cambridge, Mass.: MIT Press, 1963), 28.

52. Only minimization occurs because no filter—whether analog or digital—can foresee the future without itself consuming time. A gifted engineer formulated the aporia as follows:

> Another limitation is that filters cannot be expected to predict the future! While this may seem obvious, a low-pass filter specification with zero phase shift at all passband frequencies is asking exactly that. For example, if the filter were presented the first three samples of a low-frequency yet high-amplitude wave, it would have no way of "knowing" whether it really was part of a low-frequency cycle or part of a high-frequency but low-amplitude cycle without further data. Zero phase shift implies that such a decision is made immediately and the samples either pass to the output or are blocked. (H. Chamberlin, *Musical Applications of Microprocessors* [Rochelle Park, N.J.: 1980], 433f.)

On this uncertainty principle [*Unschärferelation*] of communications technology, which ever since the work of Dennis Gabor has confronted the measurement of frequency and time in the same way that quantum physics faces waves and corpuscles, cf. also, Bell, *Information Theory*.

Chapter 13

1. Thus a letter from an Australian listener to *Der Deutsche Kurzwellensender* (a radio program for foreign audiences during the Third Reich) in 1936; quoted in Heinz Pohle, *Der Rundfunk als Instrument der Politik. Zur Geschichte des deutschen Rundfunks von 1923/38* (Hamburg: Hans-Bredow-Institut, 1955), 437: "England may rule the waves, but Germany rules the air."

2. Hans Magnus Enzensberger, *Mausoleum: Thirty-Seven Ballads for the History of Progress*, trans. Joachim Neugroschel (New York: Urizen, 1976), 140.

3. Cf. John von Neumann, "General and Logical Theory of Automata," *Collected Works* (Oxford: Pergamon, 1963), V, 312ff.

4. Jacques Lacan, *Écrits: The First Complete Edition in English*, trans. Bruce Fink (New York: Norton, 2007), 245f.

5. Norbert Wiener, *Cybernetics or Control and Communication in the Animal and the Machine* (Cambridge: MIT Press, 1963), 28.

6. Neumann, "General and Logical Theory of Automata," 296.

7. Erich Murawski, *Der deutsche Wehrmachtbericht 1939–1945. Ein Beitrag zur Untersuchung der geistigen Kriegführung* (Boppard: Boldt, 1962), 113.

8. Cf. Rolf Oberliesen, *Information, Daten und Signale. Geschichte technischer Informationsverarbeitung* (Reinbek: Rowohlt, 1982), 109, 128.

9. Cf. Enzensberger, *Mausoleum*, 74.

10. Cf. Richard Hennig, "Telegraphie und Seekriegsrecht," *Preußische Jahrbücher* 144 (1911): 30.

11. Cf. ibid., 36f.:

In questions of undersea cables and the law of war, the theory of international law, notwithstanding all particular differences, could at least theoretically determine the most important considerations on the main point with some certainty, even if, for now, its suggestions have failed to meet practical recognition. The matter is different with wireless telegraphy. Here, the theory and practice of international law, indeed, the whole of juristic research, stands for the most part before entirely new concepts, unknown until now, before a *terra nova*, which must first be laboriously explored on paths that are as yet untrodden and must first be created. A single question may elucidate the entire difficulty of these legal problems: may a neutral station on land or at sea [*eine neutrale Land- oder Schiffsstation*] for wireless communication exchange dispatches with a blockaded harbor or not? Without a doubt, a break of the blockade is not the case, for according to juridical nomenclature until now, a blockade can only be physically broken—by ships, human beings, carrier pigeons, cables, air balloons, and so on; all the same, everyone will admit without reservation that the exchange of wireless dispatches with the blockaded coast runs directly counter to the interests of the blockade. The juridical concept of blockade will therefore need to undergo an expansion and reinterpretation, similar to what occurred with the concept of theft, when it was determined that incorporeal and invisible electrical power could be stolen.

12. Quoted in Orrin E. Dunlap Jr., *Marconi. The Man and His Wireless* (New York: Macmillan, 1941), 353.

13. Cf. Winfried B. Lerg, *Die Entstehung des Rundfunks in Deutschland. Herkunft und Entwicklung eines publizistischen Mittels* (Frankfurt a.M.: Knecht, 1970); and

Notes to Chapter 13

Wolfgang Hagen, "Die verlorene Schrift. Skizzen zu einer Theorie der Computer," *Arsenale der Seele*, ed. Friedrich Kittler and Georg Christoph Tholen (Munich: Fink, 1989), 211–29.

14. Cf. Paul Virilio, *War and Cinema*, trans. Patrick Camiller (New York: Verso, 1989).

15. Cf. Ernst Jünger, *Der Kampf als inneres Erlebnis* (Berlin: Mittler & Sohn, 1922), 8 and passim.

16. Cf. Ernst Volckheim, *Die deutschen Kampfwagen im Weltkriege* (=2. Beiheft zum 107. Jg. des Militär-Wochenblattes) (Berlin: Miller, 1923), 14; and also, Dermot Bradley, *Generaloberst Heinz Guderian und die Entstehungsgeschichte des modernen Blitzkrieges* (Osnabrück: Biblio, 1978), 156ff.

17. Cf. Karl Heinz Wildhagen (ed.), *Erich Fellgiebel. Meister operativer Nachrichtenverbindungen. Ein Beitrag zur Geschichte der Nachrichtentruppe* (Wennigsen: [no pub.], 1970), 31:

> In 1934, professional opinion held that ultra-short wave (between 10 and 1 meters) spreads in a straight line and is not to be used on the battlefield, because it cannot overcome the hindrances of the landscape. Colonel Gimmler, then the head of Testing Division 7 in the Army Weapons Office, went with his technicians into the Harz mountains and determined, without onlookers, that—contrary to what was taught—ultra-short waves adapt to elevations of the land when appropriate antennas are used. Accordingly, many more radio devices could be accommodated in its frequency band than before. The atmospheric disturbances of devices until this point ceased—Colonel Guderian, the creator of the German tank force, then demanded radio equipment for every tank. [...] It was Gimmler's discovery that first gave Fellgiebel the possibility to fulfill the hitherto unusual, far-reaching demands of this ingenious tactician. Thus, in 1939, every German tank, equipped with an ultra-shortwave radio device, departed for Poland, France, and Russia.

18. Cf. ibid., 181ff.

19. Andrew Hodges, *Alan Turing: The Enigma* (New York: Simon & Schuster, 1983), 30.

20. Ibid., 14.

21. Alan M. Turing, *The Essential Turing*, ed. B. Jack Copeland (Oxford: Oxford University Press, 2004), 421.

22. Quoted in Hodges, *Turing*, 120.

23. Ibid., 148.

24. Ibid.,168.

25. Jürgen Rohwer and Eberhard Jäckel (eds.), *Die Funkaufklärung und ihre*

Rolle im Zweiten Weltkrieg. Eine internationale Tagung in Bonn-Bad Godesberg und Stuttgart vom 15. bis 18.10.1978 (Stuttgart: Motorbuch, 1979), 336.

26. Hodges, *Turing*, 192.

27. Ibid., 192.

28. Cf. B. Randell, "The Colossus," *A History of Computing in the Twentieth Century*, ed. N. Metropolis, J. Howlett, and G.-C. Rota (New York: Academic Press, 1980), 72–75.

29. Cf. Rohwer and Jäckel, *Funkaufklärung*, 110–12.

30. Cf. Rudolf Brauner, *Die Schreibmaschine in technischer, kultureller und wirtschaftlicher Bedeutung* (Prague: Deutscher Verein zur Verbreitung gemeinnütziger Kenntnisse, 1925), 40f.: "A further stage in the applicability of the typewriter for calculation- and recording operations was achieved in 1907 when the Wahl Adding Machine Company in Chicago brought out a device for addition and subtraction, with which Remington equipped a calculating model in 1910."

31. Cf. Hodges, *Turing*, 277.

32. Cf. Konrad Zuse, *Der Computer. Mein Lebenswerk* (Berlin, Heidelberg, New York, Tokyo: Springer, 1984), 52f.

33. Cf. Oberliesen, *Information*, 205.

34. Cf. Zuse, *Computer*, 80–83 (the V-2 facilities at Nordhausen as mere storage space for his V4 computers); and in contrast, Hodges, *Turing*, 299: "Too late for Germany, as with all its scientific effort, Zuse calculators were used in the engineering of V-2 rockets, and in 1945 Zuse himself was installed in the Dora underground factory, a place where there were no jokes about 'slaves.'"

35. Erik Bergaust, *Wernher von Braun. Ein unglaubliches Leben* (Düsseldorf, Vienna: Econ, 1976), 95. The account differs in Zuse, *Computer*, 83.

36. Hodges, *Turing*, 335, 301, 304, and 413. Cf. Oberliesen, *Information*, 297f.

37. Robert Jungk, *Heller als tausend Sonnen. Das Schicksal der Atomforscher* (Bern: Scherz, 1956), 314. For the details on von Neumann's ICBM designs, see Steve J. Heims, *John von Neumann and Norbert Wiener. From Mathematics to the Technologies of Life and Death* (Cambridge, Mass.: MIT Press, 1982), 272f.

38. Hodges, *Turing*, 362.

39. Cf. József Garlinski, *The Enigma War* (New York: Scribner, 1979), 119–44. For a partial confirmation, cf. F. H. Hinsley, *Intelligence in the Second World War. Its Influence on Strategy and Operations* (London: Stationery Office, 1981), II, 59f.

40. Cf. Virilio, *War and Cinema*.

41. Hodges, *Turing*, 337.

42. Cipher A. Deavours and Louis Kruh, *Machine Cryptography and Modern Cryptanalysis* (Norwood, Mass.: Artech, 1985), xi.

43. James Bamford, *The Puzzle Palace: Inside the National Security Agency, America's Most Secret Intelligence Organization* (New York: Penguin, 1983), 434.

Notes to Chapters 13 and 14

44. Deavours and Kruh, *Cryptography*, 28. Cf. Zuse, *Computer*, 96: "The edge that, in some respects, we had after the war in Germany was largely lost in the years that followed. In the United States, building could continue unabated and on a broad scale. Even though industry was reserved even there, scientific institutes furthered development. As a rule, military instances provided the most support. From them, computer research in the United States received its strongest impulses."

45. Deavours and Kruh, *Cryptography*, 29.

Chapter 14

1. Raymond Cartier, *Der Zweite Weltkrieg* (Munich, Zurich: Piper, 1985), 603.

2. Casablanca was half liberated because, pursuant to the diplomatic policy of the United States that sought to avoid offending either Vichy's General Giraud or de Gaulle, the Nuremberg Race Laws remained in effect for half a year after the Allied landing. (Information shared by Jacques Derrida).

3. Cf. K. R. Greenfield, "Die amerikanische Luftkriegführung in Europa und Ostasien 1942–1945," *Probleme des Zweiten Weltkrieges*, ed. Andreas Hillgruber (Cologne and Berlin: Kiepenhauer & Witsch, 1967), 292–311; 296.

4. On the thought-provoking connections between Mussolini's effort to suppress the Mafia, German U-boat strategy in the western Atlantic, American dockworkers' unions, and finally, the choice of Sicily as the first site for Allied landings in Europe, see Andreas Schäfer, "Drogen und Krieg," *Arsenale der Seele. Literatur- und Medienanalyse seit 1870*, ed. Friedrich A. Kittler and Georg Christoph Tholen (Munich: Fink, 1989), 151–69; 162–64.

5. Günter Moltmann, "Die Genesis der Unconditional-Surrender-Forderung," *Probleme des Zweiten Weltkrieges*, 171–202; 172.

6. Cf. ibid., 185.

7. Cartier, *Der Zweite Weltkrieg*, 606f.

8. Cf. Walter Baum, "Regierung Dönitz und deutsche Kapitulation," *Probleme des Zweiten Weltkrieges*, 347–78; 362.

9. Thomas Pynchon, *Gravity's Rainbow*, 520–21.

10. Cf. Karl-Heinz Wildhagen (ed.), *Erich Fellgiebel. Meister operativer Nachrichtenverbindungen. Ein Beitrag zur Geschichte der Nachrichtentruppe* (Wennigsen/Hannover, 1970), 31f.

11. Cf. Ian S. Milward, "Hitlers Konzept des Blitzkrieges," *Probleme des Zweiten Weltkrieges*, 19–40; 24–27. For Jodl's confirmation, cf. Percy E. Schramm (ed.), *Kriegstagebuch des Oberkommandos der Wehrmacht (Wehrmachtsführungsstab) 1940–1945, geführt von Helmuth Greiner und Percy Ernst Schramm* (Herrsching, 1982,) IV, 1713.

12. Cf. Wildhagen, *Erich Fellgiebel*, 31f.

13. Jodl, in Schramm, *Kriegstagebuch des Oberkommandos der Wehrmacht*, IV, 1718.

14. Cf. Willi A. Boelcke (ed.), *Deutschlands Rüstung im Zweiten Weltkrieg. Hitlers Konferenzen mit Albert Speer 1942–1945* (Frankfurt a.M.: Athenaion, 1969), 37:

> [Hitler's] interests one-sidedly concerned the traditional weapons of the army. Here, he hardly abandoned the ground of macrophysics, that is, laws concerning mechanics, the strength of materials, and structural engineering, with which he was probably familiar. Whether he ever commanded differential calculus, so as to be able to intervene decisively in matters of discussion, remains questionable. It is certain that he, as an autodidact, lacked the foundations of microphysics and modern chemistry.

15. Quoted in Schramm, *Kriegstagebuch des Oberkommandos der Wehrmacht*, IV, 1721.

16. Albert Speer, *Inside the Third Reich* (New York: Simon & Schuster, 1997), 275.

17. Cf. Friedrich Wilhelm Hagemeyer, *Die Entstehung von Informationskonzepten in der Nachrichtentechnik. Eine Fallstudie zur Theoriebildung in der Technik in Industrie- und Kriegsforschung*, Typescript, Diss. phil. (FU Berlin, 1979), 340.

18. Cf. Michael Geyer, *Deutsche Rüstungspolitik 1860–1980* (Frankfurt a.M.: Suhrkamp, 1984), 101–8.

19. Cf. ibid., 159f.

20. Cf. Bernd Ruland, *Wernher von Braun. Mein Leben für die Raumfahrt* (Offenburg: Burda, 1969), 123.

21. Cf. Karl-Heinz Ludwig, *Technik und Ingenieure im Dritten Reich* (Düsseldorf: Droste, 1979), 350.

22. Cf. Rudolf Lusar, *Die deutschen Waffen und Geheimwaffen des Zweiten Weltkrieges und ihre Weiterentwicklung* (Munich: Graefe, 1971), 115–17.

23. Cf. Paul Spremberg, *Entwicklungsgeschichte des Staustrahltriebwerkes* (Mainz: Krausskopf-Flugwelt Verlag, 1963).

24. Cf. Ludwig, *Technik und Ingenieure*, 200.

25. Cf. Geyer, *Deutsche Rüstungspolitik*, 166.

26. Quoted in Fritz Hahn, *Deutsche Geheimwaffen 1939–1945* (Heidenheim: Hoffmann, 1963), 11.

27. Edward L. Homze, *Foreign Labour in Nazi Germany* (Princeton: Princeton University Press, 1967), 232.

28. Cf. Geyer, *Deutsche Rüstungspolitik*, 162–66.

29. Cf. Ludolf Herbst, *Der Totale Krieg und die Neuordnung der Wirtschaft. Die Kriegswissenschaft im Spannungsfeld von Politik, Ideologie und Propaganda 1939–1945* (Stuttgart: Deutsche Verlags-Anstalt, 1982), 261–66.

30. Cf. ibid., 383–97 and 443, n.394.

31. Cf. William H. McNeill, *The Pursuit of Power. Technology, Armed Forces, and Society since A.D. 1000* (Chicago: University of Chicago Press, 1982), 360–65.

32. Cf. Werner Durth, *Deutsche Architekten. Biographische Verflechtungen 1900–1970* (Braunschweig: Vieweg, 1987), 209–21.

33. Ibid., 15.

34. Quoted in Schramm, *Kriegstagebuch des Oberkommandos der Wehrmacht*, IV, 1705.

35. Ibid., 1580f.

36. Quoted in Speer, *Inside the Third Reich*, 451.

37. Cf. Jorge Luis Borges, *Obras Completas* (Buenos Aires: Emecé, 1964–66), III, 131f.

38. *OKW-Lagebuch*, 10 April 1945, reproduced in Schramm, *Kriegstagebuch des Oberkommandos der Wehrmacht*, IV, 1233.

39. Cf. Hugh R. Trevor-Roper, *The Last Days of Hitler* (London: Macmillan, 1947), 185.

40. Cf. Ludwig, *Technik und Ingenieure im Dritten Reich*, 506–14.

41. Speer, *Inside the Third Reich*, 474.

42. Felix Steiner, *Die Armee der Geächteten* (Göttingen: Plesse, 1963), 225.

43. Ibid., 228ff.

44. Cf. Speer, *Inside the Third Reich*, 300; and Trevor-Roper, *Last Days of Hitler*, 63.

45. Cf. Wassilij Tschuikow, *Das Ende des Dritten Reiches* (Munich: Goldmann, 1966), 213.

46. Ludwig, *Technik und Ingenieure im Dritten Reich*, 514.

47. Schramm, *Kriegstagebuch des Oberkommandos der Wehrmacht*, IV, 1679ff.

48. Cf. Claude R. Sasso, *Soviet Night Operations in World War II* (Fort Leavenworth, Kan.: Combat Studies Institute, US Army Command and General Staff College, 1982), 21f.

49. Thomas Pynchon, *Gravity's Rainbow* (New York: Viking, 1983), 449.

50. Cf. Tom Bower, *The Paperclip Conspiracy: The Hunt for Nazi Scientists* (Boston: Little Brown, 1987), 14.

51. Cf. ibid., 206.

52. Cf. Brian Southall, *Abbey Road. The Story of the World's Most Famous Recording Studio* (Cambridge: Haynes Manuals, 1982), 137.

53. Cf. Bower, *Paperclip Conspiracy*, 190–94.

54. Cf. Paul Virilio, *Guerre et cinéma I: Logistique de la perception* (Paris: Galilée, 1984), 106.

55. Cf. Jürgen Rohwer and Eberhard Jäckel (eds.), *Die Funkaufklärung und ihre*

Rolle im Zweiten Weltkrieg. Eine internationale Tagung in Bonn-Bad Godesberg und Stuttgart vom 15. bis 18. 9. 1978 (Stuttgart: Motorbuch, 1979).

56. Cf. Andrew Hodges, *Alan Turing: The Enigma* (New York: Simon & Schuster, 1983), 312.

57. Cf. Bower, *Paperclip Conspiracy*, 150f.

58. Ibid., 335.

59. Quoted in ibid., 255.

60. Ludwig, *Technik und Ingenieure im Dritten Reich*, 513.

61. Cf. Ruland, *Wernher von Braun*, 253.

62. Cf. Cartier, *Der Zweite Weltkrieg*, 1023.

63. Cf. McNeill, *Pursuit of Power*, 353.

64. Robert Jungk, *Heller als tausend Sonnen. Das Schicksal der Atomforscher* (Bern: Scherz, 1956), 414.

65. Cf. Steve J. Heims, *John von Neumann and Norbert Wiener. From Mathematics to the Technologies of Life and Death* (Cambridge, Mass.: MIT Press, 1982), 230–90.

66. Cf. Gordon Welchman, *The Hut 6 Story: Breaking the Enigma Codes* (New York: McGraw-Hill, 1982), 264.

67. Compare Walter Bruch, *Kleine Geschichte des deutschen Fernsehens* (Berlin: Haude und Spener, 1967), 74–77; and Walter Dornberger, *V2—Der Schuß ins Weltall* (Esslingen: Bechtle, 1953), 10, to see that the television screen at Test Stand VII—like almost all simulations ever since—had the purpose of preventing deadly results for the experimenters themselves.

68. Bruch, *Kleine Geschichte des deutschen Fernsehens*, 73ff.

69. Cf. Hagemeyer, *Die Entstehung von Informationskonzepten*, 278–87.

70. Cf. Norbert Wiener, *Cybernetics or Control and Communication in the Animal and the Machine* (Cambridge, Mass.: MIT Press, 1961), 28–30.

71. Cf. Hagemeyer, *Die Entstehung von Informationskonzepten*, 338–45. For example, the "Würzburg Primer" [*Würzburg-Fibel*] published in two parts in October 1943 by the Supreme Command of the Navy, Office for Technical Communications [*Amtsgruppe Technisches Nachrichtenwesen*], Reich Aviation Ministry, and Army Communications [*Wehrmachtnachrichtenverbindungswesen*], gives up on all mathematical elegance where rectangular pulses are concerned.

72. Speer, *Inside the Third Reich*, 524.

73. Ibid., 520f.

74. Pynchon, *Gravity's Rainbow*, 105.

75. Ruland, *Wernher von Braun*, 141.

76. Cf. Bernhard Siegert, "The Fall of the Roman Empire," *Materialities of Communication*, ed. Hans Ulrich Gumbrecht and K. Ludwig Pfeiffer (Stanford: Stanford University Press, 1994), 303–18.

77. Cf. vol. II of Karl Otto Hoffmann, *Die Geschichte der Luftnachrichtentruppe* (Neckargmünd: Vowinckel, 1965).

78. Ludwig, *Technik und Ingenieure im Dritten Reich*, 513ff.

79. Quoted in Takushiro Hattori, "Japans Weg aus dem Weltkrieg," *Probleme des Zweiten Weltkrieges*, 389–436; 389f.

Chapter 15

1. Cf. Andrew Hodges, *Alan Turing, The Enigma* (New York: Simon & Schuster, 1983), 399.

2. Cf. Alan M. Turing, "On Computable Numbers, with an Application to the Entscheidungsproblem," *Proceedings of the London Mathematical Society* 42 (1937): 230–65.

3. Cf., for example, *Microsoft Macro Assembler 5.1, Reference* (1987), 115: "This section provides an alphabetic reference to instructions of the 8087, 80287, and 80387 coprocessors. The format is the same as for the processor instructions except that encodings are not provided."

4. *TOOL Praxis: Assembler-Programmierung auf dem PC, Ausgabe 1* (Würzburg, 1989), 9.

5. Ibid., 39.

6. On one-way functions in mathematics and cryptology, cf. Patrick Horster, *Kryptologie* (Mannheim, Vienna, Zurich: Bibliographisches Institut Wissenschaftsverlag, 1985), 23–27.

7. Personal communication from Hartley to Friedrich-Wilhelm Hagemeyer (Berlin).

8. Klaus-Dieter Thies, *Das 80186-Handbuch* (Düsseldorf, Berkeley, Paris, 1986), 319.

9. For the details of the Pentagon's instructions, see D. Curtis Schleher, *Introduction to Electronic Warfare* (Norwood, Mass.: Artech, 1987).

10. Cf. Gerry Kane, *68000 Microprocessor Book* (Berkeley: Adam Osborne & Associates, 1978), 8ff.

11. Cf. Friedrich A. Kittler, *Gramophone, Film, Typewriter*, trans. Geoffrey Winthrop-Young and Michael Wutz (Stanford: Stanford University Press, 1999), 253–63.

12. Cf. Michael Löwe, "VHS1C: Ultraschnelle Schaltkreise frisch vom Band ins Pentagon," *Militarisierte Informatik*, ed. Joachim Bickenbach, Reinhard Keil-Slawik, Michael Löwe, and Rudolf Wilhelm (Münster: Marbuch, 1985), 64.

13. Harald Albrecht, "MSDOS in a Box," *c't* 3 (1990), 258.

14. Cf. Carl Schmitt, *Gespräch über die Macht und den Zugang zum Machthaber* (Pfullingen: Neske, 1954).

15. This occurred under QEMM386 by Quarterdeck, when attempting to execute the commands LGDT and SGDT within CODEVIEW.

Notes to Chapter 15

16. Cf. Arne Schäpers, *Turbo Pascal 5.0* (Munich: Addison-Wesley, 1989), 1: "The processor's address space is divided by the system architecture into RAM- and ROM areas, and RAM is subdivided by the operating system into storage blocks for recording programs. A given program consists of individual segments, some of which contain procedures and functions that, for their part, can define further routines locally." One is almost reminded of a picture by Escher (or more in keeping with the times, of the Mandelbrot set [*Apfelmännchen*]).

17. Cf. Jacques Lacan, *Écrits: The First Complete Edition in English*, trans. Bruce Fink (New York: Norton, 2007), 691.

18. Cf. Intel Corporation, *80386 Programmer's Reference Manual* (Santa Clara, Calif., 1988), ch.17, p. 145: "The DS, ES, FS, and GS segment registers can be set to 0 by the RET instruction during an interlevel transfer. If these registers refer to segments that cannot be used by the new privilege level, they are set to 0 to prevent unauthorized access from the new privilege level." In the CS code segment, however, these sledgehammer methods of protection are inoperative, so as not to block the entire system.

19. Cf. ibid., ch. 14, p. 1 and passim.

20. Ibid., ch. 9, p. 16.

21. Cf. Harald Albrecht, "Grenzenlos. Vier Gigabyte im Real Mode des 80386 adressieren," *c't* 1 (1990), 212:

> However, the segment boundary of 64 KB is in no way set as firmly in the 80386 operating in Real Mode as appears, for example, in Intel's documents about the 386DX. If one duly continues the steps necessary for bringing the 80386 back out of Protected Mode into Real Mode, the whole address space of 4 GB promptly opens for Real Mode (whereby the grins of Motorola aficionados ought to visibly diminish in width).

What follows largely occurs thanks to this altogether ingenious suggestion.

22. For an exception (albeit—and tellingly—without any commentary), cf. Klaus-Dieter Thies, *PC XT AT Numerik Buch. Hochgenaue Gleitpunkt-Arithmetik mit 8087.. 80 287.. 80 387 ... Nutzung mathematischer Bibliotheksfunktionen in "Assembler" und "C"* (Munich: te-wi Verlag, 1989), 638. In contrast, Ed Strauss—even though he "has seen the full range of system issues and devised many practical solutions during his work for Intel" (according to the foreword authored by Robert Childs, the architect of the 80286)—has managed to write an authoritative handbook that says nothing at all about nondocumented spaces of leeway [*Spielräume*]. Cf. Edmund Strauss, *80386 Technical Reference. The Guide for Getting the Most from Intel's 80386* (New York: Brady, 1987), passim.

23. Cf., for example, Andreas Stiller, "Bitter für 32-Bitter," *c't* 8 (1990), 202. For more detail on the command LOADALL (including the dubious assertion that only

the 80286 recognizes it), cf. Norbert Juffa and Peter Siering, "Wege über die Mauer. Loadall—Extended Memory im Real Mode des 80286," *c't* 11 (1990), 262–66.

24. Löwe, *VHSIC*, 70.

Chapter 16

1. Thanks to Wolfgang Hagen (Radio Bremen), who performed the textual comparison of the two versions of "Eve of Destruction" for listeners in a live broadcast.

2. Cf. Thomas Hafki, *Franklin—Frankenstein. Zum Verhältnis von Elektrizität und Literatur*, Master's thesis [*Magisterarbeit*], Bochum, 1993.

3. In 1978, when the Intel 8086 processor was being designed, the blueprints are said to have filled sixty-four square meters of graph paper. Cf. Klaus Schrödl, "Quantensprung," *DOS* 12 (1990): 102f.

4. Cf. Alan M. Turing, "On Computable Numbers, with an Application to the Entscheidungsproblem," *Proceedings of the London Mathematical Society* 42 (1937): 230–65.

5. The extent of such repression is demonstrated by the authors (or typesetters) of Intel's reference manuals: for example, the floating point instruction *f2xm1*—that is, an input quantity squared, minus one—does not become, once translated into everyday English, "Compute 2^x-1," but rather "Computer 2^x-1." Cf. Intel, *387 DX User's Manual. Programmer's Reference* (Santa Clara, 1989), 4–9; as well as Intel, *i486 Microprocessor. Programmer's Reference Manual* (Santa Clara, 1990), 26–72.

6. Cf. Wolfgang Hagen, "Die verlorene Schrift. Skizzen zu einer Theorie der Computer," *Arsenale der Seele. Literatur- und Medienanalyse seit 1870*, ed. Friedrich A. Kittler and Georg Christoph Tholen (Munich: Fink, 1989), 221:

> Therefore, the linguistic structure of von-Neumann machine logic already establishes, in principle, the divergence of software and software manuals, and so, from 1945 on, a Babylonian tower of computer performances has piled ever higher; using them has nothing more to do with the meaningful arrangement of a machine language. A tower of software with undocumented errors, hopelessly confused dialects, and a mass of linguistic acts that no one at all can follow.

In an image that for being less precise is all the more desperate, a UNIX expert has written: "Almost all operating systems are marked, after a certain age, by a high 'degree of pollution.' They grow wild in all directions and give the impression of being ruins that can only be held together with great effort" (Horst Drees, *UNIX. Ein umfassendes Kompendium für Anwender und Systemspezialisten* [Haar: Markt & Technik, 1988], 19). The UNIX expert is too polite to weave the proper name of a company such as Microsoft into the welter.

7. It is no accident that the only counterexample of which I am aware comes

Notes to Chapter 16

from Richard Stallman's Free Software Foundation, which has declared war—a struggle that is just as heroic as it is doomed—on software copyright in general. This counterexample goes: "When we say that '*C-n* moves down vertically one line' we are glossing over a distinction that is irrelevant in ordinary use but is vital in understanding how to customize Emacs. It is the function *next-line* that is programmed to move down vertically. *C-n* has this effect because it is bound to that function. If you rebind *C-n* to the function *forward-word* then *C-n* will move forward by words instead" (Richard Stallman, *GNU Emacs Manual* [Cambridge, Mass., 1988], 19).

8. May this be accepted as a free translation for "booting."

9. Stephen C. Kleene, quoted in Robert Rosen, "Effective Processes and Natural Law," *The Universal Turing Machine. A Half-Century Survey*, ed. Rolf Herken (New York: Springer, 1995), 489.

10. Cf. Johannes Lohmann, "Die Geburt der Tragödie aus dem Geiste der Musik," *Archiv für Musikwissenschaft* (1980), 174.

11. Cf. Andrew Hodges, *Alan Turing: The Enigma* (New York: Simon & Schuster, 1983), 399.

12. Cf. *TOOL Praxis: Assembler-Programmierung auf dem PC, Ausgabe 1* (Würzburg, 1989), 9.

13. Nahajyoti Barkalati, *The Waite Group's Macroassembler Bible* (Carmel, Ind.: Howard H. Sams, 1989), 528.

14. Cf. James D. Foley, Andries van Dam, Steven K. Feiner, and John F. Hughes, *Computer Graphics. Principles and Practice* (Reading/Mass., 1990), 398.

15. On the connection between the Pentagon, Ada, and Intel's iAPX 432, the first microprocessor in Protected Mode—which, when it failed economically, gave rise to the industry standard from the 80286 up to the 80486—cf. Glenford J. Myers, "Overview of the Intel iAPX 432 Microprocessor," *Advances in Computer Architecture* (New York: John Wiley & Sons, 1982), 335–44 (with thanks to Ingo Ruhmann, Bonn). Whoever wishes to understand the failure should meditate the following: "The 432 can be characterized as a three-address-storage-to-storage architecture, there are no registers visible to programs" (342).

16. Cf. Chapter 15.

17. On the following, cf. Patrick Horster, *Kryptologie* (Mannheim, Vienna, Zurich: Bibliographisches Institut Wissenschaftsverlag, 1985), 23–27.

18. See Chapter 12.

19. Charles H. Bennett, "Logical Depth and Physical Complexity," *Universal Turing Machine*, 210.

20. Thanks to Oswald Wiener, Dawson City.

21. Cf. M. Michael König, "Sachlich sehen. Probleme bei der Überlassung von Software," *c't* 3 (1990): 73.

Notes to Chapter 16

22. One might sooner—as it stands in a letter from Dirk Baecker (15 April 1991)—

suppose that the distinction between hardware and software is a distinction that is supposed to escort [*betreuen*] the reentry of the distinction between programmability and non-programmability back into the realm of what can be programmed. It stands, so to speak, for the calculability of technology, and in this sense, for technology itself. This can be only because the "unity" of the program can only be achieved if equation and calculation are, respectively, distributed on two sides, so that only one of the sides is ever operationally available and the other one can be kept constant.

23. Consequently, I am at a loss to explain how Turing's famous paper could declare in the first sentence that "the 'computable numbers' may be described briefly as the real numbers whose expressions as a decimal are calculable by finite means" (Turing, "On Computable Numbers," 230), then define the set of calculable numbers as countable, and finally, call π, as "the limit of a computably convergent sequence," a computable number (256).

24. Brosl Hasslacher, "Beyond the Turing Machine," *Universal Turing Machine*, 391.

25. Ibid., 389.

26. Friedrich-Wilhelm Hagemeyer, *Die Entstehung von Informationskonzepten in der Nachrichtentechnik. Eine Fallstudie zur Theoriebildung in der Technik in Industrie- und Kriegsforschung*, Diss. phil. (typescript), Berlin, 1979, 432.

27. Alan Turing, "Intelligent Machinery: A Heretical Theory," *The Essential Turing: Writings in Computing, Logic, Philosophy, Artificial Intelligence, and Artificial Life*, ed. B. Jack Copeland (Oxford: Oxford University Press, 2004), 475.

28. Michael Conrad, "The Price of Programmability," *Universal Turing Machine*, 264–65.

29. Cf. John von Neumann, "General and Logical Theory of Automata," *Collected Works* (Oxford: Pergamon, 1963), V, 296ff.

30. Conrad, "Price of Programmability," 268.

31. Ibid., 265.

32. Ibid., 279.

33. Thus, the first integrated neural network—from, of all places, Intel's discrete chip empire (and as far as I can see, for the second time in the entire history of the firm after the relatively hybrid i2920 signal processor)—went back to straightforward analog operational amplifiers.

Chapter 17

1. Martin Heidegger, "The Origin of the Work of Art," *Basic Writings*, ed. David Farrell Krell (New York: HarperCollins, 2008), 187.

2. Cf. Reginaldo Giuliani, *Gli arditi* (Milan: Treves, 1934), 1.

3. Cf. Ferdinando Cordova, *Arditi e legionari dannunziani* (Padua: Marsilio, 1969), 1ff.

4. Cf. Graf von Schwerin, "Das Sturmbataillon Rohr," *Das Ehrenbuch der deutschen Pioniere*, ed. Paul Heinrici (Berlin: Tradition Wilhelm Kolk, 1932), 559.

5. Cf. Hellmuth Gruss, *Die deutschen Sturmbataillone im Weltkrieg. Aufbau und Verwendung (Schriften der kriegsgeschichtlichen Abteilung im Historischen Seminar der Friedrich-Wilhelms-Universität Berlin, Heft 26)* (Berlin: Junker & Dünnhaupt, 1939), 152.

6. Cf. the unsurpassed study by Paul Fussell, *The Great War and Modern Memory* (Oxford: Oxford University Press, 1975).

7. Ernst Jünger, *Der Arbeiter. Herrschaft und Gestalt* (Hamburg: Hanseatische Verlagsanstalt, 1932), 143.

8. Ibid., 105.

9. Cf. John Ellis, *The Social History of the Machine Gun* (New York: Pantheon, 1975).

10. Cf. Hans Linnenkohl, *Vom Einzelschuß zur Feuerwalze. Der Wettlauf zwischen Technik und Taktik im Ersten Weltkrieg* (Koblenz: Bernard & Graefe, 1990), 272.

11. Jünger, *Der Arbeiter*, 107.

12. Cf. Gruss, *Die deutschen Sturmbataillone*, 20.

13. Cf. Hugo Ott, *Martin Heidegger. Unterwegs zu einer Biographie* (Frankfurt a.M., New York: Campus, 1988), 85–87. Ott goes so far as to suppose that Heidegger's secret reading of all correspondence made it impossible for professors in Freiburg to halt his career (i.e., his return from Marburg to Freiburg).

14. Franz Schauwecker, *Im Todesrachen* (Halle: Diekmann, 1921), 282. This and other violations performed by assault battalions were later made into the norm for the infantry; as the *Truppenamt der Reichswehr* declared (presumably in words chosen by Lieutenant Ernst Jünger): "In open order, the rifle is slung over the shoulder if necessary, but not carried on it. For the most part, the strap is extended. The bearing and movement of gunmen are free and unforced" (*Ausbildungsvorschrift für die Infanterie [=Heeresdienstvorschrift Nr. 130, Heft II]*; reprint: [Berlin, 1934], 50). Finally, in 1942, the Waffen-SS made the small-caliber Sturmgewehr 44, the direct precursor of the Kalashnikov, the standard of today (Felix Steiner, *Die Armee der Geächteten* [Göttingen: Plesse, 1963], 134ff.).

15. What is more, Major Bernhard Reddemann—who came from the Leipzig

fire department, logically enough—"is supposed to have been the actual creator of the word 'shock troop'" (Gruss, *Die deutschen Sturmbataillone*, 24).

16. Cf. Linnenkohl, *Vom Einzelschuß zur Feuerwalze*, 179ff.

17. Cf. ibid., 209.

18. This kind of specialization went even further among the shock troops of other states: the French Army did not even expect general knowledge of rifles and hand grenades (Linnenkohl, *Vom Einzelschuß zur Feuerwalze*, 211).

19. Gruss, *Die deutschen Sturmbataillone*, 21. This matched the organigram of the Italian Arditi: a battalion of one thousand men consisted of four companies which each included four infantry platoons, one machine-gun division, and one flamethrower division (Cordova, *Arditi e legionari dannunziani*, 5).

20. Cf. Christoph Albrecht, *Geopolitik und Geschichtsphilosophie 1748–1798*, Diss. (typescript), Bochum, 1994, 179.

21. Cf. Gruss, *Die deutschen Sturmbataillone*, 165, 181.

22. Barbed wire—this measure taken by American ranchers in the Wild West against their herds of cattle—had already been used for prisoners since the Boer War; in the Russo-Japanese War, it was employed against the enemy in combat.

23. Cf. Gruss, *Die deutschen Sturmbataillone*, 80: "Following the suggestion of Mackensen's Supreme Command, the troops [of assault battalions] were called 'grenadiers' (hand-grenade launchers). Here, too, the exceptions were Assault Battalion No. 5 (Rohr) and Rifle (Assault) Battalion No. 3."

24. Cf. Max Bauer, *Der große Krieg in Feld und Heimat* (Tübingen: Osiandersche Buchhandlung, 1921), 87f.: "The novelty was that attack no longer occurred in a broad line of fire, but rather in narrow and deeply articulated squads. Troops were armed in keeping with the task awaiting them. The principle of uniformly arming all infantrymen was broken for the first time."

25. Linnenkohl, *Vom Einzelschuß zur Feuerwalze*, 171.

26. Gruss, *Die deutschen Sturmbataillone*, 26ff.

27. Ibid., 152.

28. Quoted in ibid., 48; cf. 121.

29. Michael Geyer, *Deutsche Rüstungspolitik 1860–1980* (Frankfurt a.M.: Suhrkamp, 1984), 102. Geyer, incidentally, stresses that he is unrelated to Captain Hermann Geyer.

30. Cf. Timothy T. Lupfer, *The Dynamics of Doctrine: The Changes in German Tactical Doctrine during the First World War* (Leavenworth Papers, No. 4, Fort Leavenworth/Kansas, 1981), 9: "Besides personal visits, the telephone was another major means of communication with the front line during the First World War. Ludendorff used it extensively and thought that it was good to use when personal visits could not be conducted. He also felt that the telephone had some value as a counter to the drawbacks of personal visits, such as false personal impressions." Accordingly,

the war was lost at the precise historical moment when military psychiatry induced Ludendorff to "detach himself from the telephone" and "put flowers in the room" (Dr. Hochheimer, quoted in Wolfgang Foerster, *Der Feldherr Ludendorff im Unglück. Eine Studie über seine seelische Haltung in der Endphase des Ersten Weltkrieges* [Wiesbaden: Limes, 1952], 77f.).

31. Cf. Geyer, *Deutsche Rüstungspolitik*, 102f.; and Linnenkohl, *Vom Einzelschuß zur Feuerwalze*, 194f.

32. Cf. Peter Berz, *08/15. Ein Standard des 20. Jahrhunderts* (Munich: Fink, 2001).

33. Hermann Cron, *Die Organisation des deutschen Heeres im Weltkriege (Forschungen und Darstellungen aus dem Reichsarchiv, Heft 5)* (Berlin: Siegismund, 1923), 49ff.

34. Geyer, *Deutsche Rüstungspolitik*, 101.

35. Bauer, *Der große Krieg*, 87.

36. Cf. Lupfer, *Dynamics of Doctrine*, 27.

37. Erich Ludendorff, *Meine Kriegserinnerungen* (Berlin: Mittler, 1919), I, 208.

38. Cf. Cron, *Die Organisation des deutschen Heeres*, 47.

39. Cf. Gruss, *Die deutschen Sturmbataillone*, 28–31.

40. Bauer, *Der große Krieg*, 88.

41. Cf. *Comando dell corpo d'armata d'assalto*, "Norme per l'impiego tattico delle Grandi Unità d'assalto, 1. 7. 1918," quoted in Cordova, *Arditi e legionari dannunziani*, 3.

42. Captain Rohr, 27 May 1916, quoted in Gruss, *Die deutschen Sturmbataillone*, 45.

43. Oberpostdirektor Nehrkorn (1937), quoted in Gruss, *Die deutschen Sturmbataillone*, 88.

44. Cf. Curzio Malaparte, *Die Wolga entspringt in Europa, mit einem Vorwort von Heiner Müller* (Cologne: Stahlberg, 1989), 23 and –1 [sic].

45. Cf. Cron, *Die Organisation des deutschen Heeres*, 141.

46. Cf. Martin Van Creveld, *Supplying War. Logistics from Wallenstein to Patton* (Cambridge: Cambridge University Press, 1977), 143.

47. Cf. Gruss, *Die deutschen Sturmbataillone*, 45.

48. Ibid., 101. The numbering of the battalions did not follow the chronology according to which they had been established, but rather reflected the numbering of the armies to which they were assigned.

49. Ibid., 135. Cf. also Schwerin, "Das Sturmbataillon Rohr," 562.

50. Cf. also, in the context of the history of the word "storm troopers," Wotan's historical enjoinder to his Valkyries:

Daß stark zum Streit	That mighty for battle
uns fände der Feind,	the foe would find us,
hieß ich euch Helden mir schaffen;	I ordered you fetch me heroes;

die herrisch wir sonst	such as once, masterfully,
in Gesetzen hielten,	we subjected to laws,
die Männer, denen	men whose courage
den Mut wir gewehrt,	we checked,
die durch trüber Verträge	and through dark treaties'
trügende Bande	deceptive bond
zu blindem Gehorsam	held in blind obedience
wir uns gebunden—	to ourselves—
die solltet zu Sturm	them, now, to storm
und Streit ihr nun stacheln,	and strife, should you spur,
ihre Kraft reizen	goad their forces
zu rauhem Krieg,	into rugged war,
daß kühner Kämpfer Scharen	that I may assemble hordes
ich sammle in Walhalls Saal!	of bold warriors in Valhalla's hall!

(Richard Wagner, *Die Musikdramen* [Munich: DTV, 1978], 613 [Act II, Scene 2])

The line-by-line commentary this passage calls for is reserved for later studies.

51. Cf. Cordova, *Arditi e legionari dannunziani*, 2.
52. Ibid., 22.
53. Cf. Giuliani, *Gli arditi*, 21.
54. Cf. Cordova, *Arditi e legionari dannunziani*, 4f.
55. Cf. Pamela Ballinger, "Blutopfer und Feuertaufe," *Der Dichter als Kommandant. D'Annunzio erobert Fiume*, ed. Hans Ulrich Gumbrecht, Friedrich Kittler, and Bernhard Siegert (Munich: Fink, 1996), 189.
56. Mario Carli, quoted in Cordova, *Arditi e legionari dannunziani*, 5. Following the same logic, Guido Keller—who incarnated *Arditismo* in Fiume like no one else—died in a "tragic automobile accident [*disgrazia*]" in 1929 (Atlantico Ferrari, *L'asso di cuori. Guido Keller* [Rome: Cremonese, 1933], 163). Ways of death [*Todesarten*], especially after Ingeborg Bachmann's own, remain a desideratum for scholarship.
57. Cf. Filippo Tommaso Marinetti, letter to Benito Mussolini, quoted in Cordova, *Arditi e legionari dannunziani*, 51.
58. Michael A. Ledeen, *The First Duce. D'Annunzio at Fiume* (Baltimore: Johns Hopkins University Press, 1977), 66. According to Bettina Vogel ("Guido Keller—Mystiker des Futurismus," *Der Dichter als Kommandant*, 117–32; 124), the number of vehicles was forty.
59. Cf. Ferrari, *L'asso di cuori*, 110.
60. Cf. ibid., 161.
61. Elena Ledda, "L'esercito liberatore," *Fiume. Rivista di studi fiumani*, NF 19 (1990), 21.

62. Gabriele d'Annunzio, *Altri Taccuini*, ed. Enrica Bianchetti (Milan: Mondadori, 1976), 377.
63. Cf. Ledda, "L'esercito liberatore," 3.
64. Cf. Gabriele d'Annunzio and Giuseppe Pfiffer, "Entwurf einer neuen Ordnung des Befreiungsheeres, §49," *Der Dichter als Kommandant*, 76.
65. Ibid., 79.
66. Cf. also D'Annunzio, Letter to Captain Nino Host-Venturi, 16 September 1919, quoted in Ledda, "L'esercito liberatore," 4: "Athletic activities should expect a great development."
67. Cf. Ibid., 8.
68. Ledda, ibid., reports contests for 100- and 21,000-meter races that legionaries in Fiume conducted.
69. Futurist Arditi added "flying" to "running and swimming" (*L'Ardito Futurista—Manifesto* [November 1919], cited in Cordova, *Arditi e legionari dannunziani*, 215).
70. Nino Host-Venturi, Letter to Gabriele d'Annunzio, 28 April 1920, quoted in Ledda, "L'esercito liberatore," 9.
71. Cf. Horst Bredekamp, *Florentiner Fußball. Die Renaissance der Spiele: Calcio als Fest der Medici* (Berlin: Wagenbach, 1993).
72. Gruss, *Die deutschen Sturmbataillone*, 130.
73. VIe Armée, État majeur, 2e bureau, Annexe au Bulletin de Renseignements, *Le "Sturmbataillon" Rohr à la date du 7 Août 1916*, cited in Gruss, *Die deutschen Sturmbataillone*, 180. In the mythos of frontline soldiers [*Frontschweine*], the teamwork continued: on Christmas Eve, 1914, British and German (or more precisely, Saxon) soldiers are said to have faced each other in a soccer match. Cf. Modris Eksteins, *Rites of Spring: The Great War and the Birth of the Modern Age* (New York: Mariner, 2000), 113.
74. Gruss, *Die deutschen Sturmbataillone*, 140.
75. Geyer, *Deutsche Rüstungspolitik*, 100.
76. Gruss, *Die deutschen Sturmbataillone*, 131.
77. Steiner, *Armee der Geächteten*, 24.
78. Ibid., 25.
79. Ibid., 91–97.
80. Cf. ibid., 24, 351.
81. *Völkischer Beobachter*, 21 May 1939, quoted in Steiner, *Armee der Geächteten*, 108.
82. Cf. Geyer, *Deutsche Rüstungspolitik*, 101: "[The Third Supreme Army Command under Ludendorff and Hindenburg] changed the tactics and principles of deployment in the German army . . . so radically and decisively (especially in con-

Notes to Chapter 17

trast to the French army) that it basically took until 1941–1942 for all the consequences to be drawn."

83. Steiner, *Armee der Geächteten*, 106.

84. Cf. Walter Bloem, *Seele des Lichtspiels. Ein Bekenntnis zum Film* (Leipzig: Grethlein, 1922). Incidentally, on 7 October 1918—that is, just in time for World War N+1—Assault Battalion No. 7 made its first training film (Gruss, *Die deutschen Sturmbataillone*, 120).

85. Gruss's malicious commentary on the passage ("Captain Rohr was not of noble birth" [*Die deutschen Sturmbataillone*, 56n.]) simply elucidates how the criteria for recruiting the elite changed between 1914 and 1939—that is, between the imperial army and the Waffen-SS.

86. Walter Bloem, *Das Ganze—halt!* (Leipzig: Grethlein, 1934), 229.

87. Cf. Linnenkohl, *Vom Einzelschuß zur Feuerwalze*, 272.

88. Cf. Cordova, *Arditi e legionari dannunziani*, 3.

89. Cf. Linnenkohl, *Vom Einzelschuß zur Feuerwalze*, 39.

90. Cf. Lupfer, *Dynamics of Doctrine*, 45.

91. Cf. ibid., 44: "The Germans wanted to avoid any prolonged artillery fire, for surprise would be lost and an artillery duel would develop in which the Allies, with greater amounts of munitions, would eventually prevail. Therefore, German fire had to be fast and accurate, and its mission was neutralization, rather than elusive and costly destruction."

92. Cf. Hermann Geyer, "Der Angriff im Stellungskriege," *Urkunden der Obersten Heeresleitung über ihre Tätigkeit 1916/1918*, ed. Erich Ludendorff (Berlin: Mittler, 1921), 648, 659, 671.

93. Geyer, "Angriff im Stellungskriege," 672, cf. also 652. The Ludendorff Offensive shortened the distance between artillery and infantry to between fifty and thirty meters, and it accelerated the barrage of fire to 1.5 kilometers/hour. That is exactly what the Red Army did in March 1945 at the Oder (Linnenkohl, *Vom Einzelschuß zur Feuerwalze*, 272–74).

94. Geyer, "Angriff im Stellungskriege," 657. The same principle also governed the training guidelines of the Reichswehr (*Ausbildungsvorschrift für die Infanterie*, 50).

95. "Something like that," Hitler had appreciatively remarked of Steiner's maneuvers at the Munster Training Area, "can only be done with select people!" (Steiner, *Armee der Geächteten*, 106).

96. Lupfer, *Dynamics of Doctrine*, 46.

97. Von Schwerin, "Das Sturmbataillon Rohr," 560.

98. Cf. Ernst Jünger, *Der Kampf als inneres Erlebnis* (Berlin: Mittler, 1922), 101–16.

99. Ott, *Martin Heidegger*, 104ff.

100. Martin Heidegger, Letter to Karl Jaspers, 19 June 1923, quoted in Ott, *Martin Heidegger*, 122.

101. All that follows simply reproduces thoughts that Hans Ulrich Gumbrecht, during the winter semester of 1988–89, shared as we sat between the palm trees of a lobby and iced-over highways.

102. Cf. Martin Heidegger, *Being and Time*, trans. John Macquarrie and Edward Robinson (New York: Harper & Row, 1962), 76.

103. Ibid., 303.

104. Ibid., 304, 311.

105. Ibid., 303.

106. Ibid., 307.

107. Ibid., 307f.

108. Ibid., 385.

109. Ibid. 303. The thesis that the concept of death in *Being and Time* derives from the First World War was proposed by Domenico Losurdo and defended against unhistorical readings (if one concedes that Habermas is capable of reading at all): *Die Gemeinschaft, der Tod, das Abendland. Heidegger und die Kriegsideologie* (Stuttgart, Weimar: Metzler, 1995). I would simply suggest replacing the non-concept of "war ideology" with documented plans of attack and exchanging Heidegger's later invocation of the "comradeship of front soldiers" for "anticipating death" [*Vorlaufen zum Tode*] in the well-defined military sense.

110. Cf. Lupfer, *Dynamics of Doctrine*, 10n.: "One important characteristic of the Imperial German Army was its extreme stinginess in promotions during the war. This army could not be accused of inflation of rank, unlike its World War II counterpart, the Wehrmacht."

Chapter 18

1. Unless otherwise noted, references follow Plato, *Collected Works*, ed. Edith Hamilton and Huntington Cairns (Princeton: Princeton University Press, 1994); occasionally, the translation has been modified.

2. Cf. Friedrich Nietzsche, "Ueber die Zukunft unserer Bildungsanstalten," *Gesammelte Werke*, ed. Giorgio Colli and Mazzino Montinari (Berlin: de Gruyter, 1973), III/2, 231ff.

3. Cf. Jacques Lacan, *Le séminaire, livre VIII: Le transfert*, ed. Jacques-Alain Miller (Paris: Seuil, 1991), 147.

4. Wilhelm Heinrich Roscher, *Nektar und Ambrosia. Mit einem Anhang über die Grundbedeutung der Aphrodite und Athene* (Leipzig: Teubner, 1883), 75.

5. Cf. Harry Neumann, "Diotima's Concept of Love," *American Journal of Philology* 86 (1965): 50.

6. Lacan, *Le séminaire, livre VIII*, 147.

Notes to Chapter 18

7. Diotima's epithet for Poros is *euporos*, and the one she applies to Penia *aporē*. Cf. Lacan, *Le séminaire, livre VIII*, 147: "Voilà donc l'Aporia femelle en face du Poros, l'Expédient."

8. Cf. Lacan, *Le séminaire, livre VIII*, 148.

9. Cf. ibid., 141.

10. Roscher, *Nektar und Ambrosia*, 36. Literally translated, the fragment reads roughly as follows:

> According to Orpheus, Kronos is pursued by Zeus by means of honey. Filled up with honey, he is intoxicated; it grows dark before his eyes, as if by wine, and he sleeps. But it was no wine. Namely, according to Orpheus, Night gives advice to Zeus, whereby she tells him the treacherous trick with honey: "When you see him under the broad-canopied trees, intoxicated by the works of the buzzing bees, bind him straightaway."

11. For Ouranos-Kronos, cf. Hesiod, *Theogony*, l. 175.

12. Hesiod, *Theogony*, l. 177.

13. Cf. the "Homeric Hymn to Aphrodite," available in many editions.

14. Cf. Pausanias, *Description of Greece*, VIII, 12, 8.

15. Ibid., VIII, 12, 9. On the age of this temple, cf. Walter Immerwahr, *Die Kulte und Mythen Arkadiens, Bd. I, Die arkadischen Kulte* (Leipzig: Teubner, 1891), 170–72.

16. Felix Bölte, in *Paulys Real-Encyclopaedie der classischen Altertumswissenschaft* (Stuttgart: Metzler), s.v. "Mantineia," col. 1320f.

17. Cf. Aristides, *Complete Works: Orations I-XVI*, ed. C. A. Behr (Leiden: Brill, 1997), 158.

18. Cf. Plato, *Protagoras*, 347 c-348a.

19. Cf. Thucydides, *The History of the Peloponnesian War*, trans. Richard Crawley (London: Longmans, Green, & Co., 1874), 385 (V, 66).

20. Cf. ibid., 390 (V, 76).

21. Michel Foucault, *The History of Sexuality, Vol. 2: The Use of Pleasure*, trans. Robert Hurley (New York: Vintage, 1990), 241.

22. Ibid., 245: "One would be missing the crucial point if one imagined that the love of boys gave rise to its own interdiction, or that an ambiguity peculiar to philosophy accepted its reality only by demanding its supercession."

23. Xenophon, *Memorabilia, Oeconomicus, Symposium, Apologia*, trans. E. C. Marchant and O. J. Todd (Cambridge, Mass.: Loeb Classical Library [Harvard], 1997), 567.

24. Plato, "Gedicht XXXIII," *Sämtliche Werke* (Cologne: Hegner, 1967), III, 784.

Chapter 19

1. Hermann Diels and Walther Kranz (eds.), *Fragmente der Vorsokratiker* (Berlin: Weidmann, 1951), cited here and in the following as "DK," followed by section division; DK 58, B 40.
2. DK 58, C 2.
3. DK 58, C 2.
4. DK 58, C 1.
5. DK 58, C 2.
6. Julian Jaynes, *The Origin of Consciousness in the Breakdown of the Bicameral Mind* (New York: Mariner, 2000), 75.
7. DK 58, C 4.
8. DK 58, C 4.
9. Martin Vogel, *Onos Iyras, Der Esel mit der Leier*, 2 vols. (Düsseldorf, 1973) (=*Orpheus- Schriftenreihe zu Grundfragen der Musik, Bd. 13 und 14*): Hesychios, 386.
10. DK 58, C 4.
11. DK 18, 13.
12. Johannes Lohmann, *Musiké und Logos. Aufsätze zur griechischen Philosophie und Musiktheorie*, ed. Anastasios Giannarás (Stuttgart: Musikwissenschaftliche Verlagsgesellschaft, 1970), 32.
13. DK 47, B 2; cf. DK 18, 15.
14. DK 44, B 6.
15. DK 28, B 12.
16. DK 44, B 6.
17. DK 45, 1.

Chapter 21

1. Cf. Barry B. Powell, *Homer and the Origin of the Greek Alphabet* (Cambridge: Cambridge University Press, 1991).
2. Cf. Richard Bentley, *Remarks upon a Late Discourse on Free-thinking*, quoted in Alfred Heubeck, *Schrift* (Göttingen: Vandenhoeck & Ruprecht, 1979), 170.
3. *Odyssey*, XII, l. 184.
4. Joachim Latacz, *Troia und Homer. Der Weg zur Lösung eines alten Rätsels* (Munich: Koehler & Amelang, 2003), 25.
5. *Theogony*, ll. 1011–20.
6. *Aeneid*, VI, ll. 851–53.
7. *Elegies*, II 34, 65f.
8. *Inferno*, III, ll. 1–6.
9. Stefan George, *Gesamt-Ausgabe der Werke. Endgültige Fassung*, 18 vols. (Berlin: Bondi, 1927–34), XI, 59 [*Inferno*, XXVI, ll. 91–99].

Notes to Chapters 21 and 22

10. *Paradiso*, XV, l. 26.
11. *Purgatorio*, XIX, l. 33.
12. Gottfried von Straßburg, *Tristan [und Isolde!]*, ed. Karl Marold (Berlin: de Gruyter, 1969), l. 4860ff.
13. Suetonius, *Tiberius*, LII, 3.
14. *Aeneid*, IV, 25.
15. Jacques Lacan, *The Seminar: On Feminine Sexuality, the Limits of Love and Knowledge (Encore) (Bk. 20)*, trans. Bruce Fink (New York: Norton, 1999), 89.
16. Martin Heidegger, *Gesamtausgabe* (Frankfurt a.M., 2000), III, 75, 89, 115.
17. William S. Burroughs, *The Electronic Revolution* (Bonn: Expanded Media Editions, 1976), 4.
18. *Politics*, 1253b25–1254a1, trans. Jowett (New York: Viking, 1966).
19. It is hardly known that Turing, already in 1923, was drawn by his mother, as he contemplated daisies (*Bellis perennis*) instead of playing hockey; in 1952, he authored a concise text on the topic. Cf. Alan M. Turing, "Outline of the Development of a Daisy," *Morphogenesis*, ed. P. T. Saunders (Amsterdam, London: Elsevier, 1992).

Chapter 22

1. Martin Heidegger, "Das abendländische Gespräch," *Gesamtausgabe* (Frankfurt a.M.: Klostermann, 2000), LXXV (*Zu Hölderlin/Griechenlandreisen*), 57–196; 115. References provided without the name of an author are to Heidegger.
2. *Being and Time*, trans. John Macquarrie and Edward Robinson (New York: Harper & Row, 1962), 95–122 (§§ 15–18).
3. "The Thing," *Poetry Language, Thought*, trans. Albert Hofstadter (New York: HarperCollins, 2001), 161–84; 164.
4. "The Origin of the Work of Art," *Basic Writings*, ed. David Farrell Krell (New York: HarperCollins, 2008), 159.
5. *Being and Time*, 142.
6. *Being and Time*, 139f.
7. "The Age of the World Picture," *The Question Concerning Technology, and Other Essays*, trans. William Lovitt (New York: Harper, 1982), 135.
8. Ibid., 149.
9. "The Question Concerning Technology," *The Question Concerning Technology*, 1–35; 22.
10. Ibid., 24.
11. "The End of Philosophy and the Task of Thinking," *Basic Writings*, 427–49; 435.
12. "Das Ding," VII (*Vorträge und Aufsätze*), 165–87.
13. "Question Concerning Technology," 17.

14. Cf. Martin Heidegger and Eugen Fink, *Heraklit. Seminar Wintersemester 1966/1967* (Frankfurt a.M.: Klostermann, 1970), chs. I–III.

15. Cf. Heinrich Wiegand Petzet, *Auf einen Stern zugehen. Begegnungen und Gespräche mit Martin Heidegger 1929–1976* (Frankfurt a.M.: Societäts-Verlag, 1983), 219f. On the Greek gods as athletes, cf. Hans Ulrich Gumbrecht, *In Praise of Athletic Beauty* (Cambridge: Harvard University Press, 2006).

Chapter 23

1. Martin Heidegger, *Being and Time*, trans. John Macquarrie and Edward Robinson (New York: Harper & Row, 1962), 172 (§ 29).
2. Aristotle, *Metaphysics*, A 1.
3. Aristotle, *Poetics*, 4.
4. Aristotle, *Politics*, I 1.
5. Aristotle, *Historia animalium*, IV 9; cf. Plutarch, *Gryllos*, 9.
6. Aristotle, *De interpretatione*, 1. On the basis of the Greek vocalic alphabet there developed, in twofold recursion, first a numerical system and then a system of musical notation for voice and instruments.
7. Aristotle, *Poetics*, 4.
8. Aristotle, *De interpretatione*, 1.
9. Aristotle, *Rhetoric*, I 11.
10. Plato, *Phaidros*, 259b.
11. Homer, *Odyssey*, I, 10.
12. Barry B. Powell, *Homer and the Origin of the Greek Alphabet* (Cambridge: Cambridge University Press, 1991).
13. Bruno Snell, *Die Entdeckung des Geistes. Studien zur Entstehung des europäischen Denkens bei den Griechen* (Hamburg: Claasen Goverts, 1948), 19.
14. Hesiod, *Works and Days*, l. 137.
15. *Charaktēr* in Greek, by the way, seems to be borrowed from Akkadian, where it refers to stamping (engraving) money. See Walter Burkert, "Die orientalisierende Epoche in der griechischen Religion und Literatur," *Sitzungsberichte der Heidelberger Akademie der Wissenschaften* 1 (1984): 1–135; 39.
16. Heraclitus, B 119 (Diels/Kranz).
17. Heraclitus, B 32, 45, 101, 115 (Diels/Kranz).
18. Aristotle, *Rhetoric*, III 3.
19. Aristotle, *Poetics*, 6.
20. Sappho, Fragment 1 L-P.
21. Sophocles, *Antigone*, l. 1142.
22. Aristotle, *Poetics*, 1.
23. Archilochos, Fragment 67d.

24. Aristotle, *Politics*, V III 5 to Il. I 601–4.

25. Aristotle, *De anima*, III 8.

26. Niklas Luhmann, "Sinn als Grundbegriff der Soziologie," in Jürgen Habermas and Niklas Luhmann, *Theorie der Gesellschaft oder Sozialtechnologie—Was leistet die Systemforschung?* (Frankfurt a.M.: Suhrkamp, 1971), 25–100; 77.

Chapter 24

Manuel Rodriguez, an undergraduate majoring in economics at Stanford University, established the original German text for this collection of essays and checked all quotes and historical references.

1. Friedrich Kittler, *Gramophone, Film, Typewriter*, trans. Geoffrey Winthrop-Young and Michael Wutz (Stanford: Stanford University Press, 1999), 243.

2. P. 81 in this volume; further references to the essays collected here appear parenthetically.

3. In going through the original versions of Kittler's essays, a number of passages were detected that appeared to reference texts imprecisely and/or to contain inaccurate historical references. Such instances contradict the author's frequently emphasized insistence on philological and historical "facticity." At the same time, however, they underscore the strong lines of interpretation and shaping of the past that I am calling "mythographic" here. Clearly, reading Kittler's work as a contemporary mode of mythology alters the status of its claim to facticity; instead of being purely positivistic, it highlights an understanding of history that the author considered necessary and pertinent in the present of his lifetime.

4. Friedrich Kittler, *Eine Kulturgeschichte der Kulturwissenschaft* (Munich: Fink, 2000).

5. Martin Heidegger, "Nur noch ein Gott kann uns retten," *Der Spiegel* 23/1976 (31 May 1976), 193–219.

Credits

"Der Dichter, die Mutter, das Kind. Zur romantischen Erfindung der Sexualität." In *Romantik in Deutschland, Deutsche Vierteljahresschrift für Literaturwissenschaft und Geistesgeschichte, Sonderband*. Ed. Richard Brinkmann. Stuttgart: Metzler, 1978. 102–14. © 1978 J. B. Metzlersche Verlagsbuchhandlung and Carl Ernst Poeschel Verlag in Stuttgart.

"Nietzsche (1844–1900)." In *Klassiker der Literaturtheorie. Von Boileau bis Barthes*. Ed. Horst Turk. Munich: C. H. Beck, 1979. 191–205. By kind permission of C. H. Beck Verlag.

"Lullaby of Birdland." Expanded reprint in *Dichter—Mutter—Kind. Deutsche Literatur im Familiensystem 1760–1820*. Munich: Wilhelm Fink 1991. 103–18. By kind permission of Wilhelm Fink Verlag. [Originally published in *Der Wunderblock. Zeitschrift für Psychoanalyse* 3 (1979): 2–16.]

"Der Gott der Ohren." Expanded reprint in *Das Schwinden der Sinne*. Ed. Dietmar Kamper and Christoph Wulf. Frankfurt a.M.: Suhrkamp 1984. 140–55. [Originally published in *europaLyrik 1775–heute. Gedichte und Interpretationen*. Ed. Klaus Lindemann. Paderborn: Schöningh, 1982. 467–77.]

"Flechsig/Schreber/Freud. Ein Nachrichtennetzwerk der Jahrhundertwende." *Der Wunderblock. Zeitschrift für Psychoanalyse* 11/12 (1984): 56–68.

"Romantik—Psychoanalyse—Film: Eine Doppelgängergeschichte." In *Eingebildete Texte. Affairen zwischen Psychoanalyse und Literaturwissenschaft*. Ed. Jochen Hörisch and Georg Christoph Tholen. Munich: Wilhelm Fink, 1985. 118–35. By kind permission of Wilhelm Fink Verlag.

"Medien und Drogen in Pynchons Zweitem Weltkrieg." Reprinted in *Die unvollendete Vernunft: Moderne versus Postmoderne*. Ed. Dietmar Kamper and Willem von Reijen. Frankfurt a. M.: Suhrkamp, 1987. 240–59. [Originally published in *Narrativität in den Medien*. Ed. Rolf Kloepfer, Karl-Dietmar Mueller, Münster: MAkS Publikationen 1985. 231–52.]

Credits

"'Heinrich von Ofterdingen' als Nachrichtenfluß." In *Novalis. Beiträge zu Werk und Persönlichkeit Friedrich von Hardenbergs (Wege der Forschung,* Vol. 248.). Ed. Gerhard Schulz. Darmstadt: Wissenschaftliche Buchgesellschaft, 1986. 480–508.

"Weltatem. Über Wagners Medientechnologie." In *Diskursanalysen 1.* Ed. Friedrich A. Kittler, Manfred Schneider, and Samuel Weber. Opladen: Westdeutscher Verlag, 1986. 94–107. By kind permission of VS Verlag für Sozialwissenschaften.

"Die Stadt ist ein Medium." Reprinted in *Mythos Metropole.* Ed. Gotthard Fuchs, Bernhard Moltmann, and Walter Prigge. Frankfurt a.M.: Suhrkamp, 1995. 228–44. [Originally published in *Geburt einer Hauptstadt. Bd. 3: Am Horizont,* Vienna: Edition BuchQuadrat, 1988. 507–31.]

"Rock Musik—ein Mißbrauch von Heeresgerät." Expanded reprint in *Medien und Maschinen. Literatur im technischen Zeitalter.* Ed. Theo Elm and Hans H. Hiebel. Freiburg: Rombach, 1991. 245–57. By kind permission of Rombach Verlag. [First published in *Appareils et machines à représentation.* Ed. Charles Grivel. Mannheim: *Mannheimer Analytica* 8 (1988): 87–101.]

"Signal-Rausch-Abstand." In *Materialität der Kommunikation.* Ed. Hans Ulrich Gumbrecht and K. Ludwig Pfeiffer. Frankfurt a.M.: Suhrkamp, 1988. 342–59.

"Die künstliche Intelligenz des Weltkriegs: Alan Turing." In *Das Subjekt der Dichtung. Festschrift für Gerhard Kaiser* (with Gerhardt Buhr, Hort Turk). Würzburg: Königshausen & Neumann, 1990. 187–202. By kind permission of Königshausen und Neumann.

"Unconditional Surrender." In *Paradoxien, Dissonanzen, Zusammenbrüche. Situationen offener Epistemologie.* Ed. Hans Ulrich Gumbrecht and K. Ludwig Pfeiffer. Frankfurt a.M.: Suhrkamp: 1991. 515–33.

"Protected Mode." Reprinted in *Strategien des Scheins. Kunst—Computer—Medien.* Ed. Florian Rötzer and Peter Weibel. Munich: Boer, 1991. 256–67. [First published in *Computer, Macht und Gegenwehr. InformatikerInnen für eine andere Informatik.* Ed. Ute Bernhardt and Ingo Ruhmann, Bonn: FIFF 1991. 34–44.]

"Es gibt keine Software." First published in German in *Writing/Ecriture/Schrift.* Ed. Hans Ulrich Gumbrecht. Munich: Wilhelm Fink, 1993. 367–78. By kind permission of Wilhelm Fink Verlag. [Originally published as "There is no software," *Stanford Literature Review,* 9.1 (1992): 81–90.]

"Il fiore delle truppe scelte." In *Der Dichter als Kommandant. D'Annunzio erobert Fiume.* Ed. Hans Ulrich Gumbrecht, Friedrich Kittler, and Bernhard Siegert. Munich: Wilhelm Fink, 1996. 205–25. By kind permission of Wilhelm Fink Verlag.

"Eros und Aphrodite." In *Maskeraden. Geschlechterdifferenz in der literarischen Inszenierung*. Ed. Elfi Bettinger, Julika Funk. Berlin: Erich Schmidt, 1995. 31–39.

"Homeros und die Schrift." Reprinted in *Die Geburt des Vokalalphabets aus dem Geist der Poesie. Schrift, Zahl und Ton im Medienverbund*. Ed. Wolfgang Ernst and Friedrich Kittler. Munich: Wilhelm Fink, 2006. 47–53. By kind permission of Wilhelm Fink Verlag. [Originally published in *Claude Elwood Shannon, Aus/Ein. Ausgewählte Schriften zur Kommunikations- und Nachrichtentheorie* (with Peter Berz, David Hauptmann, and Axel Roch). Berlin: Brinkmann und Bose, 2000. 47–59.]

"Das Alphabet der Griechen. Zur Archäologie der Schrift." In *Die Aktualität des Archäologischen in Wissenschaften, Medien und Künsten*. Ed. Knut Ebeling and Stefan Altekamp. Frankfurt a.M.: Fischer, 2004. 252–60.

"Im Kielwasser der Odyssee." In *Odysseen. Mosse-Lectures 2007*. Ed. Elisabeth Wagner and Burkhardt Wolf. Berlin: Vorwerk 8, 2008. 96–120. By kind permission of Vorwerk 8 Verlag.

"Martin Heidegger, Medien und die Götter Griechenlands. Ent-fernen heißt die Götter nähern." In *Philosophie in der Medientheorie. Von Adorno bis Žižek*. Ed. Alexander Roesler and Bernd Stiegler. Munich: Wilhelm Fink, 2008. 133–43. By kind permission of Wilhelm Fink Verlag.

"Pathos und Ethos. Eine aristotelische Betrachtung." In *Passionen. Objekte—Schauplätze—Denkstile. Festschrift für Sigrid Weigel*. Ed. Corina Caduff, Anne-Kathrin Reulecke, and Ulrike Vedder. Munich: Wilhelm Fink, 2010. 27–32. By kind permission of Wilhelm Fink Verlag.

When the initial version of an essay has not been used for this volume, reference is provided in brackets.

The authorized representative in the EU for product safety and compliance is:
Mare Nostrum Group
B.V Doelen 72
4831 GR Breda
The Netherlands

www.ingramcontent.com/pod-product-compliance
Lightning Source LLC
Chambersburg PA
CBHW031750220426
43662CB00007B/351